Population Policies in the People's Republic of China

Xueyuan Tian

Population Policies in the People's Republic of China

Reviews and Prospects

福建人民出版社
FUJIAN PEOPLE'S PUBLISHING HOUSE

Springer

Xueyuan Tian
Beijing, China

Translated by
Yingmei Luo
Henan University
Zhengzhou, Henan, China

Ge Zhang
Henan University
Zhengzhou, China

Wenjia Zhu
Henan University
Zhengzhou, China

ISBN 978-981-96-9657-4 ISBN 978-981-96-9658-1 (eBook)
https://doi.org/10.1007/978-981-96-9658-1

Jointly published with Fujian People's Publishing House
The Print edition is not for sale in China Mainland. Customers from China Mainland, please order the book from Fujian People's Publishing House.

Sponsored by Chinese Fund for the Humanities and Social Sciences (本书由中华社会科学基金资助).

This Springer imprint is published by the registered company Springer Nature Singapore Pte Ltd.
The registered company address is: 152 Beach Road, #21-01/04 Gateway East, Singapore 189721, Singapore

I am not blindly opposed to population growth, yet I am opposed to blind population growth. I advocate for an all-round improvement in population capacities. And I am against any theory, behaviour or preaching that impedes or obstructs the promotion of population capacity.

History is recorded in words, people create history with their actions, and scholars elucidate history with their expertise and theory. Yet history is history and with the passing of time, history becomes increasingly clear and truths reveal themselves.

China is a nation that favours the births of many children in a family. The reasons for this preference lie in the deep-rooted natural economy underpinned by small peasants that has lasted for thousands of years. The views of population and fertility that have formed on the foundation of this natural economy should not be underrated. Their influence is discernible even today.

How do we evaluate the one child per couple policy that was advocated from 1980 to the twenty-first century when adjustments were made to it? The only criterion is practice: the influence of population policy practices on the development of the population, the economy, the society, natural resources, and the environment.

Population policies have specific purposes and ultimately, are constrained by economic development. The "three-step" population strategy in China constitutes the key to understanding the implementation, adjustment, and future changes to China's population policy. We are presently in a period of implementing the second step of the population strategy, which highlights the enhancement of population capacities in an all-round way. This situation closely accords with the high-quality economic development in today's China.

Family planning in the Western world is an inevitable outcome of population change and socio-economic development. With China transitioning to the second step of its population strategy and rapidly moving to the post-demographic transition, it must make the inevitable choice to continue family planning with Chinese characteristics.

Acknowledgments

It took more than two years to complete the project of translating a seminal work on the formulation, application, and transformation of population policies throughout China's history, with a particular focus on the evolution of family planning policy in the People's Republic of China since the 1970s. I thank all those who directly and indirectly guided us in this endeavor.

Professor Xueyuan Tian (Member of the Chinese Academy of Social Sciences), a leading expert on demographic studies in China, has been a source of inspiration and encouragement throughout the entire process. The significant differences between Chinese and English languages and cultures posed substantial challenges in fulfilling the daunting task of rendering the voluminous Chinese work—*Da Guo Zhi Ce: Xin Zhong Guo Ren Kou Zheng Ce Hui Gu Yu Zhan Wang (Policy in a Big Country—Population Policies in the People's Republic of China: Reviews and Prospects)*—into English.

I am sincerely indebted to Ge Zhang (an English teacher at the International Business School, Henan University, Zhengzhou, China), a devoted friend and supporter. With her tremendous courage, sound training in translation, and enduring passion for translation studies, we overcame all the challenges encountered during this journey. She was highly supportive both in the early stages of the project and throughout the translation process. Conscientiously, she translated all the chapters of the original text into English except Chap.8 and the first two sections of Chap.10. Her contribution to the translation exceeded 230,000 Chinese characters. Honestly, the translation of this book would not have been possible without her commitment and persistence.

My thanks also go to Wenjia Zhu (PhD in English Literature, Henan University), who translated Chap.8 and the second section of Chap.10 (approximately 40,000 Chinese characters). I am deeply grateful to Rick Qi, a renowned translation scholar from Monash University, Australia, for his valuable translation advice. I would like to express my heartfelt gratitude to Kathleen Weekley (PhD, an Australian editor) for her meticulous proofreading of our translation manuscript.

I am grateful for the support of Professor Ling Jiang (a professor specialized in translation studies) from Henan University, Zhengzhou, China, for her professional advice on translation. Special thanks are also extended to Fiona Anderson

(Associate professor, Victoria University, Australia) and Associate Professor Rod Neilsen from Deakin University, Australia, for their friendship, support, and encouragement. I also appreciate the support and professional advice of Professor Tongping Li (China University of Geosciences, Wuhan, China) and Associate Professor Wei Zhang (Henan University, Zhengzhou, China), both experts in demographic studies.

Finally, my team members and I are deeply grateful to Jintuan Wang (Director of the Editorial Department, Fujian People's Publishing House, Fuzhou, China) and Leana Li from Springer (Editorial Director, Humanities and Social Sciences, China), whose cooperation and continuous efforts made the publication of our translation work possible.

Yingmei Luo

Zhengzhou, China
November 2024

Introduction

By 2009, the population policy encouraging each couple to have only one child had been in place for nearly 30 years so it was time to adjust it. But at the time it seemed that a change to the emphasis on a low fertility rate was out of the question.

As one who participated in creating the policy, I could not help fidgeting because I saw no sign of any coming policy amendments though the time suggested for adjustment had arrived. Before 2009, some experts, including myself, had appealed to the authorities through internal channels for changes to the population policy but we received no responses to our written proposals. I had no choice but to make our ideas public and in 2009, I published *China's Population Policies in 60 Years.* I also wrote an article titled *Review and Prospects of Population Policies in People's Republic of China.* The book was published by Social Sciences Literature Press and was included in the series "Glorious History: Key Books in Celebration of the 60th Anniversary of People's Republic of China". The article was printed on the front page of the theoretical edition of *People's Daily*, on December 4, 2009. These publications, especially the paper, were reprinted and commented on by the press both at home and abroad, sparking wide public interest and much speculation about China's population policy. A leader of a certain government department commissioned an official to phone the newspaper office to enquire into my background. Some people remarked that Professor Tian was once again in the midst of a storm. But since I had chosen to make my views public in this way, I did not care much about how dangerous the storm was; I was prepared for the potential consequences. Fortunately, the Third Plenary Session of the 18th Central Committee of the Communist Party of China (CPC) in 2013 published the *Decision of the Central Committee of the CPC on Some Major Issues Concerning Comprehensively Deepening the Reform.* This document launched the policy *that allows married couples to have two children if one parent is a single child*, signaling the first change to the fertility policy. Two years later, the Central Committee decided to allow all couples to have two children. The across-the-board implementation of the two-child policy indicated that the fertility policy adjustment was in place. With that, the reproach of me disappeared.

Since the 18th CPC National Congress, adjustments to the population policy have been smooth, which is very encouraging to me, an old scholar in his seventies. I was aware of inadequacies in my previously published work on population policy. For one thing, my earlier published theoretical explorations were inadequate, and I had failed to cite some important literature, which affected the power of my arguments. When I was pondering how to address these inadequacies, Fujian People's Publishing House invited me to write a new book on population policy. Initially, I presumed that the task would not be easy, because I had already elaborated on all the background to the one child policy, the process of its creation and its application, in my earlier writings. It made no sense to repeat my own views. Later, I tried to approach the subject from a more theoretical perspective. With the help of more theoretical resources and new literature, together with the new situation and the post-18th CPC National Congress policy amendments, I came to believe that it was possible to write a new book with new material and new ideas, so I accepted the offer. Against the background of deepening reform and expanded opening-up, I started writing, with the emphasis on what I had witnessed of the policy history.

Policy is defined here as a set of guides to action developed by a state or a political party in order to accomplish certain political aims or tasks in certain periods. Where do these aims and tasks come from? From the general plans and strategies that guide the social era, and from certain specific strategies. Thus, research on population policy cannot be divorced from a population strategy, because it is the application of guidelines, tactics, and principles of strategy in different historical moments. This provides a logical clue for policy studies: we can start from the general design of the strategy, including its goals, tasks, and phases and investigate the policy focus and orientation during different periods, along with the tactics, regulations and roadmaps used for policy application. Accordingly, if we investigate Chinese population policies since 1949 and possible future policy directions, we can begin by analysing the "three-step" population strategy. More than 30 years ago, I explained China's population development strategy. Though slight amendments were later made to my analysis, they related only to the wording. The key theme and the main content remained the same. The basic points of the "three-step" population strategy were as follows:

The first step involved population size control, population capacity improvement and population structure adjustment, with the focus on size control but also taking the other two aspects into consideration. The chief goal was to bring the fertility rate under replacement level and complete the transition from a pattern of high birth rate, low mortality rate and high increase rate to one of low birth rate, low mortality and low increase rate. The total fertility rate (TFR) dropped to 2.01 in 1991, below replacement level (2.10), signifying that the first step of population strategy had been completed.

In the second phase, we continued to combine size control, skills improvement and structural changes in the period before zero population growth and after the fertility rate had dropped below replacement level. In this interim, the population strategy focus would gradually shift from size control to all-round capacity improvement. Meanwhile, size control and structural adjustments, and the sustainable development of population and economy, society, resources, and environment were also to be considered. Some believed that this second step should emphasise changes to population structures because the focus on size control created serious problems, including an aging population. So, how to shift the population structure? There seems no choice but to raise the fertility rate and even to adopt a policy that stimulates fertility. A substantial increase in fertility and birth rates could temporarily alleviate the pressure of the aging problem. But when these new-borns become old, the aging pressures might be even greater. Such a measure is neither a comprehensive solution to the population question nor does it help to settle the aging problem fundamentally. On the contrary, when the population strategy focus is shifted to an all-round improvement in the population's capacities, and we give equal consideration to population size and structure, there will be several advantages. For one thing, problems of population structure created by attempts to control its size will be limited to a reasonable range. The aging problem will not become too serious, for example, and for another, greater population capacities will be conducive to settling population problems more comprehensively. All-round improvement in population capacities is conducive to transformation of the mode of economic development, adjustments to economic structure, and high-quality economic development. Moreover, all-round capacity improvement and high-quality economic development are favourable to saving natural resources, protecting the environment and promoting sustainable development.

The Chinese population is expected to achieve zero growth in 2030. After that it will experience negative growth. According to the third step of the population strategy, we should achieve an all-round optimum population conducive to sustainable development thanks to the changes in population size, its capacities and structure and to the changes to the economy, society, natural resources, and the environment. An all-round optimum population is one of adequate size, high skills, and reasonable distribution, of age, genders, the urban rural divide, and regional distribution. There will also be harmonious and sustainable relationships among the population and resources, environment, economy, and society.

The focus of the three-step population strategy in China has shifted from size control to all-around improvement of the people's capacities, and focus will be shifted to all-round optimum population. Policy adjustments must adapt to shifts in the strategic focus. Policy is the trail breaker, it must ensure the smooth implementation of strategy. Since strategy and policy are different in goals, tasks, and time spans, each strategic step can have more than one corresponding policy. For example, the second step of the population strategy has seen the focus from size control to capacity improvement, so a suitable range of policies must be developed. As the strategy proceeds, policies must be adjusted and their focus shifted accordingly. One of the themes running through this book is that the adoption and implementation of the three-step population strategy has required that population policies be adjusted accordingly, as they are essential to its success.

With the unification of logic and history, this Preface summarises this theme and the history of the population policy changes. The book contains three parts, and ten chapters.

Part one, The Development of Population Policy despite Controversies contains Chap. 1: A Historical Overview of Chinese Population Policies, Chap. 2: Debates on the Population Problem in the People's Republic of China (1949–1979), and Chap. 3: Growing Concerns about Population Policy. In these three chapters, I present a general overview of Chinese population policies since ancient times and especially in the first 30 years after the founding of the PRC (1949–1979). I analyse the policies from a theoretical perspective and explain how they worked in practice. I explain the pre-1949 policies in relation to people's inclination to have many children because of the small-scale peasant economy in place for thousands of years, and the arguments in favour of population growth in Chinese history. My post-1949 analysis is divided into development periods. In the period of national economic recovery (1950–1952), reproduction patterns shifted from high birth rate, high mortality rate and low increase rate to high birth rate, low mortality rate and high increase rate. Two baby booms (1953–1957 and 1962–1973) demonstrated how the child cost—benefit theory worked. This was why fertility and birth rates remained high and the extreme difficulty we had in restraining population growth. In the period 1949 to 1979, population policies shifted from encouraging growth to restraining it. The restraining policy was intensified gradually. There were many reasons for this and it was difficult to reach consensus on the question. My research has found that the change of population policies in the first 30 years of the PRC basically followed Mao Zedong's thoughts and judgements on the question, and the policy directions he suggested.

During the national economic recovery period (1950–1952), there was no explicit population policy and birth control practices were limited, mainly through regulations that restricted abortion. During the first five-year plan (1953–1957), Mao Zedong approved population control proposals made by Shao Lizi and Ma Yinchu, suggesting that we advocate for birth control and that people have children in a planned way. Thus, the notion of family planning was proposed in the PRC for the first time.[1] In 1957, during the Anti-Rightist Campaign,[2] all sociologists in favour of restrained population growth were labelled Rightists. One of the accusations against them was that "they exploited the population question to launch attacks on the CPC and socialism". In 1958, the "Great Leap Forward Movement" was started. With the surge of enthusiasm among society at large, Mao made some concluding remarks, "Is it good to have a large population or a small population? In my opinion, it is good to have a large population at present. I am afraid the current population needs to

[1] The National Population and Family Planning Commission. (Ed.). (2007). *Population and Birth Control History in China.* China Population Publishing House. p. 25.

[2] The Anti-Rightist Campaign was launched to fight the bourgeois conservative faction in June 1957. When the CPC initiated a "rectification" movement in May 1957, many intellectuals and non-communist political figures offered valuable suggestions about the work of the Party. However, a minority of "bourgeois rightists" were considered to be exploiting the situation to attack the CPC and the socialism. To stabilize the newly established socialist system, an Anti-Rightist Campaign was launched across the country. Though it was necessary to fight against the attacks of the Bourgeois Rightists, the Central Committee of CPC had committed the mistake of overestimating the situation which had led to a considerable number of intellectuals, members and officials of CPC, as well as democratic patriots, being mistakenly labelled Rightists. From 1959 to 1964, most of those wrongly accused were rehabilitated. In April 1978, the CPC Central Committee determined to rehabilitate all remaining"Rightists"and by the end of 1981, the job had been basically completed.

continue to increase a little".[3] Then in the 1960s, he began to express concern about controlling population growth. He repeatedly told the mobile medical teams assigned to provide health care in the rural areas to carry out family planning work.[4]On the surface, his remarks on the population question and policies seemed contradictory. But he had his own thoughts on China's population development, and the strategic and policy considerations required for it.

Mao's basic views on the population question are reflected in two of his statements. One appeared in "On the Correct Handling of Contradictions among the People" in 1957, in which he said, "It is good to have a large population in our country, but there are also difficulties with it".[5] The second appeared in a speech delivered to a Supreme State Conference in January 1958, where he said, "When the population in China reaches 700 or 800 million, it will be the time for Mr. Shao Lizi's proposals to be extensively applied."[6]The first statement implied that we should see China's large population dialectically. He affirmed that it was good to have a large population but was admitting at the same time that it created difficulties. The second statement meant that having a large population was a problem but the problem could not be solved until it was fully exposed and recognized. Only when the population reached 700 or 800 million would the resultant problems be fully exposed. Then Mr. Shao Lizi's proposal of population control could be widely applied.

Mao's first observation was more important than the second, for it embodied his general view on the Chinese population question. When he said it was good to have a large population, he was speaking from personal experience. During the Autumn Harvest Uprising of 1927, in the War of Resistance against Japanese Aggression (1937–1945), the War of Liberation (1946–1950) and in the War to Resist US Aggression and Aid Korea (1950–1953), victory was based on the support of the people and soldiers. Relying on the people, informing them, mobilizing them, and organizing them, the Communist Party fought together with them and won. Finally, the "three big mountains" (imperialism, feudalism, and bureaucrat-capitalism) were removed from the backs of the Chinese people. During war time, the aim is to centralize power and fight a war of annihilation, transforming strategic disadvantage into the tactical advantage provided by more people. Likewise, in the period of economic construction, we launched many mass campaigns, such as the people's commune and the Great Leap Forward Movement. Yet at the same time, Mao saw the difficulties created by a large population. In this context, he suggested some policies concerning population control and "family planning".

How then were these two different, even contradictory views reconciled? The speech at the Supreme State Conference in 1958 provided a clear answer. Mao said:

---

[3] The National Population and Family Planning Commission. (Ed.). (2007). *Population and Birth Control History in China*. China Population Publishing House. p. 44.

[4] Idem, p. 52.

[5] Literature Research Office of the CPC Central Committee. (Ed.). (1999). *Selected Writings of Mao Zedong* (Vol. 7). People's Publishing House. p. 228.

[6] The National Population and Family Planning Commission. (Ed.). (2007). *Population and Birth Control History in China*. China Population Publishing House. p. 44.

Is it good to have a large population or a small population? In my opinion, it is good to have a large population at present. I am afraid the current population needs to continue to increase a little. If we want people to practice birth control, there are two disadvantages before us. First, there are not enough contraceptives. Second, birth control methods are not propagated widely enough. Most peasants are illiterate and accustomed to marrying at an early age. We can't force them to practice birth control and they cannot control themselves. Human beings are still unable to control their own reproduction. Neither are several hundred million people in China able to control their own actions. In my opinion, the situation will become tough when the population in China reaches 700 million. When it reaches 700 or 800 million, then will be the time for Mr. Shao Lizi's proposals to be extensively applied. When we talk about birth control now, people will turn a deaf ear to it... A large population can make the nation look more powerful. At the same time, we should also note the problems that ensue. Yet there is no necessity to fear the problems. Anyway, I don't fear them. If there is an increase of 200 million people, the population problem will be solved soon. Why? Because things will develop in the opposite direction when they become extreme. Now I think the population problem still hasn't reached its most serious point.[7]

At the previous Supreme State Conference in 1957, Mao had said,

human beings should learn to control themselves. Population change should fluctuate. In other words, it might sometimes increase, sometimes stay at the same level, and at other times it could decrease a bit. From my point of view, birth control should be carried out, so that people will have children in a planned manner. Government should create some offices to take charge of such matters. There should be some work measures. For example, we could launch a pilot project to promote birth control for three years … then apply the experience more broadly for four years ... Let's make a ten-year plan ... We shouldn't promote birth control in the regions inhabited by the minority ethnic groups. Neither should we promote it in the places with fewer people. Even in the populous areas, we should also first run pilot programmes, and then gradually … apply family planning extensively step by step.[8]

We see here that Mao Zedong not only mentions the notion of family planning, but also envisaged a rough "ten-year plan". He did not expect people to practice birth control right away. What he expected was to advocate birth control, to set up offices and to take steps such as launching three-year pilot projects. "Now" he wanted the population to grow a little, then people would be aware of the situation. When it reached 700 or 800 million, the problems would be exposed and only after that it would be the right time to apply more widely the birth control proposals made by Shao Lizi and Ma Yinchu. History confirmed Mao's views. By the early 1970s the population had surpassed 800 million, 200 million more than in 1958. Accordingly, the government quickened its pace to tighten population policy.

---

[7] The National Population and Family Planning Commission. (Ed.). (2007). *Population and Birth Control History in China*. China Population Publishing House. p. 44.

[8] Idem, p. 25.

Part two, The Whole Story of the One Child Policy contains four chapters, Chap. 4: Setting Things Right in Demographic Theory, Chap. 5: Context for the One-child Policy, Chap. 6: Population Control Policy in 2000 and Chap. 7: Population Strategy Supported by the Basic National Policy.

In 1980 began the general advocacy of the one child per couple policy. I participated in discussions about this policy and drafting the relevant documents; later I hosted the project "China's Population and Employment in 2000" and wrote the report with the same title (as a lead author). I also published some other pieces under my own name. All efforts were made to break the doctrine "the more people, the better", which had been manipulated to criticize Ma Yinchu's population theory in the 1960s. My writings pointed out that a shortage economy coexisted with a surplus population, hence it was urgent for us to accelerate social and economic development. They also elaborated on the necessity and feasibility of intensifying efforts to control population growth, the optional population development strategy, and the relevant policies. Years later, in an interview, I said that the population policy had been formulated by the CPC Central Committee, and that I had a clear conscience as a participant discussing the policy and as a drafter of some of the related documents. I meant that it was a fact that the policy had been formulated by the Central Committee and I had a "clear conscience" because I participated in the forming of that state policy and published some scholarly pieces from my own convictions and my own will. My work was based on facts and not in response to someone else's interests or pressure. The documents I drafted and the writings I published sufficed to illustrate what I said.

My publications elaborated on the population questions both theoretically and practically. My predictions about the major problems and development trends were confirmed. The decision in 1980 to advocate only one child per couple, created a knot in my heart that could not be spoken of in public at the time. I participated in the 1980 Symposia on the Population Question convened by the CPC Central Committee and assumed the task of drafting the Report on Symposia on the Population Question (the first draft for discussion, henceforth, the Report). The Report focused on whether it was feasible for each couple to have only one child, what problems might be encountered, and how to solve them. I think it is safe to say that all the major problems we meet today were covered in those discussions and analysed as clearly as possible. Why was the analysis in the Report not adequately publicized, remaining unknown to many people? Because when we introduced the policy, the primary task was to propagate the policy and explain its necessity and significance, and to encourage people to practice it. It was essential to raise the rate of only children among all new births and to reduce the fertility rate quickly. If we talked too much about the possible pursuant problems (and solutions), we would be suspected of disrupting the implementation of the policy. We had to decide whether we had made the potential problems sufficiently known to the public and whether the Report had paid due attention to them. In truth, the Report's authors had made all possible efforts to provide adequate analysis of the possible problems and available solutions. Time proved their policy suggestions to be basically correct and sensible. The Report on Symposia on the Population Question (the first draft for discussion) was submitted to the Secretariat of the CPC Central Committee in 1980. Based on the copy of the report that I kept, I discuss here some of its major content.

There are three sections in the Report.

The first pointed out that strongly encouraging one child per couple was the key to solve the current population problems in China. At the time, China's population was increasing at a fast pace. The population was large and young, predicting strong growth in the future. According to the Report, "encouraging each couple to have only one child is the policy we must persist with for a certain time".

Section two reviews discussions that were had about questions that might arise in the practice of the one child policy. The discussions revolved around six issues:

(1) The work force and the dependent population. Trends in the working-age population over the next five decades were analysed. It was found that it would first increase and then decrease. At that time, we need not worry about labour power or troop shortages, or the increase in the dependent population. Since population development is progressive and long term, we had to consider potential future problems and develop preventive measures in population planning and policies.
(2) Population capacity. With the one child policy, we should pay special attention to the improvement of skills and other capabilities. Research into genetics and eugenics should be reinforced to improve the collective capacities of the Chinese nation.
(3) Old-age social security. According to the Report, an effect of the policy would be that the elderly parents of one party of a married couple would often be left unattended, which necessitated planning for retirement pensions and some other associated services. For example, each relevant government department should carefully study important matters such as housing and medical services, and deal with them in a systematic manner. We should address people's concerns.
(4) The education of the only children. Only children might be wilful, lack respect for their teachers and elders, or be too dependent on their parents.
(5) Birth control methods and contraceptive devices. It was noted that we should strengthen research into safer, more reliable, more convenient and effective birth control pills and contraceptive devices as rapidly as possible.
(6) The gender ratio. Under the one child policy, some people might still be deeply attached to traditional ideas about the importance of having a male heir to carry on the family line or boys being more valuable than girls. In extreme cases, parents might even kill or abandon their baby girl in order to have a boy. We should be attentive to such appalling phenomena and the law should explicitly prohibit such behaviour.

In section three, seven policy proposals were made:

(1) We should reinforce education about family planning and draw public attention to the issue. Encouraging couples to have only one child each constituted a real revolution in ideas about fertility, so we must do a good job in propagating the policy. It was suggested that special columns be opened in *People's Daily* and *Guangming Daily* to disseminate demographic theory and family planning

knowledge; a Population Research Institute be established in Chinese Academy of Social Sciences, and journals on population research be created; courses such as physiology, population, and sex education be introduced in middle schools, demographic theory be brought into political economy departments in universities; and more research personnel be hired by the Family Planning Work Office of the State Council to strengthen research and carry out the work more professionally.

(2) The Family Planning Law should be enacted as soon as possible. Presently, the standards for rewarding those who comply with the family planning guidelines and punishing those who do not have not been consistent across the country. As well, there was no guaranteed prize money. Family planning regulations lacked legal grounding and many contradictions emerged as they were carried out. When certain family planning provisions conflicted with other regulations, they were usually branded "local rules" and not implemented conscientiously. Symposia participants strongly hoped that the unified family planning law could be promulgated as quickly as possible.

(3) A national population commission should be established. It was noted that in China, a big country with around one billion people, population problems affected many sectors of society, and there were also many international exchanges on population issues. The lack of a national population commission was hampering China's development. To improve implementation of the national population policy, promote population development, and fulfill the strategic goal of population control, the Report recommended the establishment of a national population commission that would take charge of all population-related issues.

(4) An old-age social security system should be established step by step, starting now. Three social insurance schemes suggested by insurance companies were submitted along with the Report. In the units under collective ownership where the retirement system had not been adopted, pension insurance funds should be established. It was emphasized that the government should ensure unified planning to provide decent living facilities for those who had no children to take care of them in their old age.

(5) The timetable for slowing population growth was covered in the Report. It expressed no approval or disapproval of the target submitted for discussion at the Symposia to drop the increase rate down to five per 1,000 people in 1985 and to zero in 2000. Instead, it said, "Even if the one couple, one child policy is strictly applied in 1985, the natural increase rate will approach zero by the end of the 20th century but will not actually reach zero until 2004." The Standing Committee of the National People's Congress proposed that the legal age of marriage be lowered and the Symposia responded that since it was not consistent with the recommendations for late marriage and childbearing; the Report suggested that "the Central Committee should reconsider the suggestion of lowering the marriage age. If the change is inevitable, some relevant measures should be adopted."

(6) In encouraging each couple to have only one child, we must provide good health care to women and good education to children. It was recommended that the health and education departments strengthen women's health care, popularize new methods of delivering babies, and promote the development of relevant medical services, scientific research, and education so that we might reach the goals of fewer births, taking good care of children, and offering them good education.
(7) There were suggestions related to rewarding parents who had only one child. The positive influence of such rewards had been acknowledged in all parts of the country in the previous year. Considering that some enterprises and public institutions, administrative units, jobless urban citizens, rural communes and brigades were short of funds, it was suggested that the central and local government financial departments should supply them with the reward funds.

The main Report was only about 5600 words. Since it was not long, many issues could not be elaborated within it, so five appendices were attached: Current Population Problems in China; Proposal of the Plan that Each Couple Have Only One Child; Family Planning Law in the People's Republic of China (fifth draft for discussion); A Survey of Population Agencies in the Philippines, Thailand, Malaysia, and Singapore, and Exploration of Family Planning Insurance in the Rural Areas. Among these, Appendix Two was most directly related to advocacy of the one child policy; in fact, it was a further explanation of the main Report on this subject. I was the author of this second Appendix, so the leader of the symposia told me to sign it before submitting it.

Appendix Two contained three parts. Part One noted that it was imperative to implement this policy, emphasising the analysis of the current population situation and future population trends.

Part Two suggested that "the advocacy of the policy of only one child per couple might last until around 2010" and that we must "tell the people explicitly that the 'one child per couple' policy would not be in place permanently. It would be applied for only a certain period of time ... about one reproduction cycle" by which time "there will no longer be any serious problem with the general population, the working-age population, the dependency rate, or the level of aging."

Part Three suggested that adjustments be made to the fertility rate after 2010. According to the analysis, although in 2010 the total population would be higher than the current one, its growth tendency will have slowed. While the working-age population would also be greater than currently, its rate of decrease will have quickened. Though the dependency rate would be lower than the current level over the long term, radical changes would occur to the dependent population and the aging index would rise rapidly. "Therefore, to avoid and alleviate such circumstances, the 'only one child per couple' plan can only be implemented until around 2010 … After that, the average fertility rate should be raised slightly, so that the above-mentioned problems will not become too serious and the entire population size will be stabilized at a near-ideal level." These suggestions, and the analysis behind them, made 40 years ago, can be considered reasonable, practical, and farsighted.

In sum, as I see it, the one couple, one child policy was introduced in 1980 against a special historical background of contradiction between a shortage economy and a surplus population. Double measures were taken to resolve this contradiction: for one thing, economic development was boosted and supply increased. For another, population growth was restrained to lighten the consumption load. It was essential to solve the contradiction from both production and consumption sides. Practice proved that such approaches were effective, enabling the national economy to develop rapidly and continuously over a long period. We reaped abundant demographic profits and demographic dividends in the "golden age" brought about by population structure change. It is undeniable that the conditions of people's lives improved massively, from starvation to subsistence then to moderately prosperous. Now, a well-off society in an all-round way is around the corner. These achievements are chiefly the fruit of the reform and opening-up policies, but we should not forget that population policy and management were part of that broad reform project. The one-child policy and adjustments made since the 2010s were elements of the reforms. In other words, had there been no reform and opening-up, there would be no one-child policy, nor the adjustments made to it in the 2010s. In general terms, we can observe that population policies since the 1980s were basically correct and brought mainly positive influences and benefits. Was there any negative side? Yes, there were some conspicuous drawbacks that had a negative impact. For example, the problems confronted by the families with only one child were not dealt with adequately. The old-age social security system plan was not fully established. The regulation permitting rural families to have a second child if their first was a girl helped to raise the gender ratio. Although it was time to adjust the population policy, there was no signal of policy change until several years had elapsed. As mentioned, these issues had been anticipated when the policy was introduced in 1980 and measures were also suggested to deal with them, yet these were not completely carried out and the policy changes were somewhat belated. Only the negative outcomes were realistically assessed and investigated could some practical measures be implemented for an optimal remedy for the situation. In particular, the individuals and groups that have been harmed during the application of the policy should be compensated both economically and emotionally.

For 30 years, the one couple, one child policy remained in place. In 2013, it was altered to allow married couples to have two children if one parent is a single child. In 2015, the two-child policy was introduced for all. Behind all the policy adaptions were changes to the marginal costs and benefits of having children. Why were rewards given to parents with only one child? The purpose was to increase the benefit that an only child could bring to their parents. Why were parents fined if they violated the policy? In order to increase the costs of additional children. The aim was to induce people to comply with the policy, to increase the rate of only children among new births and to restrain population growth effectively through regulating the costs and benefits of having children. In the 2010s, the post-demographic transition proceeded to the middle stage and the time to adjust the population policy had arrived. At that time, China had reached the middle income level, and the costs and benefits of children had been through a fundamental transformation, from stimulating fertility to inhibiting it. It was the right time to relax the fertility policy because it would no longer spark a new baby boom. Along with the introduction of the cost-benefit theory of children, in this book, I propose the concept of "social elements in the cost–benefit theory of having children" and elaborate on it both theoretically and practically to facilitate a theoretical understanding of the actual population policy.

Part Three of this book, Prospects for Population Policy Adjustments, has three chapters; Chap. 8: General Population Trends and Developments in the 21st Century, Chap. 9: Population Policy and the Population Development Strategy, and Chap. 10: Population Policy in the Future.

Chap. 8 is about the new trends in population change.

After the one child policy was implemented, the percentage of people between 15 and 64 reached a peak in 2010 and the proportion of the total dependent population (old people aged 65 and above and children aged 0 to 14) dropped to its lowest point. The demographic profits and dividends brought by the "golden age" of population age structure change have peaked. Then, the ratio of the working-age population showed a decline. After 2013, the absolute numbers of working-age people also started to decline. In contrast, the proportion of dependents, especially old people, rose rapidly. The reduction of demographic profits and dividends accelerates and a new population development appears.

The size of the population is new: though at present the population is increasing slowly, both the birth rate and natural increase rate have decreased to very low levels. It is predicted that population will reach zero growth around 2030. After 2013, a couple that included one parent who was an only child was permitted to have two children and in 2015, the two-child policy was applicable to all; both changes influenced the birth and natural increase rates. They first dropped and then rose slightly, dipping to 12.07 and 4.96 per 1,000 people in 2015, rising to 12.43 and 5.32 per 1,000 people respectively in 2017,[9] and then dropping again slightly to 10.94 and 3.81 per 1000 in 2018.[10]

Population capacity, or quality, is also new. For some time, people have interpreted and defined the term in different ways. What is important is that a concept have clear connotations and denotations. A definition should contain quantitative indexes that can be used for assessment. The connotations and denotations of certain terms, such as a population's cultural capacity or its political and moral awareness are arbitrarily defined. My definition of population capacity combines three elements: physical fitness and good health, intelligence and scientific training, and cultural and moral fitness. What I mean is that population capacity is part of humankind's special quality, showing different physical, intellectual, cultural and moral potential. This being the case, there should be quantitative indexes to demonstrate that population capacity can be measured. It is now generally agreed that physical fitness, health, and education levels have improved substantially. Yet compared with these capacities in the developed countries, China is still some way behind. Comprehensive population

---

[9] National Statistics Bureau of the PRC. (Ed.). (2016). *China Statistical Yearbook of 2016*. China Statistical Publishing House. p. 34.

[10] [EB/OL] https://data.stats.gov.cn/easyquery.htm?cn=C01.

capacity enhancement is an urgent matter. The solution lies in deepening the reforms across the board. In other words, we need to deepen our understanding and intensify efforts to promote increased population capacity. To promote good health, medical reforms should focus on providing high-quality medical services. Sporting reforms should focus on improving physical fitness levels. To improve education attainments, the focus should be on reforming the education and technology sector, specifically, breaking the examination-oriented system. Opinion is divided about whether people's cultural and moral fitness has improved. The emphasis of future reforms should be on helping people develop a correct outlook on life and correct values, oriented towards the ideal of prosperity for all.

Changes to the population's natural structure are also new. With the one-child policy in place for more than 30 years and control of population growth intensified, big changes have taken place in population structure, especially in age and gender structures. The key to deepening reforms comprehensively is to understand the role of population structure adjustments in the population strategy and policy. I do not approve of insisting on population size control while regarding age and gender structures as insignificant, thereby ignoring imbalances in them. But neither are age and gender structures our primary concern. If we pay no attention to adjusting population structures then there will be imbalances, exerting irreversible adverse influence and creating further contradictions between the population and socio-economic development. If we take population age and gender structure adjustment as our first concern, we will lose control of other aspects of the population question. For example, to avoid aging, we encourage new births and raise the fertility rate. When the new generation becomes old, there will be a new wave of aging. Then the plan to resolve the population problem reasonably and comprehensively will fall through. Thus, we should not take stopgap measures to settle the aging of population age structure. Instead, we can only settle it *within* the general population development strategy. A solution can be secured only through concentrating on the key point while simultaneously maintaining an overall balance and coordinating other elements at the same time.

Chap. 9 is about the shift in focus of the population development strategy and adjustments to the population policy.

Since the years of reform and opening up, I have elaborated on China's three-step population development strategy that involves size control, capacity improvement and structural adjustment, with its different focus in different periods. The first focus was on population size control, with some consideration for the other two aspects. The second focus was on improving population capacity in an all-round way, while also considering size control, structural adjustments, and the sustainable development of population and resources, environment, economy and society. The goal of the third step is to achieve an all-round optimum population, meaning optimal size, high capacity, and reasonable structural distribution, along with the harmonious and sustainable development of the population's own change, as well as in relationship to the natural and social environments.

At time of writing, the PRC has existed for more than 70 years and the reform and opening-up policies have been in place for more than 40 years, the same period of the strengthening of the population policy. In the dramatic changes to the economy, politics and society, the population development strategy and policy have gradually matured, showing different characteristics in different periods. Presently, it is in the middle period of the second step. With the two-child policy in place for some years now, we are only one step away from the international community's concept of family planning. We are closer and closer to a population development strategy and policy that focuses on the ultimate goal of achieving an all-round optimum population. This second stage is a decisive step that connects the first and the third steps, and it will exert a deep influence on the future population development strategy and policy.

The second step is focused on comprehensive improvement of the population's capacities. We should have a clear understanding of this point. Why should the focus of population development strategy and the current population policy be shifted to comprehensive improvement of population capacity rather than to distribution adjustments? Why should we not continue to focus on restraining population growth?

There are two reasons. First, the shift in focus is determined by the natural laws of population change and development. In 1991, the fertility rate dropped below replacement level, then the intrinsic natural increase rate became negative, and its increase potential has weakened. If we maintain the current fertility rate over a long time, the total population will gradually achieve zero growth or experience negative growth. Thus, to shift the strategic focus from size control is an objective requirement of a natural law of population reproduction. Where should it be shifted to? The academic community is divided on this question.

Some scholars think that the focus should now be on population structure change, especially the age structure, on the following grounds: 30 years of the one-child policy saw the birth rate decline, resulting in a shrinking working-age population, labour shortages and serious aging. So, they argued, "it will be too late if we do not start to adopt policy that promotes reproduction now". Yet this view still relies on raising the fertility rate via policy; they want the population policy to be shifted from birth control to rewarding fertility. The suggestion has some merit. Adjustments to fertility policy in 2013 and 2015 took the situation into consideration. The difference is, however, that the existing policy adjustments were designed to relax the one-child policy rather than to encourage and reward fertility. We need a clear answer to the question: is now the time to stimulate population growth because we are already short of population and labour? I do not endorse encouraging reproduction right now for I think the time has not come yet for the view "it will be too late if we do not start to adopt policy that promotes reproduction now". Instead, I think the aim of current policy adjustments is to let natural persons have the right to decide how many children they want and the spacing of their children. I would like to ask the supporters of pronatalist policies what size the Chinese population and the labour force should be and why? It is true that a certain size population is required if we are to make population age structure adjustments. Yet how should we decide on the quantity? According to what standards? In 2017, people over 65 in China made up 11.4% of the total population, which took till 2019 to grow to 12%, when China reached the middle stage of aging. In these circumstances, it will not do to overstate the aging problem.

Aging is a global demographic trend, and China is no exception. It is the outcome of a decreased birth rate and increased life expectancy. What we want is to control the degree of aging rather than eliminate it, and to hope it does not break through the level that the society can support. Thus, when adjusting population policy, we should avoid repeating a historical mistake. We raised the fertility and birth rates to keep the pace of aging under control, not to boost the size of the total population. We should first recognize aging as inevitable and then control the degree of aging to keep it within a reasonable range. Objectively, the present changes to the general and the working-age populations are natural changes, a new normal. Current and future trends have not gone beyond earlier predictions and they basically follow an expected trajectory. Aging is an essential stage we must go through to complete the second step of our population strategy and move on to the ultimate goal of the third step. What is a reasonable level of aging? Aging is a constantly developing concept. With the rise in life expectancy, the standards for measuring aging are constantly being raised. It is a fact that aging is positively correlated with socio-economic development. China's Two Centenary Goals suggest that by the mid-21st century China will reach the level of the moderately developed countries. So, with the general aging level of the developed countries as the upper limit, it is reasonable if we do not exceed it by then. It will be both safe and feasible to take that level as the limit and to adjust our fertility and birth rates policy accordingly. The strength and the rhythm of the policy adjustments should have in mind the end goal at a certain stage of population development strategy.

Second, changing the focus of population development strategy and policy to comprehensive improvement of the population's capacities is an objective prerequisite to high-quality economic and social development. There are two main ways of developing production: increasing labour and improving productivity. In the early reform and opening-up period (1978–2012), China's economy was developed mainly through labour increases, following the extensive growth model that depended on increasing investment and quantity.

In 2013, the Third Plenary Session of the 18th CPC Central Committee decided to deepen reform and opening-up across the board, ushering in a new era of high-quality, all-round development that depended mainly on improving productivity. The key to this improvement was scientific and technological improvement and the basis of the development was the promotion of education. Finally, it also depends on all-round improvement of the population's skills and capacities in various areas. We should integrate population capacity improvement with deep reforms in health care, science and technology, education, sports, and social programs. Continuous reinforcement of human capital accumulation will promote further, stable improvements in labour productivity and high quality economic and social development.

Now, the Chinese population development strategy and population policy is in its second step of development with its focus on improving population capacity in an all-round way. I have always been convinced that socio-economic development is the foundation for population change and development. Population policy provides the external conditions that facilitate or hinder change and development, and its role should not be neglected. Yet, policy is also limited, so we must draw lessons both positive and negative from a careful review of the population strategy and policies implemented in the past and presently. More importantly, we should look to the future and search for strategies and policies that are scientific and realistic rather than imagined, and integrate them into the goal of the great rejuvenation of the Chinese nation.

Chap. 10 is about choosing Chinese-style family planning.

I summarize the third step of population development strategy as creating an all-round optimum population in terms of size, physical fitness and health, education levels, cultural and moral fitness, and reasonable age and gender distribution. In addition, it means sustainable development among population and natural resources, environment, economy, and society. Starting from the realities of Chinese population history and the current situation, especially since the introduction of the two-child policy, I make the following suggestions:

First, population policy should not be abandoned. In recent years, thc past policy has been questioned and criticised. Some suggest that we should completely abolish the population and fertility policy and allow people to have children when and how they want. I agree that people should be permitted to have children autonomously and freely, but not that the policy should be abolished. Our 30+ years of family planning policy offer lessons we can learn from, and some remaining problems that need to be settled. More importantly, we should look to the future of population development. We are presently in the second stage of the population development strategy and we should advance the third, whose goal is to achieve an all-round optimum population. In the process, we should pay attention to the shifts in focus at the different stages of development. Population policy is needed to guide, facilitate, and guarantee the development. Besides its function of restraining population growth and encouraging family planning, there are other functions to the population policy. It serves the strategy, the route and the guidelines for population development and functions as a tool to achieve population development goals in the various periods.

Second, restrictions on the number of births should be lifted gradually. The focus of population policy so far has been on population size control. The number of births constitutes the major index for measuring the application of the policy. Now that we have a two-child policy, lifting all restrictions on the number of births is around the corner. Recent history has shown that the two-child policy has not caused a big rebound in the fertility rate. According to several surveys of people's fertility desires, most families prefer to have only two children, and only a small portion choose to have three. We are justified in saying then that lifting all restrictions will not trigger a big leap in the fertility rate. This is a revolutionary turning point—population policy no longer needs to revolve around controlling the number of births and naturally shifts to providing services to people.

Third, Chinese-style family planning should be carried forward, step by step. Chinese-style family planning should restore reproductive rights to people, end restrictions on birth numbers and allow people to have children of their own free will. What then is the task of a population policy? It is to provide services, in an all-round way: first to pregnant women and women in labour, and newborn babies, and to facilitate sound child rearing and reproductive health. It should also help to improve children's physical fitness and health, their education levels, and their cultural and moral fitness. Further, it should also serve to settle social problems associated with the population question. For example, it should provide relevant services for certain groups of people when they go to see a doctor, go to school, or seek employment. It should facilitate urbanization and regional economic development. It should also help to provide social security services for the elderly and cater to old people's needs in comprehensive ways. The services should adapt to the characteristics of aging in China, such as its fast pace, its high prevalence, and the regional imbalances. Traditionally, family members take care their elders at home, and this is the main way of caring for the elderly in China. Chinese-style services to the elderly are likely to have both traditional and modern features: a combination of home care and social support, for example. With the promotion and deepening of reform and opening-up, Chinese-style family planning will blaze a bright new path from its own experience, and incorporation of other successful experiences and practices around the world.

References

Literature Research Office of the CPC Central Committee. (Ed.). (1999). *Selected Writings of Mao Zedong* (Vol. 7). People's Publishing House. p. 228.

National Statistics Bureau of the PRC. (Ed.). (2016). *China Statistical Yearbook of 2016*. China Statistical Publishing House. p. 34.

The National Population and Family Planning Commission. (Ed.). (2007). *Population and Birth Control History in China*. China Population Publishing House. p. 44.

[EB/OL] https://data.stats.gov.cn/easyquery.htm?cn=C01.

Contents

Abbreviations

CPC	Communist Party of China
GDP	Gross Domestic Product
TFR	Total Fertility Rate
the PRC	the People's Republic of China
UN	United Nations

List of Figures

List of Tables

List of Pictures

Part I
The Development of Population Policy despite Controversies

Chapter 1
A Historical Overview of Chinese Population Policies

As a popular Chinese saying goes, "If the name is not correct, the words will not ring true." This book is about historical changes to population policy in the People's Republic of China (hereafter the PRC), so the first thing we need is some sound explanation of key terms relating to policy. With reference to other scholarly research, together with my understanding of relevant concepts from my own research experience, policy is defined here as a set of action guidelines developed by a state or a political party in order to accomplish certain political aims or tasks, in certain political periods. The following sections present an overview analysis of population policies throughout Chinese history.

1.1 Encouragement and Restraint in Population Policy

Population policy generally refers to policies concerning population quantity and quality, changes to population structure, and population-related development. Since fertility directly affects population numbers and its influence is particularly noticeable, fertility policy is sometimes also referred to as population policy. Yet, it should be noted that fertility policy is population policy in its narrowest sense, and cannot represent it in the broad sense. Policy is usually interdisciplinary and overlapping. Population policy in its broadest sense involves social and economic development, various sectors and categories like resources, environment, and so on. In its narrow sense, fertility policy not only determines the tendency and process of population change, but also involves certain aspects of the general population policy. The former and the latter interact and affect each other (which must be kept in mind in research activity).

Population change and population policy under certain historical circumstances often reflect prevailing social and economic conditions, the level of productivity and

X. Tian, *Population Policies in the People's Republic of China*,
https://doi.org/10.1007/978-981-96-9658-1_1

the nature of production relations at the time. China is one of the four ancient civilizations of the world. Chinese people began to reproduce along the Yellow River basin more than 4000 years ago, nourishing a celebrated Chinese culture. Among the numerous treasures of this traditional Chinese culture, a deep belief in "fertility and fortune"—the core of a Chinese understanding of population production—has survived and remains popular. It influenced not only the formulation of feudal ruling class population policies but also common people's views about procreation, marriage, and family. This belief was a unique feature of human reproduction in China, and we cannot understand the population policies during the ancient feudal dynasties unless we know the ins and outs of this characteristic.

1.1.1 Population and Reproduction Worship

More than two decades ago, I visited Cologne Cathedral together with a local professor while on an academic exchange in Europe. After the visit, my companion asked me: "Professor Tian, have you noticed some architectural features associated with demography during your visit?" I was so confused that I could not answer his question, so I asked him for the answer. He asked: "Did you notice anything special about the design of the cathedral's gate and the building in general?" I was still at loss, so I pressed for more details. Finally, the professor said, "The gate of a cathedral is different from those in China. Usually, a gate in China is square, with two posts supporting a ledger strip. But the gate of a cathedral here and in many other European countries takes the shape of a herringbone, with several layers of engravings within, looking like a cracked walnut without its kernel." He slowed his speech and further explained, "The gate also looks like a female's cunnus, with its upper part resembling the clitoris and the vaginal orifice underneath, like a gate of life, and a gate of holiness. Such an interpretation is anything but vulgar. The design of such a gate, I guess, evolves from reproduction worship." I was totally astonished by his remarks. I have seen reproduction worship sculptures in Italy, France, Austria and elsewhere in Europe, most of which are physically explicit. I have also visited dozens of well-known European cathedrals, but this was the first time I had heard such a particular interpretation. However, when I closed my eyes and recalled the churches I have visited, especially the cathedral gates and huge stained-glass windows, I realized that most of them have similar features to the gate of the Cologne Cathedral. Few visitors would recognize the implicit reproduction worship in the elaborate gate designs.

Reproduction worship has existed in China for thousands of years, more than European countries like Germany, France, and Italy. Evidence of reproduction worship and reproduction worship culture goes back to the prehistoric times of Three Sovereigns and Five Emperors. According to the legends, the Eight Diagrams of the Bagua, which contains connotations of reproduction worship, was created by Fuxi, one of the Three Sovereigns, to reveal the laws that govern changes in all things and the interdependence of heaven, earth, and man. In particular, it reveals the laws of human reproduction, and of the development of the relationships between humans and nature. Human reproduction determines the laws of development and changes thus the image of reproduction worship is at the centre of the Bagua.[1] Guo Muoruo (1892–1978), a prominent Chinese scholar, maintained that the arrangement of Yin and Yang in the form of the Yin Yang fish in the middle of the Eight Diagrams symbolizes the union of man and woman, reflecting reproduction worship. The Yin Yang Fish at the center of the Bagua, the evolution of Eight Diagrams and numerous variations of Eight Diagrams such as Twenty-Four Hexagrams, Sixty-Four Hexagrams, Five Elements and Eight Diagrams, Yin-Yang and Eight Diagrams and so on, are all based on the interactions of the Yin Yang Fish with the different arrangements of Yin and Yang. Compared with other reproduction worship forms that explicitly display human reproductive organs and behaviour in sculptures, engravings and portraits, the reproduction worship represented in the Eight Diagrams seems more dynamic and essential, being an abstract and refined form that dispenses with physical objects, and suggests an integration of implicitness, profundity, comprehensiveness and science.

Another legend relates the custom of keeping a piece of deer skin as a token for the control of matrimony, also initiated by Fuxi, who had begun the notion of a practice of high-quality procreation. Fuxi lived in the transitional period between a matriarchal and a patriarchal society, when group marriage was practiced within tribes. When he saw the drawbacks of consanguinity, which had hindered population reproduction and harmed people's health, he determined to reform matrimony. According to extant research, Fuxi and his tribe lived in a place called Tianshui (in Gansu province today), an open forest with abundant deer at that time. They stipulated that a deer skin would be cut in half when a baby was born. While the left half would remain with the tribe, the right half would be given to the new-born baby. Later, when a couple wished to marry, they would be obliged to produce their piece of deer skin. If it matched the skin half held by the tribe, then consent would not be given to their marriage because the matching skins indicated that they were "cousins". If their deer-skin pieces did not match, or one party did not own the deer skin, then they would be permitted to marry. Gradually, the rite and custom of extra-tribal marriage was established based on this practice of control through deer skin tokens. As the custom spread, it evolved into a system of exogenous marriage. Though not a system of monogamy, it enabled a tribe to thrive and laid solid foundations for conquering and annexing neighbouring tribes. Fuxi was the first person to advocate idea that "marriage between a man and a woman within the family clan will produce no strong and healthy offspring". Reproduction worship, together with some of the earliest instances of producing sound offspring, contributed to power and prosperity of Fuxi's tribe.

Why in primeval society, in particular the Neolithic period, did reproduction worship occur and why was it displayed with such enthusiasm? It is generally believed that reproduction worship began with the admiration of the female: people started to adore women and their reproductive organs when they saw that babies come from their bellies; they were amazed by female reproductive power of fertility because

[1] For more information about the Eight Diagrams see *The Book of Changes* (*Yi Jing*).

they were ignorant and knew no science at the time. But while such an explanation makes some sense, it is unconvincing. Did reproduction worship disappear following scientific advances? No, reproduction worship continues to exist. What then is the explanation? We take up this question again later in the book; for now let us say that the cultural significance of fertility must be investigated in the context of human production, and contemporary social and economic development, if we are to understand the long history of reproduction worship.

Though tools of production and weapons used in wars in the Neolithic period had significantly progressed since the Paleolithic period, the reality that human strength still constituted the basic power had not changed. Human strength was determined on the one hand by the number of people available, so a tribe was stronger the more heads it counted. On the other hand, human power was determined by the quality of the population, so a tribe was stronger if its members had strong constitutions. When a tribe happened to be more numerous and those people happened to be stronger, it would become powerful enough to conquer other tribes.

Fuxi lived in a transition period between a matriarchal society and a patriarchal one, and after his death, reproduction worship shifted to focus on men. This shift had a tremendous impact on the marriage system, and fertility activities, as well as population changes in feudal China. In various places, a growing number of sculptures, engravings, and paintings expressing male reproduction worship have been discovered. Apparent reproduction worship of males has also been traced in the history of Chinese characters, astronomy, legislation, and so on. Take the Chinese character "祖"[2] for example. Originally, only the right-hand element ("且") of the character represented the currently used character "祖". The right part ("且") looks like an erect penis, doesn't it? People in many Western countries believe that God is the human ancestor and that "He" created humans. As China became patriarchal, reproduction worship also evolved and patriarchy was reflected in the duty to worship and honour male ancestors and to father as many male heirs as possible, a tradition in line with ethics in feudal China.

1.1.2 Arguments in Favour of Population Growth in Chinese History

When talking about feudal ethics in China, we cannot avoid Confucius and Confucianism. His views on population questions have been studied by scholars in the fields of demography, history, sociology, and economics. In general, much extant research maintains that he was a pioneer, who had carried forward the tradition of reproduction worship, and advocated both natural population growth and encouraging immigrants from other states. However, there is a lack of convincing elaboration on the status of Confucius's views on population within Confucianism and how they were integrated with the broader system of his thought. I believe that his population views are

[2] The character "祖" literally means "one's ancestor".

the starting point and foundation of his teaching practices and his political ideal of establishing peace and stability through a strong state, ideas that he developed while traveling around other states outside his native state of Lu. His views on population play an essential role in Confucianism.

The core of Confucianism is the idea of "benevolence". How is the concept connected with people and population? What role do Confucius's views on population play in this core concept of benevolence? In *the Analects of Confucius*,[3] Yan Yuan, an accomplished disciple of Confucius, asked him about benevolence. The Master said, "To return to the observance of the rites through overcoming the self constitutes benevolence. If for a single day people could return to the observance of the rites through overcoming themselves, then the whole Empire would be considered to be benevolent." This means that if one wants to be benevolent then one must first be strict with oneself in observing the rites. Only when someone overcomes himself could he really observe the rites. And what are "the rites"? In the same chapter of *the Analects*, Duke Jing of Qi asked Confucius about government, to which Confucius answered, "Let the ruler be a ruler, the subject be a subject, the father a father, the son a son." In other words, rulers, subjects, fathers, and sons should behave as required by morals; only then can they observe the rites. "The rites" consist of rules for how to treat parents before and after their death. As Confucius told one of his disciples in the *Analects*, "When your parents are alive, comply with the rites in serving them; when they die, comply with the rites in burying them and in offering sacrifices to them." Thus Confucius not only attached importance to complying with the rites in serving our parents when they are alive, but also called particular attention to complying with the rites when our parents die. This is Confucius's interpretation of filial duty. How does a family manage to offer sacrifices to their ancestors and keep the incense burning continuously before the ancestral tablets? The answer lies in securing the line of succession; thus can we pass down the mission of offering timely sacrifices to ancestors unfailingly.

Confucius lived more than 2500 years ago, when both fertility and death rates were high, and the population growth rate was low. In addition, wars were commonplace among the dukes of rival states so the death rate was consistently very high during the Autumn and Spring period (770 BC–476 BC). In order to observe the rite of offering sacrifices to the ancestors, and to keep the incense burning before the ancestral tablets, a family had no choice but to have as many children as possible. The more children were born the greater the chance that some male heirs would survive the wars. Confucius's view of the population question then was quite clear. He had once been impressed by a teeming population in another state that he had visited, and wished to see such "a teeming population" elsewhere. He also advocated implementing benevolent policies, to "ensure that those who are near are pleased and those who are far away are attracted", suggesting that benevolent policies make it possible to please the common people and to restore peace and stability to the state. These circumstances were likewise conducive to population growth, including through attracting migrants from neighbouring states. Although Confucius knew

[3] Hereafter, the quotes are from the book below. Confucius. (2008).

nothing of demography, his views on population coincide with the push–pull theory of migration, proposed 2,000 years later.

Now, I venture to suggest a formula that could clarify Confucius's view on population:

Benevolence—Rites—Filial Duty—A Growing Population—Benevolence

Here, "benevolence", both the start and end point of Confucianism, is simultaneously the core of Confucianism. How do we achieve the goal of "being benevolent"? Only through strict observation of the rites. Rites within a family and across generations are embodied through the performance of filial duty both while parents are alive and when they are dead. How can the practice of filial duty be performed generation after generation? "A growing population" is required. "A growing population" is the result of ethics and the implementation of benevolent policies both within and outside the state—starting with "benevolence" and finishing with "benevolence".

Mencius's development of Confucius's view on population is chiefly reflected in two ideas, one of which still sounds very familiar to Chinese people, that "there are three forms of unfilial conduct, of which the worst is to have no descendants". To fulfill the filial duty embodied in the observation of rites whether one's parents are alive or dead, as demanded by Confucius, the precondition is to have descendants. Therefore, to have descendants is of primary importance in the fulfilment of one's filial duty. Consequently, men who do not marry, or married men who do not have children and thus cannot perform the rite of making offerings to the ancestors are seen as not devoted to their parents. In ancient China, a married man had a sound reason to divorce his wife when she did not produce a male heir. In these circumstances, an individual, trapped in the web of family ethics, was reduced to a tool for performing filial duties, and had to have a son in order to continue the family line. Mencius' idea deeply affected traditional Chinese society and reminded families to think of their deceased ancestors. As the tradition in China goes, a male is entitled to an entry in the family tree, with his descendants following him, symbolizing the continuation of the family blood. Mencius' remarks also aroused yearning for boys, and bearing a male heir is celebrated as bringing honour to the family. By contrast, a family with no male heir will be despised by their neighbours. Their pathetic circumstance will be regarded as some kind of retribution, as if they have done something immoral. Even today, surveys about unplanned births show that a high percentage of families hope to have a second boy and a higher percentage of families have a strong wish for a boy when they already have a girl.

Another aspect of Mencius' thought is also well-known: "the time is not as important as the terrain; but the terrain is not as important as unity with the people." Starting from the idea of "benevolence", Confucius wanted to achieve the goal of a teeming population through "pleasing the ones who are near and attracting those who are far away". Mencius upheld Confucius' ideal by putting special emphasis on "the unity with the people" among the three favourable conditions—the time, the environment, and the human relationships—for achieving success. Upon seeing vast tracts of uncultivated land during the Warring States period (475 BC—221 BC), Mencius wrote: "When a vast piece of land is sparsely populated, it faces the danger of being annexed by another state. So, it is urgent for a Duke to attract people to settle in his state. When scholars settle in the land as his subjects, they become talents that can serve the government. When peasants settle there, they can reclaim the land. When merchants settle there, they can restore the prosperity of business. Hence unity among the people marks the strong power of a state and is also the condition of paramount importance that determines its power". Henceforward, the influence of Mencius' idea of "the unity of the people" is far-reaching and his remarks on how to achieve it are still commended, celebrated, and practiced today.

During the Warring States period, there were also thinkers who were against encouraging population growth. For example, Han Feizi (280 BC–233 BC), a philosopher and a leading scholar of the Legalism School wrote in his article "The Five Vermin"[4]:

> In ancient times husbands did not have to till the fields, for the seeds of grass and the fruit of the trees were enough for people to eat. Wives did not have to weave, for the feathers of birds and skins of beasts provided sufficient clothing. No one had to struggle to keep himself supplied. The people were few, but there was an abundance of goods, and so no one quarrelled. Therefore, no rich rewards were doled out, no harsh punishments were administered, and yet the people of themselves are orderly. But nowadays no one regards five sons as a large number, and these five sons in turn have five sons each, so that before the grandfather has died, he has twenty-five grandchildren. Hence the number of people increases, goods grow scarce, and men have to struggle and slave for a meagre living. Therefore, they fall to quarrelling, and though rewards are doubled and punishments are piled on, they cannot be prevented from growing disorderly.

This passage implies that "scarce goods", "a meagre living" and "the disorderly situation" all resulted from "the increasing number of people". However, those who discouraged population growth were by far a minority with few supporters in Chinese history compared with those who have encouraged population growth.

Confucius's and Mencius's encouragement of population growth and the idea of regarding men as superior to women thrived in ancient China and have been passed down because of two opportunities. The first came when Confucianism was secured as the orthodox political system of ancient China through a proposal by Dong Zhongshu (179 BC–104 BC), a philosopher and scholar versed in Confucianism, who said "The emperor should discard the hundred schools and respect only Confucianism".[5] Another opportunity was the establishment of the ethical principles of Neo-Confucianism in the Song dynasty (960–1279 AD).

During the rule of Emperor Wu Di (156 BC–87 BC) of the Western Han dynasty (202 BC–8 AD), the social economy went through a period of great development following a long time of peace and stability. Generally, Emperor Wu Di's empire was prosperous and powerful, equipped with strong military forces. However, his reign was under threat, from the strong feudal separatist dukedoms and from harassment and invasion of the border regions by the Huns, as well as the grave situation of

---

[4] The quotes are from the book below.

Han (2003).

[5] Guo and Wang (2002).

peasant landlessness as a result of land being annexed by landlords and merchants. It became urgent to strengthen central rule. Under these circumstances, Dong Zhongshu made his famous proposal:

> The current political principles advocated by hundreds of schools are different from one another and they hinder the implementation of unified policies and a nation's rise. Thus, they should be superseded by Confucianism, which advocates the rites of establishing the authorities of the sovereign, father and husband, and supports an emperor's centralized power and the theory of great unification. What's more, government officials should be selected from the capable scholars who practice Confucianism.

Following his advice, Emperor Wu Di established Confucianism as the official political ideology, while the popular doctrine of the School of the Yellow Emperor and Master Lao, which advocated government inaction, and other non-Confucian political views proposed by philosophical schools were ruled out. This was the historical event known as "discarding hundreds of schools and respecting Confucianism only". Henceforth, Confucianism formed the orthodox political principles of ancient China, its views on population included.

It is generally believed that Zhou Dunyi (1017–1073 AD) of the Northern Song dynasty (960–1127 AD) inaugurated the establishment of the ethical principles of Neo-Confucianism. His two disciples, the brothers Cheng Hao (1032–1085 AD) and Cheng Yi (1033–1107 AD) inherited and spread their master's thought and became known as the founders of Neo-Confucianism. Scores of years later, Zhu Xi (1130–1200 AD), a disciple of one of the Cheng brothers' disciples, who had synthesized the main achievements of Neo-Confucianism, created a comparatively inclusive system of Neo-Confucianism. According to Zhu Xi, changes to heaven, earth and all things under heaven were determined by the immutable and inherent "heavenly principles". Humans should "eliminate their desires and preserve the principle of Heaven".[6] People should adhere to ethical principles such as the "Three Cardinal Guides and the Five Constant Virtues".[7] The principle of the Three Guides is that "the sovereign is the guide of his subjects, the father is the guide of his sons, and the husband is the guide of his wife". The principle of the Five Constant Virtues refers to "humanity, righteousness, propriety, wisdom, and fidelity". Later, a specific ethical code called the "Three Obediences and Four Virtues"[8] was formulated about women. The Three Obediences include "obedience to Father before marriage, to Husband after marriage, and to Son after her husband's death". And the Four Virtues include "morality, proper speech, modest manner, and diligent work". These feudal ethical principles have influenced the social, political, and cultural development of China tremendously, reflected in their intangible social binding power operating through peer pressure and public opinion. They have also had a profound and far-reaching influence on population size, gender structures, education, and population policies.

After "discarding the hundred schools and respecting only Confucianism" and the creation of Neo-Confucianism, Confucianism became the orthodox ideology of

[6] Guo and Wang (2002).

[7] Guo and Wang (2002).

[8] Idem., p. 23.

feudal China. Confucius' and Mencius' encouragement of population growth also evolved into an aspiration to fertility and fortune. Indeed, these two notions are basically the same, both expressing a wish for a populous nation. Through more than 2000 years of feudal Chinese history, the aspiration to fertility and fortune was deepened and modified, and spread widely, gradually solidifying into popular conviction. The refinement and development of this view was mainly manifested in the following aspects.

First, fertility is associated with fortune, indicating the purpose of a growing population. Why do people want to have more sons? Ultimately, to enjoy more happiness. The causal link between happiness and more sons is deeply rooted, and accepted by great numbers of people for some objective economic reasons. The feudal society of China that has lasted more than twenty centuries was built on an agricultural economy. Within the span of those 2000 years, there were advances in production tools and there were new inventions too. However, the tools remained basically hand-tools and there were no technological breakthroughs that saw machines being produced. Tools were mainly manipulated by people and domesticated animals, and a few were driven by natural energy such as water and wind. Hand tools can be seen as the extension, externalization, and materialization of human physical power. Among the factors of production, labour power plays the dominant role. In the agricultural economy of ancient China, the labour force was chiefly constituted by males. The number of male family members and the condition of their physical strength determined the wealth of a family. To build an empire on the basis of an agricultural economy in a feudal society, the quantity and quality of the population and its labour force were the decisive elements of a country's power. Therefore, a flourishing family on a micro scale and a flourishing nation on a macro scale are the ultimate and decisive factors that guarantee a prosperous and powerful country. In short, more sons are closely associated with more happiness. To some extent, more sons mean more happiness.

Second, the notion of "fertility and fortune" was incorporated into feudal ethical principles. That more happiness comes from having more sons became a highly influential belief among the ancient Chinese. Confucius and Mencius's view of encouraging population growth became an essential element of the ideological system "Benevolence—Rites—Filial Duty—A Growing Population—Benevolence", with its focus on benevolence. These doctrines were established as the orthodox ethical principles of ancient China after being reorganized, supplemented and systematized by Dong Zhongshu during the Western Han dynasty, and the Cheng Brothers and Zhu Xi during the Song dynasty. Popular views like "more happiness comes from more sons" became an inviolable norm after being incorporated into prevailing feudal ethical codes such as the "Three Cardinal Guides and the Five Constant Virtues", and the "Three Obediences and Four Virtues". The belief that more sons bring more happiness became contagious, being propagated and reinforced by practices such as updating the family tree, building memorial temples to one's ancestors, erecting arches in honour of virtuous widows, and telling stories on this theme in novels and operas. As time passed, the norm evolved into a quasi-religious belief about population. "More happiness" here not only refers to the material benefits brought by children but also spiritual happiness. In ancient China, it was popularly believed that only those who were kind and virtuous would be blessed with more sons. By contrast, evil doers would be cursed. They would not have even a single son and thus would be despised for failing to perform their filial duty. This belief reproduced an internalized strong desire for procreation, especially for having male heirs.

Third, the aspiration to fertility and fortune enlarged the idea of encouraging population growth advocated by Confucius and Mencius, incorporating additional aspects. Their propositions mainly concerned population size, gender, education, and migration. The aspiration to fertility and fortune, by contrast, involves elements of marriage, fertility, family, clan, occupation, health, ethics, morals, household registration, statistics, astronomy, geography, land, head tax, and so on. Some ancient works, including three well-known books on ancient laws and regulations (Tong Dian in the Tang dynasty, Tong Zhi in the Song dynasty and Wenxian Tongkao in the Yuan dynasty), covered topics like household registration and population management. They became classic documents of demography. Some works, though not strictly demographic studies, recorded discussions about acute contemporary problems of population, reflecting the fact that demography is interdisciplinary and promoted the role of population in the economic development of society. Population, being entwined with the fundamental contradictions of feudal society, especially land, feudal land taxes, the head tax, war, and so forth, attracted wide attention and become a prominent player in social change, including the rise and decline of dynasties through Chinese history.

In modern times, the belief that more happiness comes with more sons was challenged and underwent some changes. The eminent thinker Yan Fu (1854–1921) was inspired by Darwin's law of "the survival of the fittest". He said in his essays "Yuan Qiang" and "Baozhong Yuyi": "though China possesses a large population, the people are not well cultivated. A growing population characterized by poor quality has posed a serious problem that must now be resolved." Subsequently, he proposed that measures be taken to enlighten the people and to impose restrictions on certain marriages, in order to preserve good genes for the Chinese nation, as Western nations did. Liang Qichao (1873–1929), another celebrated thinker of modern China, also incorporated the issue of population into political reform questions. In his essay "On Banning Early Marriage", Liang Qichao explained his advocacy of later marriage and disapproval of early marriage. He argued that five harms come from early marriage: it harms the parents' health, and harms the health of next generation and their growth; third, it will harm the children's upbringing, and their education; and last, it will harm the national welfare and the people's livelihoods. Drawing a contrast between East and West, he concluded that uncivilized and backward peoples tended to marry early. Conversely, the more civilized a people, the older they were at age of marriage. He suggested improving population quality through allowing marriage only at a later age. He opposed Malthus' population theory because he thought that exponential population growth was nonsense, and believed that neither the positive nor negative measures that Malthus suggested to prevent population increase would work.

Sun Yat-sen (1866–1925), a pioneer of the democratic revolution in China, also attached great importance to changes to the Chinese population. He had once talked about why the Western colonial powers could not subjugate China. He remarked:

"One reason is that their population is too small compared with that of China. In one hundred years, if their population grows very large, and our population does not, then they are sure to conquer China with their overwhelming population. By that time, not only will China lose its sovereignty and be ruined, but also the Chinese nation will be annihilated."[9] However, Sun Yat-sen did not approve of increasing the population substantially and sometimes, he was even worried about population surplus in China.

The co-founders of Communist Party of China, namely Chen Duxiu (1879–1942) and Li Dazhao (1889–1927) also expressed concerns about population problems. Their publications on the topic included an essay "On Malthus' Theory of Population and China's Population Issue" (Chen Duxiu) and "On War and the Population Problem" (Li Dazhao). Both criticized Malthus' theory of population and refuted it by presenting the facts of population growth in both China and other countries. They pointed out that the root of Chinese population problem lay in the decadent social system rather than in the number of people, which had grown faster than the economy. The population "problem" was created by the unequal distribution of the means of livelihood, thanks to the monopoly of land by landlords and the monopoly of factories and other means of production by capitalists. They also stressed that revolution was the final solution to the population problem. Moreover, they criticized Malthus for attempting to resolve the problem of unemployment by constraining population growth instead of seeking other solutions. In their opinion, Malthus' arguments were clumsy and ridiculous.

1.1.3 Historical Arguments for Limiting Population Growth

As mentioned above, the earliest proposal for limiting population growth was made by Han Feizi, made explicit in his essay "The Five Vermin", written during the Warring States period. In the next 2000 years, several other scholars also suggested that population growth be limited. For example, Hong Liangji (1746–1809), a high official and renowned scholar in the Qing dynasty (1636–1912), concluded that population growth had overtaken the growth in the means of livelihood. His view resembled that of Malthus', so he was known as "Malthus in China". Also in the Qing dynasty, Wang Shiduo (1802–1889), a geographer and scholar, had gone so far as to suggest annihilating the surplus population by killing them and using measures like late marriage and the imposition of higher taxes on families with several children to limit procreation. His suggestions echoed the positive and negative population control measures that Malthus had proposed. However, those scholars' theories and propositions had little influence on population policy in China.

Malthus's work, *An Essay on the Principles of Population*, had been published in 1798, causing such a sensation that it even affected population research in China. What merits our attention is that China witnessed an upsurge of sociologist opinion

[9] Sun (2014).

in favor of restraining population growth since the beginning of the 1900s. For instance, Chen Changheng (1888–1987), a prominent demographer and economist, published *Research on Population in China* in 1918, and *Three People's Principles and Population Policy* in 1930. The year 1930 also saw the publication of *China's Population Problems* by Xu Shilian (1896-?), a well-known sociologist, and in 1934 another book of his entitled *Essentials of Demography.* In 1932, Chen Da (1892–1975) launched the demographic supplement to the Peiping Morning Post, and had his book *Population Problems* published in 1934. In less than two decades, so many essays and books on demographic studies were published, systematically elucidating scholarly views, propositions and theories about population questions. Such a flurry of publications had not been seen before in China's demographic studies. The main arguments and propositions of the above-mentioned scholars can be summarized as follows.

First, their assertion that there was a surplus population was based on Malthus' theory that population growth had overtaken the increase in the means of livelihood. Having assimilated this argument, the Chinese sociologists maintained that China's population growth had also overtaken the country's capacity to provide a decent livelihood to everyone, and that population density had gone beyond "moderate". Chen Changheng explained in his book *Research on Population in China*:

> The circumstances like a small population or a large population in a country, or a country with very high or very low productivity will all be detrimental to an individual's health, a nation's flourishing, a country's prosperity, and progress of civilization.... Human procreation resembles the way that a farmer sows seeds. If the seeds are sowed sparsely, the harvest won't be bumper. If they are sowed too densely, the seedlings will be too crowded to survive. The best way of sowing is to sow seeds moderately, neither too densely nor too sparsely. This done, the robust seedling is surely to appear.[10]

Chen believed that China's population density at that time had gone beyond "moderate population density" and argued that, "from the size of one's own family, one will easily find that one generation is outnumbered by the next generation at a rapid speed"[11] and concluded that the country's "population is so dense that overpopulation has emerged in China."[12]

The theory of surplus population was deeply influenced by Malthus' theory of population but some of the sociologists rejected his argument that while population increases in a geometrical ratio, food increases arithmetically. Xu Shilian argued: "Contrary to what Malthus has said, food can increase beyond the arithmetical progression because wild land can be reclaimed, and new land can be discovered. What's more, as science and technology advance, farmers can utilize new technologies to improve productivity. In the meantime, they can also tap new natural energy like hydroelectric power. More progress in other aspects has also stimulated the rapid increase of food in many civilized countries within the last two hundred years". Moreover, he said, "population does not always increase in a geometrical

[10] Chen (1918).

[11] Idem., p. 62.

[12] Chen (1918).

ratio."[13] So, while the population theories espoused by the sociological school in China were closely associated with those of Malthus, their views differed from his in some ways, and should be carefully differentiated from them.

Their second main proposition, first suggested by Chen Da, was to restrict population size and to improve its quality. "Population size and population quality are in opposition to each other."[14] A rapid growth in population would certainly result in a decline of population quality. Conversely, a decrease in birth rate and population size will contribute to improvement of the quality of the population. He stated that "a family with few children will be less burdened economically and the mother will enjoy better health. In a family with a few children, each child is likely to enjoy ample benefits in many aspects. In this case, those children are more likely to benefit the society when they become outstanding citizens after they reach adulthood."[15] To deal with the adversities of an ever-increasing population and decreasing territory in China, Chen Changheng suggested that in the first place, "government should reinforce the national defense system and be prepared for coming wars to ensure that not even a tiny piece of land be lost. Second, it should promote industrial development to take advantage of the favourable geographical conditions, and research into science and technology should be rewarded. Third, the practice of later marriage and having less children should be encouraged so as to create a nation of strong constitution and sound intelligence through mass education. Mere population increase does not necessarily create a strong nation."[16] Chen approved of the popular idea of having one son and one daughter, or two sons and two daughters per family. As a popular proverb in China goes:

> "Parents will be worried if they have only one son. It will be much better if there are two sons in a family. If there are three sons, the family will be divided into four small families after the sons marry. If there are too many children in a family, some will turn out to be ungrateful. If a family happens to have one son and one daughter in a family, the parents are fortunate. If a couple happens to have no children, they will live happily like the immortals."

In Chen Changheng's opinion, one son and one daughter are enough to care for their parents when they become old. Two sons and two daughters can also fulfill the obligation of caring for their old parents. If there are more than two sons and two daughters, then the pathetic situation may arise in which some children might want to shirk their responsibility to attend to their aging parents.

To enhance population quality, many scholars supportive of limiting population growth advocated for the development of mass education, the advance of science, the practice of eugenics, and the launching of public health campaigns. If all this basic social work were accomplished, then it was certain that population quality would improve and the number of the new-borns would decline.

Their third argument was in favour of birth control, which they regarded as the fundamental solution to the population problem. According to Xu Shilian, there were

---

[13] Xu (1930).

[14] Chen (1934).

[15] Idem., p. 267.

[16] Chen (1918).

two solutions to the over-population problem, one temporary, another fundamental. The former involved measures such as expanding agriculture and industry, increasing employment, encouraging emigration, practicing eugenics, promoting public health, and so on. The fundamental solution involves measures such as encouraging later marriage, discarding the traditional belief of "more happiness brought by more sons" and launching a nationwide campaign of birth control. Another scholar, Chen Da, endorsed Xu's propositions. For Chen Da, only birth control programs could effectively solve China's population problem, since achieving the desired effect through emigration and industrialization was too difficult and would take too long.

The above is a summary of the main theories and propositions advanced by the sociological school from the 1910s to the 1930s in China. It is not hard to see that these theories and arguments were profoundly influenced by Malthus, whom some scholars held in high esteem. However, some of their views also differed from those of Malthus. They had their doubts and criticisms, for example, about Malthus' claim that population increases geometrically and food increases arithmetically. Their "moderate population density" proposition, though perhaps unconvincingly presented, was groundbreaking. It was also significant from the perspective of demographic studies to associate the changes and developments in population with natural resources, volume of land under cultivation, agricultural development, and urbanization. It is also noteworthy that their proposal to limit population size and to enhance its quality grasped the essential element of the population problem. In the century that followed, this proposition was universally approved not only by demographers, but also by people in other sectors of the society, becoming a key topic of population debates.

1.2 Various Interventions Reflected in Population Policy

The reason for this discussion of how people thought about population in Chinese history is that social policy is always a reflection of prevailing mainstream thoughts in society. We cannot have a sense of population policy during feudal empires unless we have a thorough understanding of the idea that more happiness comes with more children, and how it was embraced in feudal China. Views had been expressed in opposition to this mainstream thought since Han Feizi's criticisms during the Warring States period and that critique went on developing, represented by Hong Liangji and Wang Shiduo in Qing dynasty and by the sociological school that advocated limiting population growth in the first half of twentieth century. But in contrast with the population policy in place for more than 2000 years, this remained a minority view. Ideas about population held by reformists and revolutionaries in modern China were only modifications of traditional beliefs, proposed as ways to save the country and the people when China's door was forced open by imperialist powers. Those views were seldom free from the constraints of traditional beliefs such as the expectation of a populous nation, and more happiness comes with more sons.

In policy terms, the proposed reforms were reflected in direct and indirect measures that influenced fertility policy. To see the whole picture, some typical policies will be illustrated chronologically.

1.2.1 Policies that Had a Direct Influence on Reproduction

In the Spring and Autumn period, Goujian (?-464 BC), Duke of Yue, had undergone self-imposed hardships so as to strengthen his resolve to obliterate the national humiliation suffered after defeat at the hands of Fuchai (?-473 BC), Duke of Wu. Goujian kept a low-profile and introduced a policy of "propagation of the population in ten years, and ten years of military training" to prepare to take his revenge. "Propagation of the population" could be achieved in two ways, through encouraging natural growth and immigration. Goujian formulated three regulations concerning natural growth; first, the ages of a bride and her bridegroom should be similar, to encourage fertility. In other words, young men should not marry old women, nor should old men marry young women. Second, the parents of young men would be punished if their sons did not marry after turning 20 years old. Similarly, the parents of young women would also be punished if their daughters did not marry after turning 17 years old. The third new regulation stated that when a boy was born, his parents would be rewarded with two bottles of liquor and a dog. When a girl was born, her parents would be rewarded with two bottles of liquor and a pig. If a couple had two children, they would receive subsidies, and if they had three children, they would enjoy some favourable policies introduced by the government. With policies in place that rewarded those who had children and punished late marriage and those who did not marry, the state of Yue became prosperous, with a flourishing population, and Fuchai, the Duke of Wu, was defeated. Those policies may have served as references for later emperors who wanted to encourage population growth in the succeeding dynasties.

1.2.1.1 Reductions and Exemptions of Citizens' Taxes

During his reign over the Western Han dynasty, Emperor Gaozu (247 BC–195 BC) decreed that any subject who had just had a baby boy would be exempt from the standard corvee labour for two years.[17] In 85 AD, Emperor Zhangdi (56–88 AD) of the Eastern Han dynasty issued an edict that a couple who had just had a baby would be exempt from the head tax for three years, and a pregnant woman would be rewarded with three *hu*[18] of grain and her husband would be exempt from the head tax

---

[17] Ban (2007). Ban Gu (32–92 AD) was a historian of the Eastern Han dynasty.

[18] "*Hu*" is a measure used in ancient China, equal to, during the Han and Tang dynasties, approximately 60 kg.

for one year.[19] Along with the rewards, here, too, there was a policy of punishment. In 189 BC, Emperor Huidi (210 BC-188 BC) of the Western Han dynasty decreed that unmarried women between the ages of 15 and 30 must pay five times the usual head tax[20] and would continue to do so until they had children. Pregnant women were not only exempted from the head tax for one year but were also rewarded for being pregnant with the payment of three hu of grain. These policies were clearly designed to increase the birth rate.

At the beginning of the Qing dynasty, the harsh land annexations carried out at the end of Ming dynasty had adversely affected the consolidation of imperial power and national development, therefore the land tax system had to be reformed. During the reign of Emperor Kangxi (1654–1722), the total amount of head tax was fixed, and the policy that new-born babies were exempt from it was introduced to lighten the load of common people. But then since some local officials and despotic gentry took advantage of certain loopholes in the regulations, the head tax had to be reformed. After many years of trial, in 1723 Emperor Yongzheng (1678–1735) decreed a new system of taxation according to which all taxes were rolled into an assessment of landholdings. In other words, the head tax became a land tax, a significant reform in feudal China because it abolished a tax that had been collected for more than 1500 years. It freed common people from the burden of more taxes if they had more children. This policy stimulated a quadrupling of the population between 1661 and 1795.

1.2.1.2 Policies that Regarded Men as Superior to Women

The view in China that men are superior to women can be dated to the Spring and Autumn period. Confucius said in *The Analects*, "In one's household, it is the women and the small men that are difficult to deal with."[21] Mencius once remarked, "There are three forms of unfilial conduct, of which the worst is to have no descendants." In the following section, I present some cases of population policy in history that favoured parents of new-born babies. In the periods of the Northern and Southern dynasties, Emperor Mingdi (452–498AD) of the Qi dynasty decreed that if a couple had a baby boy, both his parents would be exempt from both land tax and corvee labour for one year.[22] In this case, the rewards for having a boy went to both parents, and it was more generous than the previous policy. During the Han dynasty, there were some specific preconditions to rewarding parents who had just had babies: only the parents of a new-born baby boy would be rewarded. In the Tang dynasty, after Emperor Taizong (599–649AD) ascended to the throne, he decreed that a woman

---

[19] Fan (2012). Fan Ye (398–445 AD) was a Song dynasty historian of the Southern Dynasties(420–589AD).

[20] Ban (2007).

[21] Confucius. (2008).

[22] Xiao (2016).
Xiao Zixian (487–537AD) was a historian from the Liang dynasty of the Southern Dynasties.

who had a baby boy would be rewarded with a *hu* of millet.[23]In the Yuan dynasty (1271–1368), a couple who had triplets would be exempted from taxes for three years if all the babies were boys. In the Ming dynasty (1368–1644), a household blessed with triplet boys were entitled by law to receive grain until the boys turned eight. In the Qing dynasty (1636–1912), regulations included "the family that has triplets will be rewarded with ten bolts of cloth and five *dan* of rice,[24] if the triplets are all alive and are all boys." "If there are both male and female among the triplets or all of them are girls, there will be no rewards."[25] From these decrees, we can infer that the starting point of feudal emperors' encouragement of population growth was that the birth of boys rather than girls was preferred. It is safe to say, then, that gender prejudice has been a long-standing problem in China.

1.2.1.3 Policies that Encouraged Early Marriage and Childbearing

The earliest known record of a marriage regulation is found in *The Rites of Zhou*, which says, "Men should get married before 30 while women should get married before 20". According to *The Book of Master Mo,* too, sage kings introduced a law that when a young man turns 20, he should marry. When a girl turns 15, she should marry. As noted above, Goujian introduced three laws according to which the parents of young men who were not married by 20 and young women who were not married by 17 would be punished. This practice continued until the early Tang dynasty, when in 734 AD, the marriage age was lowered to 15 for men, and 13 for women.[26] During the Ming dynasty, the marriage age was once again adjusted: "When young men turn 16 and young women turn 14, they are permitted to get married". This rule corresponded with a popular saying of the period that "a young man is capable of having a child when he turns 16 and a girl is capable of bearing a child when she turns 14".[27] Alongside the law, there were also social customs that promoted early marriage. For example, at the request of both sides of parents, and through the management of a matchmaker, there were abundant forms of early marriage including childhood betrothal, which could be arranged when both parties were less than ten years old. There were also some special cases of arranged marriage, including an agreement that if two pregnant women give birth to a son and daughter respectively, the children shall be united in marriage when they grow up.

---

[23] Ouyang and Song (1975).
Ouyang Xiu (1007–1072 AD) and Song Qi (996–1061) were both historians from the Northern Song dynasty (960–1127AD).

[24] A *"dan"* is an ancient Chinese measure approximately equal to 30kg.

[25] Xiao (2007).
Xiao Shi was a historian from Qing dynasty (1636–1912).

[26] Wang (2017). Wang Pu (922–982AD) was a historian in Northern Song dynasty.

[27] Shen (2014).
Shen Shixing (1535–1614AD) was a historian in Ming dynasty.

1.2.1.4 Policy that Encouraged Big Families

The Confucian doctrines value filial duty above all other ethical principles, under the influence of which a person is likely to lose his or her own identity, being thus reduced to his or her parents' son or daughter. Within a household, filial duty is manifested in the fact that parents are central and have absolute authority. In a big family, children and grandchildren are expected to live together with parents and grandparents and to serve them wholeheartedly. Confucius said, "While one's parents are alive, one should not travel too far afield." According to Confucian teachings, it is one's duty to stay with one's parents and serve them. In 742 AD, Emperor Xuanzong (685–762 AD) of the Tang dynasty decreed, "If there are more than ten young male adults in a family, two are exempt from taxes and the corvee. If there more than five and less than ten young male adults in a family, only one is exempt from taxes and the corvee. The young men exempt from forced labour should live with their parents and grandparents and serve their elders in order to reinforce the ethical codes."[28] Here, the number of young men exempt from the corvee is in direct proportion to the size of the family. In the language of modern statistics, the rewards are weighted and progressive. There were not only rewards for preserving the custom of having a big family, but also punishments for deviating from it. According to the Tang dynasty code, "When parents and grandparents are alive in a family, their children and grandchildren should not live in different places or divide up family property and live apart. Those who break the rule will be subjected to forced labour for three years. As for adoptive sons, they will be obliged to do the corvee for two years if they desert their childless adoptive parents."[29] Over time, big families, such as those with four generations under one roof, or those that maintained their prosperity over five generations, flourished and become increasingly powerful.

1.2.1.5 Policies that Introduced Mechanical Population Change

Direct interventions that changed population mechanics mainly increased the numbers through immigration and enhanced mobility, which formed indispensable elements of feudal population policy. As in the above-mentioned case of Duke Goujian's strategy of "ten years of accumulation of population and wealth", wealth accumulation was achieved by attracting migrants into the state of Yue, effectively reinforcing its political power and military strength. Shang Yang's reform in the state of Qin was another success story of strengthening a state's national power

[28] Du (2016).

Du You (735–812AD) was a Tang dynasty historian.

[29] Zhangsun et al. (2013). This series of books was collectively edited by several historians from Tang dynasty (618–907AD).

through increased migration. When Shang Yang[30] saw that neighbouring states like Han, Wei and Zhao were densely populated despite being geographically small, he wanted people from those populous states, especially Han, Wei and Zhao, to move to Qin, which was sparsely populated with a large territory. If those migrants engaged in farming, then Qin's agriculture would be developed. At the same time, the rear would be reinforced, and military strength would be increased thanks to increased agricultural production; consequently, Qin would be better equipped in its wars against neighbouring states. Simultaneously, when people from the neighbouring states came to settle in Qin, the population and military strength of the states they came from would decrease accordingly.

So how to attract people from those other states into Qin and get them involved in farming? It was done mainly through a new system of land division. With the abolition of the ancient farmland subdivision system, Qin recruited migrants to claim uncultivated lands and provided them with mechanisms to do so, such as allowing them to buy and sell land freely. In addition, there were incentives and punishments in Shang Yang's reforms. For example, when a peasant's output of grain and cloth (including silk) had surpassed the usual yield, he was eligible to be exempted from taxes and forced labour. Conversely, those who gave up farming to engage in business, and those who were poor because they were idle (and their whole family as well) would be made slaves of the government. These "carrot and stick" policies motivated the local peasants and migrants from other states to take up farming, and thus laid the economic foundations for Qin to defeat six other states and subsequently to unify all of China.

After the Qin dynasty was overthrown, the Han, Tang, Song, Ming and Qing dynasties established centralized feudal empires, except for some small local separatist powers. Those empires implemented population policies to attract migrants from one region to another. After the Eastern Han dynasty ended, people in the northern part of China witnessed many changes of regimes and were reduced to the victims of the turmoil of war for hundreds of years thereafter. Many northern people were displaced by the wars, moving to the southern part of China, leaving the land in the Yellow River basin a wilderness. When the Ming dynasty was founded, Emperor Taizu (1328–1398) introduced policies to attract people to move back to the north. According to *The History of the Ming Dynasty*, "people are being recruited to do farm work in the north. Each migrant will be allocated 15 mu of land and two mu for planting vegetables, free from land tax for three years. Those who reclaim wild land spontaneously will be exempt from paying the land tax to the government."[31] Such encouraging policies effectively drew people from southern provinces like Jiangsu, Zhejiang, Anhui and Jiangxi back to north, contributing to a more even distribution of population. In addition to the regional population policy, there was another special policy that reflected migration patterns in the ancient feudal empires—the policy of

---

30 Shang Yang (390?–338 BC), was a statesman and one of the representative thinkers of the Legalist School during the Warring States period.

31 Zhang (1974).
Zhang Tingyü (1672–1755AD) was a Qing dynasty official.

using garrison troops and peasants to open up wasteland and to cultivate food crops in the border regions. Without doubt, agricultural production in the border regions in feudal empires was chiefly carried out by the army.

The various policies intended to attract people from the inner regions, contributed to population growth and to some extent to the consolidation of the border regions.

1.2.1.6 Policies that Improved Population Capacity

As mentioned above, 7,000 or 8,000 years ago, the tribal chieftain Fuxi decreed that before a couple got married both parties must produce their token—a piece of deer skin. If their tokens were the same, that meant they were cousins of the same tribe and were not permitted to marry. This practice began the practice of eugenics in China and contributed to the rise of Fuxi's tribe. Confucius had once remarked, "People with the same surname belong to the same clan and they are united with the same family name. When the head of the clan dispenses food to them, people will receive the same quota as long as they bear the same family name. Despite the passage of hundreds of years, men and women of the same surname are still not permitted to get married. People have conformed to such a regulation since the Zhou dynasty."[32] Here, Confucius was disapproving of marriage between men and women with the same family name for the consideration of human relations, which coincided with the biological law reflected in the saying "When men and women with the same family name marry, their descendants will not be strong and healthy." In 483 AD, Emperor Xiaowen (467–499 AD) of the Northern Wei dynasty declared:

> In the Xia and Shang dynasty, marriage between men and women of the same clan was not abhorred. Since the Zhou dynasty, marriage between men and women with the same surname has been prohibited. So, we should practice different customs as times change, and reform our administration when situations vary. Since I have recently ascended to the throne, I am now occupied with state affairs. Some primitive traditions passed down from ancient times remain unchanged and are still observed. But from now on, anyone who defies the tradition that men and women of the same surname are forbidden to marry will be penalized for being immoral.[33]

Subsequent dynasties also began to punish those who married even though they shared a family name. According to the Code in the Tang dynasty, "When those with the same family name and from the same clan get married, both husband and wife will be punished, and made to do corvee labour for two years. Those who marry their close relatives (relatives within five generations) will be punished as severely as those who commit the offense of adultery."[34] In the Ming dynasty, there was a rule that "if men and women with the same family name get married, they will both be flogged 60 times and then be compelled to divorce."[35] Likewise, consanguineous

---

[32] *The School Sayings of Confucius* (2011).

[33] Li (2013). Li Yanshou was a historian in Tang dynasty.

[34] Zhangsun et al. (2013).

[35] Shen (1985).

marriage was also prohibited in ancient times. According to the Code in the Tang dynasty, "If a man marries a half-sister or a stepdaughter, both husband and wife will be punished in the same ways as those who commit adultery. Those who marry a stepdaughter or a half-sister will be made to do corvee labour for three years. Those who rape a stepdaughter or a half-sister will be exiled to a remote place 1,000 kms away. If the raped woman is injured, the rapist will be hanged."[36] Nearly all dynasties after the Tang dynasty adopted similar laws to prevent consanguineous marriage. The cases cited here demonstrate that all feudal rulers had clear-cut attitudes and strict policies prohibiting marriage between two people with the same family name, and consanguineous marriage, thereby guaranteeing that the Chinese nation had, to a large extent, healthy descendants.

Where improvement of the population's education achievements is concerned, some remarks from Mozi (dates unclear), a thinker in the Warring States period, quoted in the ancient classic *Guangzi*, are still on the lips of Chinese people today. "If one is making a one-year plan, the best thing is to grow crops; if making a ten-year plan, the best thing is to plant trees; and if making a 100-year plan, the best thing is to edify people. In the crop-growing plan, there will be one harvest. In comparison, in the tree-planting plan there will be ten times more harvest, and edifying people will be 100 times more fruitful."[37] Confucius valued education very highly and was a great educator in his own time. It was claimed that he had about 3000 disciples, among whom 72 were very talented. After Confucius died, his thoughts and teachings were collected in *The Analects* and advocated by his disciples. Confucian doctrines gradually became the dominant ideology among Chinese people, after having been inherited, upheld, and adapted by subsequent followers, generation after generation. For example, Dong Zhongshu (179–104 BC), a devoted adherent of Confucian theories, proposed "discarding the hundred schools and respecting only Confucianism" to Emperor Wudi (156–87 BC) in the Western Han dynasty. The emperor took his advice and thus Confucianism became and remains the prevalent theory. Over 1,000 years later, the Cheng brothers and Zhu Xi in the Song dynasty remoulded and adapted Confucian doctrines. Eventually, Confucian doctrines and feudal ethic codes like "Three Cardinal Guides and the Five Constant Virtues" and "Three Obediences and Four Virtues" were internalized and observed by the general public. Confucius was revered as a sage, and Confucianism was transformed into a kind of religion. The codes of conduct including filial piety, fraternal duty, loyalty, trustworthiness, senses of shame and honour and so on, became spiritual shackles that fettered Chinese people for over 2000 years. Population views originating in those codes have had far-reaching repercussions in China. Since the imperial examination system was implemented in the Sui and Tang dynasty, Confucian classics became the exam content, and the examinees were restricted to knowing the Confucian classics. Scholars who became public servants upon passing the imperial exam applied Confucian doctrines in the performance of their official duties. The same

Shen Jiaben(1840–1913) was a celebrated jurist in the Qing dynasty.

[36] Zhangsun et al. (2013).

[37] Guan (2005).

was true even in the management of their personal business. Accordingly, Confucian doctrines and principles became *the* criteria for evaluating people's education levels. As a matter of fact, the purpose of improving the population's level of education has deviated from its original course.

1.2.2 Population Policies that Intervened Indirectly in Population Growth

Policies that intervened in population growth indirectly were those that affected population levels through economic and social policies. They incorporated population change and development into economic and social development. These policies remained in place for a long time and thus had long-lasting influences and should not be underestimated.

These "indirect" policies mainly refer to fertility programs contained in policies that made provisions for recovery from the effects of war or calamity, and land and taxation policies.

1.2.2.1 Helping the Population to Recuperate and Multiply

Policies of recuperation refer to those adopted by a new dynasty to revitalize the country through agricultural development, lightening ordinary people's loads, and maintaining social order to help people to live stable lives after surviving war and social turmoil. Investigating over 2000 years of history of frequently changing dynasties, we see that a dynasty usually went through three stages. In the first stage, it brings chaos back to order. The new dynasty created through war adopts concessionary policies to some extent, in order to resume and improve production, to build a prosperous and powerful nation, enabling people to live in happy prosperity by lightening peasants' loads through reducing both taxes and corvee labour obligations. The second stage witnesses the dynasty's transition from prosperity to decline. In this stage, with land-grabbing becoming a more serious problem, the peasant's load gradually increases, and conflicts within the ruling landlord class sharpen, resulting in social disorder. Finally, the last stage sees the collapse of the dynasty. With a growing number of peasants losing their land, government revenues fall short of expenditure despite the heavier taxes and more corvee labour. Wars and peasant up-risings break out more and more frequently, until the dynasty is overthrown. During the process, population change follows a regular pattern.

First, the population increases despite its being in decline at the beginning of the new dynasty. Next, the population increases before levelling off. In the last stage, the population decreases. In sum, as a dynasty shifts from prosperity to decline, its population will stop growing and then start to reduce. The usual pattern is that the population increases in the first half of the dynasty's life, which is closely associated with the recuperation policies. Population decreases in the latter half of the dynasty's term, which is a consequence of intensifying corruption among the ruling landlord class and incessant turmoil caused by wars. Observation of some outstanding emperors after the Qin dynasty, including Emperor Gaozu of the Western Han dynasty, Emperor Taizu of the Northern Song dynasty, and Emperor Taizu of the Ming dynasty shows that the founders of each new dynasty knew well why peasants would rise in revolt to overthrow the former dynasty, since they themselves were low-born. Therefore, after coming to power, they adopted policies that would contribute to the recuperation of society to differing degrees. For instance, Emperor Gaozu of the Western Han dynasty formulated a series of policies conducive to revitalizing the country. First, on the basis of laws of the Qin dynasty, he announced decrees to guarantee a peaceful environment for people who had suffered from the chaos caused by war. Second, he contrived to attract abundant talent to serve in government, utilizing human resources optimally. Third, he mitigated the amount of corvee labour and cut taxes. Under the new laws, peasants only had to pay the government one fifteen of their farm produce. The government called for the return of refugees, who were displaced during the wars, to their hometowns. If peasants transformed wild lands into farmland, they would be exempt from land tax for the first few years. Army veterans were allotted land and houses according to their service. They would also be exempt for several years from corvee labour. Those policies for national revitalization freed peasants of much of their burden and motivated them to engage in more agricultural production. The policies also mitigated the contradiction between peasants and landlords, creating the necessary conditions for population growth.

1.2.2.2 Land Tax Policies

Land is the foundation of agricultural production and the lifeblood of peasants' livelihood. Thus, land policy touches on every aspect of population policy. As discussed above, a dynasty's development usually followed a three-stage pattern, of prosperity, decline, and collapse. In the first stage, the most important task for a new dynasty was to solve the land annexation problem that appeared in the decline and collapse stage of the dynasty it had superseded. From the Northern Wei dynasty to the Song dynasty, feudal rulers mainly relied on equal land distribution system based on house-hold population. After the Song dynasty collapsed, people were permitted to buy and sell land.

In 485 AD, Emperor Xiaowen of the Northern Wei dynasty decreed that land would be divided into two types: *sangtian* land and *lutian* land. Sangtian land was permanent and inheritable, but could not be sold. Lutian land was allocated to a person when they reached certain age. Old people could not keep land, and it was to be returned to the government when they died. When a family's lutian land was beyond their capacity, they could sell some of it. Equally, if a family did not have enough lutian land, they could buy some. A man would be allocated 40 mu of lutian land when he turned 15, whereas a woman would be allocated 20 mu of such land. Servants and maids could also be allocated land and oxen like free subjects. During the Tang dynasty, the land policy followed the example of the Northern Wei dynasty laws, and the Zu-Yong-Diao System that contained land tax, corvée service, cloth tax was created. In 624 AD, Emperor Gaozu (566–635 AD) decreed that when a man turned 18, he could be allocated 100 mu of land, of which 20 mu was *"yongye"*[38] land and the rest was *koufen*[39] land. In a region with limited land, people would be given half the amount given to those living in a region with abundant land. Businessmen living in a region with ample land would be given half the land that peasants were allowed, while those living where there was limited land would not be given any land at all.

There were also regulations about the allocation of land to those who had recently moved to a place. Men who received land from the government had to pay a land tax called "zu" in the Tang dynasty, in the form of two dan[40] of millet per year. They also had to pay another kind of tax called "diao", which required handing over certain silk, linen or other cloth products, depending on what was produced locally. On top of paying taxes, they also had to do forced labour for 20 days per year, or 22 days in a leap year. Those who did not do the corvee had to hand in silk products as a substitute. This kind of forced labour was called "yong" during the Tang dynasty. The adoption of equal land distribution system based on house-hold population from the Northern Wei dynasty to the Tang dynasty satisfied peasants' need for land to some extent.

New forms of equal land distribution system were usually issued at the beginning of a new dynasty. At that moment, they were relatively easy to implement because the population had decreased to a large extent and a lot of farmland was deserted because of the continuous wars. Once the new dynasty had created order out of the chaos, the population grew. In the second stage, land-grabbing aggravated gradually. Then in the last stage of decline and collapse, landlords, the feudal ruling class, had monopolized farmland, resulting in the loss of the fruit of equal land distribution system.

The Zu-Yong-Diao system of land tax during the Tang dynasty was smashed by the rebellion led by An Lushan and Shi Siming (755–763 AD) in the middle of Tang dynasty and by peasant uprisings at the end of it. After the Song dynasty was overthrown, farmland transaction became an institution, gradually growing in scale. On one hand, such transactions were conducive to landlords' annexation of farmland, reinforcing the power of the ruling class, and on the other hand, it helped peasants to end their dependency on landlords, enabling them to work their own farms or to become tenant farmers in a new place.

Both the land equalization policy and the practice of farmland transaction were beneficial to population reproduction and multiplication. The land equalization policy made it possible for peasants to buy some farmland at the beginning of a new dynasty

---

[38] *"Yongye land"* is the farmland allocated to a man when he turned 18 in the Tang dynasty. This kind of farmland is inheritable and the man's family could keep it after he died.

[39] *"Koufen land"* is another type of farmland allocated to a man in the Tang dynasty. Yet after the man died, the land should be given back to the state. This type of land cannot be inherited.

[40] A *"dan"* under the Tang dynasty is approximately equal to 53 kg.

after having lost their original land. Engagement in agricultural production in a relatively stable environment was favourable to population growth. On the other hand, buying and selling land enabled landless peasants to earn a living in new ways. They could either reclaim wild land in a remote area or become small businessmen. Thus the whole society made a transition into a society with land-holding peasants in its true sense and perfected the creation of small peasant families based on an individual land-holding peasant economy. As the popular saying in China goes, "30 mu of land, an ox, a wife and several children beside a hot hearth make an ideal home for a peasant". The typical self-sustainable small family best represents the traditional belief of "more happiness brought by more sons", which provided advantageous conditions for population multiplication in feudal times.

1.2.2.3 Other Tax Policies

Investigation of tax policies in Chinese history informs us that before the Han dynasty, taxes mainly consisted of land tax and corvee labour. Though those imposts were not closely connected with population increases, neither did they impede it. In the Western Han dynasty, a head tax was levied as well as a land tax. Both the rewards granted to someone who had a son and the punishment of a woman who had not married by a certain age were related to the poll tax. Most importantly, Western Han dynasty law divided the head tax into different categories, with those between 7 and 14 paying "*koufu*",[41] those between 15 and 56 paying "*suanfu*",[42] men between 20 and 56 doing forced labour, and so on. The age groups that had to pay the head tax as koufu and suanfu varied with time. So did the amount of koufu and suanfu to be paid. The head tax was far heavier than the land tax. As main source of state revenue, it was in place for almost 1900 years until Emperor Kangxi in the Qing dynasty initiated the taxation reform that exempted new-born babies from the head tax. The scheme had long created serious problems, including that "common people tend to flee their hometowns to avoid paying it, while officials tend to conceal their family members in order to dodge paying it." Indeed, the head tax was detrimental to both population growth and feudal government revenues.

In 1723, during the Qing dynasty, Emperor Yongzheng decreed a new taxation policy: all taxes were rolled into an assessment of landholdings, thus ending the long history of associating the head tax with population size. Although this reform was not expressly a population policy, it disconnected taxation from population, implying that a bigger family did not have to pay more head taxes, thereby encouraging fertility more than any other direct population policy. The fact that the Chinese population quadrupled in about 150 years between Emperor Kangxi and Emperor Daoguang best proves the positive effect of the new taxation policy on population size.

---

[41] *"Koufu"* refers to one type of head tax that children between 7 and 14 had to pay in the Han dynasty. Its amount might vary when different emperors reigned.

[42] *"Suanfu"* refers to another type of head tax that people between 15 and 56 had to pay in the Han dynasty. Its amount might also vary when different emperors reigned.

1.3 The Formation of Traditional Population Culture

Over thousands of years, the population of China has experienced a pattern of rise-fall-rise, but always spiralling upwards. Generally, three factors have contributed to this. First, a self-sustaining natural economy has created favourable conditions for population growth. Second, all the feudal dynasties implemented direct or indirect policies conducive to population increase. And third, deep-rooted ideas about population have been constantly reproduced.

1.3.1 The Population Question in Traditional Culture

Before we talk about cultural perceptions of the population question, we need a definition of culture. We can understand culture on three levels. In a general sense, culture refers to the total physical and spiritual wealth created within a certain time span. For example, ancient Egypt created the culture of the Nile, including the pyramids, the sphinx and so on. In China, a rich culture is represented by the Great Wall, the terracotta warriors, the Palace Museum and so on. In Italy, Renaissance culture spanned two centuries, advocating the spirit of science from the 14th to the sixteenth centuries.

Culture, in its common sense, refers to an ideology or a set of concepts. The United States, first created as a nation by colonial migrants, is the product of a combination of a commodity economy, foreign trade, and colonial expansion. After the baptism of the War of Independence, US culture was born out of its history, with individualism as a core concept along with political rights and equality, freedom, and the right to private property as main features. In contrast, China, surrounded by mountains on three sides, with an ingrained ideology of feudalism, formed a closed, traditional culture, represented by Confucian creeds such as filial piety, fraternal duty, loyalty, observation of rites, trustworthiness, the sense of shame and honour.

In a narrow sense, culture refers to a person's education level, meaning how far they went in school or university.

In the term "population culture" used here, culture refers to an ideology or set of ideas about population. It is not equal to population plus culture. Instead, it is an integral entity combining population and culture. Our analyses in earlier sections have shown how popular views about population determine population policies. Conversely, certain population policies solidify and reinforce the mainstream views when society assimilates those policies as they are implemented. Over time, a unique population culture is created through the long-term interaction of population views and population policies. A comparison between Chinese and non-Chinese population cultures, especially in Western countries, shows that Chinese population culture has some distinct and prominent features. Recently, with the rising popularity of cultural studies in China, studies of population culture have also gained some popularity, and the many works published on the subject include *The General Theory of Population*

Culture, *An Introduction to Chinese Fertility Culture*, and *Studies of Fertility Culture*. We can summarize the special features of traditional Chinese population culture as follows: the high value placed on having a son to assume the duty of worshipping and honouring one's ancestors, having a son to carry on one's family name, raising a son to look after his parents when they get old, regarding girls and women as significantly inferior to boys and men, studying hard with the goal of becoming a state official, and reaping more happiness by having more sons. The last belief is the essence of traditional Chinese views on population; the rest serve as its buttresses. Together, they constitute traditional Chinese population culture.

1.3.1.1 The Expectation of More Happiness from More Sons

The belief that more sons bring more happiness conveys the means and the purpose of human production and reproduction and was the essence and spirit of population culture in Chinese feudal society. Some religious doctrines tell us that crying is the first sound a person makes when he or she is born, which means that a person is to suffer when he or she comes to the world. However, nobody wants to lead a miserable life and all people want to enjoy more happiness. But how is that to be achieved? The traditional Chinese universal answer to this question is to have more children, especially more sons. There is a world of difference between what this belief means to the working class and what it means to the feudal landlord class. Yet, in the traditional view, for a common family, there are also great differences between a family with many sons and one with few or no sons. Usually, more sons will bring more happiness, a few sons will bring less happiness, and no son will mean no happiness. This belief was universally accepted in China; the poor accepted it because they hoped to shake off poverty and enjoy more happiness through having more sons. The rich accepted it because they hoped to enhance their happiness by having more sons. Families from the exploited and oppressed classes accepted it because they hoped that their offspring would transform their adversity. Families from the ruling class accepted it because it would help them to maintain the system of exploitation and oppression: more sons would build a more bigger and more prosperous family. People of all social strata sought in different ways to improve their family situation by having more sons.

1.3.1.2 Sons and the Duty of Worshipping and Honouring the Ancestors

According to Confucian ethical codes, one should love one's parents sincerely, obey them, and look after them when they are old. There is even a saying that when a father wants his son to die, the son will be considered not devoted if he refuses to do so. Benevolence is the core of Confucianism and fulfilling one's filial duty is its quintessence. Someone who fails to perform his filial duty will be regarded as offending against the heavenly laws of morality and filial love. They will certainly be condemned by public opinion, and some will by punished by law, depending on the offence. In ancient times, when children started to learn to read and write, stories about extremely devoted sons in history and conversely about the miserable fates of those who failed in their filial duty were told to them. Thus the culture of performing one's filial duty began to be internalized during childhood. There is a reasonable and desirable aspect to the notion of filial duty, which functions as an emotional bond between parents and children. It also presses the point that each person will grow from a baby into an old person after passing through the stages of infancy, childhood, and adulthood. When someone performs their filial duty, they are setting an example for children. Exemplary behaviour will influence children naturally, which will in turn lead to a happy life in old age. The culture of fulfilling one's filial duty has permeated every aspect of Chinese life, including economic and political life, and even sporting and cultural activities. The character "孝" (which means "to perform one's filial duty")[43] in its traditional hieroglyphic form resembles an old man being supported by a young man.

1.3.1.3 The Importance of Sons to Carry on the Family Name

In order to reproduce the tradition of having sons to assume the duty of worshipping and honouring the ancestors, population reproduction must continue uninterrupted and all families should have such heirs to worship their ancestors. After matriarchal society was replaced by patriarchal society, the duty to worship ancestors was passed down to the sons and grandsons of a family. A man was considered very unfilial if he had no son or grandson. If a couple had no son, but did not want to be criticized as unfilial, how could they solve the problem? The husband had no choice but to ask for help from other people, usually his brothers or cousins. Sons of brothers and cousins were often adopted so that the family duty of worshipping the ancestors could be fulfilled. At the same time, family wealth was also passed on. Thus the practice of adopting a son was valued by Chinese people. Even now, having a son to carry on the family name is strongly adhered to in some rural areas. In the deep countryside, having no son to worship the ancestors is considered offensive and unreasonable. This is why some couples continue to try all means to have a son, even if they may incur severe punishment in doing so.

1.3.1.4 Sons to Look After Parents in Their Old Age

In Chinese feudal societies there were no social security systems. When people aged, they had to depend on their children for support. It was the only way for old people to be cared for in ancient China. The popular saying "When a boy reaches adulthood, he should get married. So should a girl" tells us that it is the natural duty of a son to take care of his parents when they get old, because a daughter will establish a new family after getting married. Having a son is preparation for one's senior years. Since

[43] The traditional form of the character "孝" (which means "to perform one's filial duty") is .

China did not experience a thorough bourgeois industrial revolution, and there is no advanced commodity economy in China, the practice of home care for the elderly has continued for a long time, with far-reaching and long-lasting effects. After the PRC was founded in 1949, with the recovery and development of national economy, a pension system was created, first in the cities, thereby breaking with the convention of home care for the elderly. Because of the rural–urban dual economic structure and corresponding dual social structure system, a dual social security system has also formed, with the elderly in urban areas predominantly depending on the social pension system, and those in rural areas mainly depending on home care. In recent years although the national government has accelerated reforms towards an integrated urban–rural old-age security system, the traditional idea of having a son for support in old age is still prevalent among some people and dominates their fertility behaviours.

1.3.1.5 Population Culture and Female Inferiority

Mencius once remarked, "There are three forms of unfilial conduct, of which the worst is to have no descendants." Here, "descendants" means sons rather than daughters. In another popular traditional idea, "more happiness will be brought by more children", but here, too, "children" means sons and not daughters. While happiness is said to follow from having many sons, daughters will only consume property. Before girls marry, they depend on their parents, who must prepare dowries for them. After marriage, women bear children for their husbands. That is why girls are considered undesirable by some parents. In feudal China, girls and women were regarded as much inferior to boys and men; gender prejudice was extreme, reflected in such ethical codes as "Three Cardinal Guides and the Five Constant Virtues" and "Three Obediences and Four Virtues", according to which, a woman is subordinate to her husband. She must observe such harsh edicts as "the husband is the guide of his wife" and "a wife should obey her husband". More punishing, a woman became subordinate to her son after her husband died. It is true that in other countries too men and boys are valued more highly than girls and women, but the devaluing of women has not been as extreme as in China, where the idea of degrading women became a spiritual shackle.

1.3.1.6 Studying Hard to Become a State Official

Feudal population culture valued improvement of the quality of the population, attaching much importance to fostering excellent sons and grandsons through hard study. Confucius was a great educator, venerated as a sage. He proposed that "When a scholar finds that he can more than cope with his studies, then he takes office", associating the pursuit of studies with becoming a state bureaucrat; this notion had a deep influence on Chinese society. The imperial examination system was implemented by the Sui and Tang dynasties; through this system, an outstanding scholar had the opportunity to become a public servant if he passed the exam. The imperial exam had two positive effects: first, it established the idea that only through study can a man have the chance to become a state bureaucrat. The exam offered him hope and incentive, a a stepping stone to success. Second, to be a public servant one had to study industriously and be knowledgeable, which encouraged bureaucrats to advance continuously in their studies.

Both effects had a positive impact on the population's education level. Among common people, tens of thousands of young men made great efforts to study diligently for more than ten years despite their adverse situations in order to become public servants, honouring their family through the imperial exam system. While most government bureaucrats were erudite, some turned out to be outstanding experts and scholars in certain fields. More importantly, the association of the imperial exam with becoming a state employee contributed to creating a social culture that valued diligent study. It also helped to create a tradition of great respect for teachers and education. Yet, what is noteworthy is that the Confucian classics were the sole content of the training. After those learned Confucian scholars became public servants, they would be committed to upholding the feudal system. They rarely accessed any scientific or technical knowledge. Consequently, the system led to an imbalance between an underdevelopment of natural sciences subjects and advanced development of social sciences like literature and history studies; this imbalance became a serious impediment to the progress of science and technology and the development of social productivity.

1.3.2 The Evolution of Traditional Population Culture

The traditional population culture with its core expectation that more sons bring more happiness also generated relatively independent thoughts, concepts, ethics, morals, traditions, customs, religions, beliefs and so on, ideologies that dominated many aspects of life, including fertility choices, upbringing of children, mobility, career choices, marriage patterns, and the development of families, clans and society. They became moral imperatives that affected population quantity, quality, and structure. Traditional population culture was an invisible net, under which population production and reproduction was free to continue. Anyone daring to try to free himself from the constraints of the net risked being condemned, deserted or even destroyed. But then after more than 2000 years, feudal rule finally collapsed in 1911 with the overthrow of the Qing dynasty. Then in 1919, the May Fourth Movement slogan "Down with Confucianism" ushered in the New Culture Movement. The traditional feudal population culture started to adapt to the new times. Since the foundation of the PRC, and especially since the economic reforms and modernisation policies begun under Deng Xiaoping, many traditional population culture beliefs have faded away, leaving a new population culture to emerge. Within the latter, traces of the traditional culture survive the times; some views and behaviours are still embraced to various degrees, and some of them even burst with new vitality.

1.3.2.1 Changes in Attitudes to Population Size

Regarding quantity, the core of traditional population culture—the expectation of more happiness brought by more sons, has largely receded. Many surveys conducted in various places indicate that an overwhelming majority of families in China have dropped the traditional idea of having many sons. Perhaps surprisingly, the first choice of about 80% of families surveyed is to have two children. This change is associated with changes in attitudes, theories, and mindsets but more importantly, it is determined by the level of economic development. People have sublated the idea of "more happiness brought by more sons", and no longer feel proud of having many children. Instead, those who do are regarded as ignorant, unambitious, and unaware of social realities. The surveys also reveal that the fertility rate is inversely proportional to parents' education level. When parents are well-educated, they tend to have less children. When they are poorly educated, they tend to have more. The fertility rate is also inversely proportional to family income. The higher the income, the lower the fertility rate. The lower the income, the higher the fertility rate. Fertility is also connected to the regions where people live. The general trend is that the fertility rate is lower in the cities than in countryside and lower in developed regions than in underdeveloped regions.

All these show that the fertility rate is ultimately restricted by economic circumstances, and primarily determined by social and economic development and family financial circumstances. Where economic development is higher, the marginal costs of raising children are higher, because of higher variable costs for schooling, health care, and so on, and there are no significant benefits for parents, even smaller than the average level in society. Conversely, in regions with lower economic development, marginal costs are lower—education, health care and on the like cost less. In this case, the benefits accruing to parents from having children are not small and are greater than the average. In these circumstances, improved socioeconomic development and family incomes, the traditional cost—benefit balance of raising children has shifted from the pattern of reaping more happiness from having more children to a pattern of reaping more happiness by having fewer children and offering them good postnatal care. In cities and highly developed regions, the shift in the costs and benefits of raising children is more or less complete. The benefits of reaping more happiness by having fewer children is accepted by a growing number of people, and the new population culture has emerged.

As attitudes to population quantity have changed, so has family culture, which is closely related. Data from two recent demographic polls show that the largest proportion of households are three-person families. While the traditional large family with four generations all living under the same roof has almost disappeared, the trend of small families advances at a steady pace. Greater changes have also taken place in the marriage culture; the traditional view that a wife must be faithful to her husband "unto death" is under serious strain and the divorce rate has surged beyond expectations.

Though traditional beliefs about family size are fading, some old-fashioned ideas still linger, and a few are vital even today. Gender discrimination, for example, remains deep-rooted in China. Some people continue to uphold traditional views such as having sons is necessary to fulfil the duties of worshipping and honouring the ancestors, to carry on the family name and to look after parents when they get old. Thus some people continue to have babies until they have a son. Adherence to these views was the primary reason why some people violated fertility policies and had more children than the law permitted in the past. There is also a significant imbalance between regions: in rural and remote mountainous areas with small economies and low income, some people still hold to the idea that having more sons will reap them more happiness. The idea of having more children, especially more sons, is very appealing to some, as is the idea of having a powerful and influential large family that belongs to a large clan.

1.3.2.2 Changes in Attitudes to Population Quality

Regarding quality, the legacy of traditional population culture that stresses enhancing people's education and cultivating moral character continues. The ideas have integrated with economic development, social progress, and population transformation, from which a new set of ideas and attitudes about population quality has gradually emerged.

First, understanding of the significance of population quality for socioeconomic development has deepened. There are four types of capital: natural, output, human, and social. There is a consensus that to boost socioeconomic development and realize modernization, is critical to develop science and technology on the basis of a robust education system, and that promoting population quality and strengthening human capital accumulation are core tasks.

Second, understanding of the relationship between elements of population quality has developed into a consciousness that population quality should be rounded: physical fitness and good health, as well as intelligent and scientific education and cultural and moral fitness are all indispensable. Among these three constituents, physical fitness and good health form the foundation, intelligent and scientific education is key, and cultural and moral fitness serves as a guarantee of the other constituents. Today, with personal income at an upper-middle level on an international scale, people have greatly deepened their understanding of the significance and application of scientific and quality education. In recent years, average educational attainments among the Chinese have grown faster than any time since the economic reforms began in the late 1970s. There is also an increasing awareness of the importance of physical fitness and good health. Health is now being seen as the basis for enhancing population quality and a goal to be promoted. As for cultural and moral fitness, China seems to be in a transition phase towards a better understanding and definition of it, and looking for ways to enhance it. With the development of practice and theory, it is time to establish values, moral standards, and a concept of civilization consistent with the development requirements of the new era.

Lastly, legal management that is conducive to population quality improvement is being promoted. In order to improve the three constituents of population quality as just discussed, people's ability to know the world and to transform both the subjective and the objective worlds must be enhanced. Furthermore, parents should attach more importance to the care and education of their children, and make sure that their children take part in social practices. Therefore, we need to strengthen social management, and instil standardized habits of self-discipline, and other mechanisms to help build personal character beneficial to individual growth, economic development, social progress, and human civilization. Successful experiences elsewhere in the world provide us with some good examples. Germany, for instance, was twice vanquished in war, its society and economy taking heavy blows each time. It is very challenging for a country to recover from such setbacks, so how did Germany achieve this national recovery? Of course, many reasons account for it, but one of them is the flourishing of science and education after both world wars. Rapid improvements in these fields, along with the promotion of good reading and learning habits, meant an accumulation of human capital which promoted scientific and technological innovation and resulted in the country becoming a leader of developments in science and technology. Meanwhile, standardized management of the state also contributed to the national rejuvenation. In Germany, the usually disciplined, law-abiding, and good-mannered citizens act in accordance with shared ideas about morality—it is a nation with an admirably high population quality. Recently, China is reinforcing legal construction. Standardized management should be incorporated into the national strategic development framework because it has proved successful when applied to promotion of comprehensive population quality.

1.3.2.3 Changes in Attitudes to Population Structures

Profound and extensive changes have taken place in population culture concerning structures, especially family structures. As the population age structure changes, so do traditional ideas such as respecting and honouring the elderly, loving one's elders and taking good care of them. The idea of having a son as preparation for old age has almost disappeared in cities and developed areas, though it is still prevalent among people in the less developed rural and remote mountainous areas. There is a notable increase in the number of households in which the elderly live alone. According to surveys, most adult children do not like living with their parents and vice versa. As a result, household size in China has been getting smaller. Family culture based on this new type of intergenerational relationship is changing significantly.

As for changes to gender structure, extreme gender prejudicial ideas about "the husband being the guide of his wife" and the exhortation that wives be faithful to their husbands "unto death" are dying out, having lost the basis of their existence. Some traces of gender prejudice can still be detected in the beliefs such as requiring a son to carry on the family name, and the gender-imbalanced allocation of educational resources among children in a household. There remains an intense preference for baby boys rather than girls, the impact of which cannot be underestimated. The continued occurrence of the murder or desertion of infant girls implies that it is difficult and time-consuming to change the traditional gender culture. To eliminate gender prejudice, relevant moral education is essential, especially education about gender equality. Meanwhile, gender equality must be promoted in aspects of policy and practice, including education and reforms in the domains of economy, politics, culture, education, employment and so on. Reforms should be equality-oriented, to achieve the goals of equal treatment of men and women, without discrimination, and with equal opportunities, and equal pay for equal work. We should persist in carrying out these reforms despite barriers and setbacks.

Regarding changes to the urban–rural population structure and regional population distribution, underlying attitudes to mobility, reflected in the sayings like "It's hard to leave one's hometown" and "While one's parents are alive, one should not travel too far afield", have changed. Nowadays, population mobility is market oriented. A commodity and market economy requires free movement of labour. The unprecedented mobility since the reform policies started is inevitable and in accord with economic development laws. Most of the 244 million "floating people" in China, are moving from rural to urban areas. The sharp rise in this migrating population has generated certain changes to the orientation and size of population flows, and to population quality. A unique floating population culture has been created, greatly promoting urbanization and socioeconomic development. Every spring festival now sees tens of thousands of migrant farm workers return to their hometown, a spectacular sight that has developed in my life time. Those who stay in the countryside, like women, children and old people, can only be united with their family members through this mass migratory return. Mass migration has become routine in their lives and has created a kind of hidden floating population culture.

References

Ban, G. (2007). *The Record of Emperor Gaozu in History of Former Han Dynasty*. Zhonghua Book Company.

Chen, C. H. (1918). *Research on Population in China*. The Commercial Press. pp. 54, 62, 63, 84

Chen, D. (1934). *Population Problems*. The Commercial Press. pp. 267, 307–308

Confucius. (2008). *The Analects of Confucius* (Lau, D. C.Trans.). Beijing: Zhonghua Book Company.

Du, Y. (2016). *A Complete Collection of Code* (*Tong Dian*) (Vol. 6). Zhonghua Book Company.

Fan, Y. (2012). *The Record of Emperor Zhangdi in History of Later Han Dynasty*. Zhonghua Book Company.

Guan, Z. (2005). *Guanzi* (Vol. 1). (Zhai, J.Y. Trans.). Guangxi Normal University Press. p. 45

Guo, S.X., & Wang C. M. (Eds.). (2002). *A Dictionary of Chinese Philosophy with English Annotations*. Henan University Press. pp.23,28, 277, 605

Han, F. Z. (2003). *Han Feizi: Basic Writings* (Watson, B. Trans.). New York: Columbia University Press. p. 98

Li, Y. S. (Ed.). (2013). *History of the Northern Dynasties*. Zhonghua Book Company.

Ouyang, X. & Song, Q. (1975). *The Record of Emperor Taizong* in *New History Book of Tang Dynasty*. Zhonghua Book Company.

Shen, J. B. (1985). *Investigation into Penal Code Throughout Chinese History(Lidai Xingfa Kao)*. Zhonghua Book Company.

Shen, S. X. (Ed.). (2014). *Code in the Great Ming Dynasty*. Zhonghua Book Company.

Sun, Y.S. (2014). *Sun Yat-sen: His Political and Social Ideals.* (Zhao, N.Trans.) Beijing: World Publishing Corporation. p. 90

The School Sayings of Confucius (Kongzi Jiayü) (Vol. 10). (2011). (Annotated by Wang, G. X., & Wang, X. M.). Zhonghua Book Company.

Xiao, S. (2007). *Yongxian Record*. Zhonghua Book Company.

Xiao, Z. X. (2016). *The Record of Emperor Mingdi* in *History of Southern Qi Dynasty*. Zhonghua Book Company.

Xu, S. L. (1930). *China's Population Problems*. The Commercial Press. p. 5

Wang, P. (2017). *Collection of Laws and Regulations in the Tang dynasty* (*Tang Hui Yao*). Zhonghua Book Company.

Zhang, T. Y. (Ed.). (1974). *History of the Ming Dynasty*. Zhonghua Book Company.

Zhangsun W.J., Chu, S.L., & Li, J. et al. (Eds.). (2013). *Code in the Tang Dynasty* (*Tanglü Shuyi*) (Vol.12 &14). Shanghai Ancient Books Publishing House.

Chapter 2
Debates on the Population Problem in the People's Republic of China (1949–1979)

With the founding of the People's Republic of China in 1949, Chinese people have stood up since then. A new long march towards national rejuvenation began. The world is generally constituted by three essential elements: people, wealth, and materials. To achieve such a formidable goal as national rejuvenation, it is critical to put the relationship between these three elements into perspective and to consider them appropriately. People, i.e., the general population, constitute the most important element. All sectors of the society are concerned about how a population changes and the direction in which it develops. After 1949, the population question caught the particular attention of a group of far-sighted scholars.

2.1 Resurgence of the Sociologists for Restrained Growth

Less than ten years after the foundation of the PRC, some sociologists began to publish articles on the population problems that were emerging from the social and economic conditions of the time. Thus, a new round of advocacy of restrained growth arose. These views were very similar in some ways to population theories and prepositions prevalent in the 1920s and 1930s, but they also differed from them significantly, mainly in their criticism of Malthusian theory. Their views are summarized below.

2.1.1 Shifts in Understandings of Malthus' Population Theory

In 1920s and 1930s China, many sociology scholars considered Malthus their mentor, approving and propagating the proposals he propounded in *An Essay on the Principle*

X. Tian, *Population Policies in the People's Republic of China*,
https://doi.org/10.1007/978-981-96-9658-1_2

of Population. However, around 1957, many of them also then wrote articles of self-criticism, one after another, explaining that they had been "misled" by Malthus and were guilty of spreading his reactionary ideas. One renowned scholar, Chen Changheng, wrote "I have published dozens of essays and books voicing my opinions on China's population problem from 1918 to 1945. Among them, some standpoints were absolutely erroneous and even reactionary, when I recall them now."[1] Another scholar, Chen Da, also expressed his regrets: "Without being equipped with ideas of classes, my views published about two decades ago did not have clear goals or a specific audience."[2] Sociologist Wu Jingchao, too, reviewed his earlier work, stating that he had once mistakenly attributed the poverty in China to its large population:

> my mistake was caused by my failure to investigate problems from the perspective of historical materialism and to apply the historical materialist approach in my research. I failed to analyse the population problem under certain social and economic circumstances. Neither had I realized that the population problem was closely associated with certain economic relationships ... Since I had not grasped the truth beneath the surface, I inevitably arrived at the wrong conclusion [which] only served to conceal the crimes committed by the reactionary classes, baffle the public's understanding of the major contradiction in our society, and obscure the targets of revolution.[3]

According to Chen Changheng, there were many fundamental differences between the kinds of population control proposed by Chinese scholars and Malthusianism. The two views could not be equated with each other. Realistically speaking, these scholars' own views and their critiques of Malthusianism were partial and rudimentary.

2.1.2 Development of the Moderate Population Density Model

While the sociology school advocated restraining population growth retreated from their positions taken under the influence of Malthus' theory, they did not abandon their belief in the need for birth control. One of their significant and valuable views targeted the new population problems that had emerged since the foundation of the PRC, including their idea of "moderate population density". Wu Jingchao commented in the book *China's Population Problems* that he used to believe that China's population could be reduced to 250 million or even 150 million through birth control but that this had turned out to be an illusion, an idealist error. He suggested that China's moderate population should be above such numbers and that the population was sure to keep increasing anyway. Fei Xiaotong, another distinguished scholar from the sociology school, said in his essay *What should population research study?* "In each society and in each period, a moderate population could have been calculated after taking

[1] Chen (1957a).

[2] Chen (1957b).

[3] Wu (1957).

into account all related conditions. If the population increases beyond a moderate number, it will retard social development."[4]

Sun Benwen, a sociologist in the geology department of Nanjing University, suggested that the optimal population for China was 800 million. After investigating population increases since 1949, he predicted that if China's population had kept increasing by 12 million people per year at the rate of 20 per thousand persons, the natural growth rate obtained from the population census conducted in 1953, then the total population would have reached 2.52 billion 72 years later, the same as the world population in 1957. Therefore, China's population cannot keep growing and must be curbed. After reviewing the relationships between population and means of livelihood, and between population and increased employment, Sun declared: "There can be no doubt that if agricultural technology keeps advancing and agricultural productivity keeps improving, grain output will absolutely continue to go up. Nevertheless, it should be noted that there is a limitation to grain output increases because the amount of farm land is limited. Hence if the population keeps growing and the total population reaches 900 million or one billion, the supply and distribution of grain will not be so optimistic." Regarding employment, he continued to voice his concern: "If the population continues to increase at the current rate, the total national population is expected to grow by 12 million people each year. Aside from the work offered by the agriculture sector, at least four million or five million people need jobs. If the population continues to increase at such a pace, there will certainly be problems in employment in the future." His analysis went on: "It is crucial that we keep our total population within the limits of the optimal number, to assure enough supply of the means of livelihood, such as grain and so forth, to resolve the employment problem, and gradually to improve people's living standards and education." On the basis of his analysis he concluded that "by the end of the third five-year-plan, that is, in 1967, our total population will probably have reached 770 million. This number is possible, even optimal … From my perspective, 800 million is probably the optimal population size for our country."[5] Though such a review is far from rigorous, and the argumentation still partial, Sun's treatment of the question of optimal population was of academic significance in terms of empirical research and his legacy was to leave the question to the coming researchers.

2.1.3 China's Rapid Population Growth After 1949 and Resultant Problems

The sociology school that advocated restraining population growth was unanimous in the opinion that China's population grew too quickly after 1949. What caused this rapid growth? After comparing the birth, death and population growth rates before and after the establishment of the PRC, Chen Da proposed five reasons. First,

[4] Fei (1957).

[5] Sun (1957).

after the overthrow of imperialist, feudalist and bureaucrat-capitalist rule in China, a popular democratic government was created, and thereby the people became masters of the country. Second, after industrial and agricultural production were restored and developed further, employment opportunities increased and living standards improved. Third, with the resumption and improvement of medical and health-care services, the death rate dropped. Fourth, scientific and cultural plans advanced rapidly and remarkable progress was made in the establishment of mass education. Fifth, as living standards rose, so did the number of people who married and had children. Although this analysis is not comprehensive, it reflects the immense impact of the political, economic and social changes on human reproduction after the success of the Chinese revolution. It shifted from a pattern of high birth rate, high death rate and low growth to one of high birth rate, low death rate and high growth.

This rapid growth generated new problems. The sociologists who advocated restrained population growth analysed the problems from many different perspectives and pointed out that if the pace of population growth did not slow down and the problems could not be resolved, then "the speed of industrialization and modernization of agriculture in our country will certainly be adversely affected. Likewise, our transitional period will certainly be greatly extended due to the rapid population increase."[6] Their work analysed population questions not only from a sociological perspective but also an economic one. According to Wu Jingchao, "the excessive high birth rate will hinder the fast increase of accumulation."[7] If the increase in accumulation was hindered, so would be the increase of investment, and the increase in technical equipment and labour productivity. Ultimately, this would impede economic development and create an imbalance between living labour and materialized labour.

2.1.4 Birth Control Policy Proposals

In accordance with their population theories such as overpopulation and the proposition of moderate population density, proper population control was the common conclusion of the scholars of the sociology school. Having compared diverse population control measures, ranging from development of agriculture and industry, rational deployment of labour power, to encouragement of migration, they found none of them ultimately capable of resolving the population problems. The problem of rapid population growth could not be finally solved unless birth control was implemented. In Wu Jingchao's opinion, to solve the problem of the imbalance between living labour and materialized labour, it was possible "to allow the coexistence of production at two different levels of technology, with one segment of production at the level of advanced technology and another at the level of relatively backward technology."[8] In considerations of investment arrangements, too, this situation should be analysed.

[6] Chen (1957a).
[7] Wu (1957).
[8] Wu (1957).

But the most effective solution was to take immediate measures to reduce the birth rate. After comparing China's birth rate with that in other countries, Wu believed it was entirely possible for the Chinese rate to drop from 37 per thousand persons to 17 per thousand persons. He declared that "conscious efforts need to be made to decrease the natural population growth rate from 20 per thousand persons to about five per thousand persons. The primary means is birth control."[9] Chen Da also expounded the merits of birth control: "Birth control is a positive measure rather than a negative one. It will not only greatly benefit individual family life, but also a country's construction and a nation's health and prosperity as well."[10] As for methods, the sociologists unanimously endorsed late marriage. Most disapproved of sterilization and abortion.

The sociology school that advocated restraining population growth in 1957 professed to have distanced themselves from Malthus' population theory, and had criticized his ideas to some extent. Yet readers will detect their enduring subtle connections to his ideas and to their own earlier work on population in the 1920s and 1930s. Nevertheless, we can distinguish their views from those of Malthus and they had indeed made significant criticisms which clarified their differences with Malthus. Unfortunately, this did not save most of them from being harshly denounced as Malthus followers after being labelled rightists in the Anti-Rightist Campaign of 1957.

2.2 Critiques of Ma Yinchu's New Population Theory

The sociologists who advocated restraining population growth did not have much influence and even this diminished after they were labelled "Bourgeois Rightists". Meanwhile, the New Population Theory, developed by Ma Yinchu, which had been praised by Mao Zedong and yet twice publicly criticized, deserves special attention, due to its deep influence on population policy in China. Ma's New Population Theory generated much controversy at the time and still does today. The story of Ma's population theory warrants retelling here.

2.2.1 *Ma Yinchu and the New Population Theory*

Ma Yinchu (1882–1982), was from Shengzhou county of Zhejiang province. After graduating from Peiyang University (Tianjin University today), he continued his studies in the United States, earning a BA in economics from Yale University, then a Master's degree and a doctorate in economics, as well as a doctorate in philosophy from Columbia University. In 1916, he returned to China and started teaching at Peking University, where he was later elected first provost. From 1929 to 1936,

[9] Idem.

[10] Chen (1957b).

he was as professor at Central University in Nanjing, Shanghai Jiaotong University, and Dongwu University in Souzhou. From 1938 to 1940, he was president of the Chongqing Business College while also a professor there. In October 1940, he was arrested under the Kuomintang government and detained in Xifeng concentration camp in Guizhou province. In August 1942, he was released, but placed under house arrest in Geleshan in Chongqing city. Thanks to the joint efforts of the Communist Party of China (CPC), including Zhou Enlai, some pro-democracy figures, and some American friends, Ma Yinchu regained his freedom after the Chinese victory in the War of Resistance against Japanese Aggression in 1945. In 1949, by special arrangement with the Shanghai branch of CPC, Ma succeeded in attending the first Chinese People's Political Consultative Conference and was elected committee member of the Central government, after reaching Beijing via a circuitous route that included stops in Hong Kong and Dalian.

After the founding of the People's Republic of China, Ma Yinchu was appointed deputy-director of the Finance and Economic Committee of the Government Administration Council of the Central People's Government of China, the vice-president of the East China Military and Political Committee, and Standing Committee Member of the first, second and fifth National People's Congresses. Besides his government positions, he served as president of Zhejiang University, Peking University, and honorary president of both Peking University and the China Population Association, and as councillor with the China Federation of Economic Groups.

Having lived for 100 years, Ma Yinchu witnessed the collapse of feudalism, the democratic revolution, the struggles against foreign invaders, and the extraordinary ups and downs of the socialist revolution and construction. Despite profound hardships and substantial setbacks, he persisted in pursuing truth and exploring ways to build a powerful country and a prosperous nation. His commitment to his country, and his moral integrity were highly esteemed. However, in recent papers about China's population policy, some have criticised Ma Yinchu's New Population Theory, and some have gone so far as to question his character and even to vilify him. History will be the judge of that question, but the New Population Theory that Ma developed was based on solid philosophical foundations and exhibits academic confidence. Ma Yinchu's theory, like the person, was sound and reliable.

As early as the 1920s, Ma had begun research in demography. His first essay in the field appeared in the Volume 4, No. 7 of the prestigious Chinese magazine *New Youth* in 1920. Yet, in the following 30 years, he focused on finance and economics, with few publications on demography. After 1949, he was excited to find socialist construction booming in every region of the country. However, whenever he returned to his hometown in Zhejiang province for research, he found that the sweets he brought back as gifts for the children were not enough, because of the rapid population growth. He was disturbed by his concern that this burgeoning population would hold back economic growth. Later, when performing his duties as a representative of the National People's Congress and as member of the Standing Committee of the National People's Congress, he would also hold meetings and discuss population issues with workers, farmers, and local officials.

In 1954–1955, he went to Zhejiang province three times for investigation. To quote his own words, "I have visited ten out of eleven prefectures[11] of Zhejiang province". During these trips, he thoroughly investigated population questions, especially the growth of the rural population and the increase in grain production. In 1955, after preparing a speech based on his survey, he decided to submit it to the National People's Congress. But before doing so, he presented it to the Zhejiang Panel of the NPC. Unexpectedly, some representatives raised objections to his ideas. Some said his research was the same old Malthusian stuff; others said that though his report differed from Malthus to some extent, their systems of thought converged, for they shared the same basic ideas. In a word, his presentation was disapproved by representatives present at the meeting. Although Ma Yinchu did not fully agree with the criticisms, he withdrew his paper, believing that the objections were made in in friendship. Then in February of 1957, Chairman Mao Zedong delivered a speech, *On the Correct Handling of Contradictions among the People*, at the Eleventh Enlarged Supreme State Conference. Mao's remarks contained a sentence about the population issue: "It is good to have a large population in our country though it also means difficulties". Ma Yinchu thought it was the right time to make his survey known to the public, so he presented the paper that he had prepared a long time earlier. Mao Zedong and other Party and state leaders considered his presentation a significant contribution.

To Ma's speech, Mao responded, "I want to talk about the population problem in China. The population is mushrooming in our country, with an annual increase of approximately 12 to 15 million people. It has become an important issue in China … To solve this problem, studies could be conducted to work out measures for people to have children in a planned way." He continued, "there is such a huge population in our country, but other countries in the world are not bothered by the same problem. Thus, we should advocate birth control, in other words, have children according to a plan. In my opinion, humans are incapable of managing themselves. In the factory, goods like steel, cloth, tables, and chairs are manufactured according to production plans. Humans do not usually have a plan when they reproduce. This is indiscipline and anarchy." At a subsequent conference in March 1957, Mao once again stressed,

> Humans should learn to control themselves. Population change should fluctuate. In other words, it might sometimes increase, sometimes stay at the same level, and at other times it could decrease a bit. From my point of view, birth control should be carried out, so that people will have children in a planned manner … This matter is worth studying. The government should set up offices to deal with it and develop measures of birth control. Do people have such a need? For my information, farmers have such needs. So do big families that have too many children. So, it is inappropriate to say there is no such need, since there are common needs of birth control both in the city and in the countryside.

Mao specifically mentioned Ma's contribution, saying, "Mr. Ma Yinchu is very insightful to put forward the suggestion of birth control."[12] Here, Chairman Mao

---

[11] A prefecture is an old term used to refer to an administrative area in feudal times in China. It is usually bigger than a county.

[12] The National Population and Family Planning Commission (2007).

explicitly mentioned "to have children in a planned way" and approved of Ma Yinchu's proposal by his compliment "Mr. Ma Yinchu is very insightful". These signals greatly encouraged Ma, and in an interview by a journalist with the *Wenhui Daily* in March 1957, he said, "Now, we can talk about the population problem openly, which indicates that our country has made rapid progress. Meanwhile, it implies that the problem is indeed serious. I will continue to speak about it at the impending National People's Congress."[13] He made some changes to his original speech and submitted the new version to the Fourth Session of the First National People's Congress held in June the same year. On July 5, the whole speech was published in the *People's Daily*, the most influential newspaper in China. This was the *New Population Theory*, Ma Yinchu's famous theoretical work.

2.2.2 The Gist of Ma's Theory

Ma's "New Population Theory", together with the articles published around 1957, his speeches, and his answers to journalists' questions explicitly expressed his basic views of the population question, analysed them in a practical way and made suggestions that suited the prevailing national conditions. In 1979, based on my collection of the above-mentioned materials, a book entitled *New Population Theory* was published by the Beijing Publishing House to showcase Ma's theory and propositions concerning population problems. His ideas can be summarized as comprising one precondition, five major problems, and three suggestions.

The precondition refers to Ma Yinchu's evaluation of population growth at the time, which was the starting point and footing of his New Population Theory. According to the First National Demographic Census conducted in 1953, the total Chinese population was 601,940,000, with a natural growth rate of 20 per thousand persons. This natural growth rate was calculated through data (birth rate 37 per thousand persons, death rate 17 per thousand persons) collected from 29 big and middle-sized cities, the entire Ningxia province, and ten counties selected from each of the other provinces. Ma Yinchu only saw the data as "a static record". According to him, though the data were correct at that time, there was no "dynamic record". "I am afraid it is inaccurate to take 20 per thousand persons, the data in 1953, as the natural population growth rate for the following years (1954-1957)." He listed reasons for his judgment from seven aspects.

First, the number of people marrying increased. Before the establishment of the PRC, young people faced unemployment after finishing school, but afterwards, the government would arrange jobs for them. Their improved economic conditions enabled them to establish their own families. Nearly everyone had a job and therefore did not need financial support from their families or friends. The lightened load contributed to the rise in marriages. The government was also concerned about the

---

[13] Ma (1979).

interests of married couples so in principle husbands and wives could now arrange to work in the same city, increasing their chances of having babies.

Second, the government was concerned about the welfare of pregnant women as well as infants. Women were entitled to 56 days of maternity leave before and after giving birth, an unprecedented entitlement. With the improvement of health services, the midwives who used to assist women with home births in the rural areas were replaced by licensed midwives in hospitals, thereby reducing new-born death rates. Nurseries were created to offer child-care to families with several children. Furthermore, school-age children could enjoy free public education.

Third, the number of elderly people increased, and they were well cared for. In the past, there had been a popular saying that it was rare to see people in their seventies, but it became common to see them. Elderly people received a pension when they retired, and the state would look after elderly people with no family. The ancient Chinese ideal that old widowed men and women, orphans, and the elderly with no family would be taken good care of was indeed realized.

Fourth, the death rate declined. In the past, people had been plagued by almost constant civil war. Innumerable people had suffered from looting by the Guomingtang army, and droughts and floods resulted in displacement and high mortality rates. There was also a decline in the number of people who die unnatural deaths, with the end of the civil war, and the disappearance of thieves and bandits, and a decrease in the murder rate.

Fifth, there were radical changes to social customs. With the transformation of the social system, more than half of the Buddhists nuns and monks resumed a secular life and chose to marry. In the future, China is likely to see few nuns and monks.

Sixth, the influence of traditional notions associated with fertility remained. With the launch of the agricultural cooperation movement, people's lives improved but older people still stuck to some traditional ideas and hoped that they will be blessed with long and happy lives. Popular traditional sayings such as "May the family prosper five generations running" and "There are three forms of unfilial conduct, of which the worst is to have no descendants" are still widespread. Under the influence of such sayings, as long as parents are rich enough, they will try to find a wife for their son and help him to settle down.

Seventh, the government adopted measures to look after multi-child families, including subsidies and rewards for the parents.

These factors and others would all contribute to the increase of birth rate and the decrease of death rate and accordingly, Ma Yinchu estimated that the population increase rate would very probably be above 20 per thousand persons.[14] In fact, the natural population growth rate was 24.8 per thousand persons in 1954, 20.3 per thousand persons in 1955, 20.5 per thousand persons in 1956, and 23.2 per thousand persons in 1957.[15]

Next, we talk about the five major "problems", that is, the contradictions between rapid population growth and the development of economy and society.

---

[14] Ma (1979).

[15] National Statistics Bureau of the PRC (1986).

First, there was a contradiction between rapid population increases and accelerated capital accumulation. Ma's main research area was financial economics. In 1914, he obtained a doctorate in economics from Columbia University with a dissertation entitled *On the Finances of New York City*, later published by Columbia University Press and used as textbook for first-year university students. No matter where he worked in universities or in government, Ma's academic approach was associated with financial economics, and his study of population problems was no exception. According to his *New Population Theory*, "the biggest contradiction in our country is between the rapid growth of population and seemingly rather slow accumulation of capital." After working people became the masters of the country with the founding of the PRC, national income was spent on meeting their material and cultural needs. The rest was accumulated in order to expand production. But consumption and accumulation should be in appropriate proportions. In the book, Ma remarked, "to better people's lives, production and reproduction must be expanded. To achieve the goal, accumulation must be increased ... Due to our country's large population, more is consumed, and less is accumulated ... My research concentrates on how to bring the population under control, so as to decrease the proportion of income used for consumption and meanwhile increase the proportion used for accumulation."[16]

Second, there was a contradiction between rapid population growth and the improvement of labour productivity. As *New Population Theory* pointed out, "the fastest way to increase capital accumulation is to improve labour productivity." To enhance labour productivity in industry, every worker's technical equipment should be updated. "At the same time, the population should be brought under control. If the population grows spontaneously, it will be difficult to accumulate capital quickly."[17] In the sector of agriculture, the reality was as follows: "In our country, there is a population as large as 640 million. However, the average arable land per person is less than three mu[18] ... The average arable land [per head] has, in fact, decreased from 2.8 mu in 1953 to 2.7 mu in 1955."[19] Hence agricultural productivity improvement depends on controlling population growth.

Third, there was a contradiction between rapid population growth and the supply of raw materials for industry. Ma pointed out that "in order to accumulate capital, it is best to develop light industry. With a relatively small amount of investment, light industry is easily constructed, yet it can reap profits within a short period of time. With capital accumulated from light industry, heavy industry can be developed faster and on a larger scale." However, most raw materials needed for the development of light industry come from agriculture. If the population multiplies unrestrainedly, then "grain production must be increased. As a result, the size of farming land for growing cash crops will decrease, which directly affects light industry and indirectly affects heavy industry. Based on this analysis, population growth means a decrease

---

[16] Ma (1979).

[17] Ma (1979).

[18] One "*mu*" is equal to 0.0667 hectare.

[19] Ma (1979).

in accumulation, resulting in the delay of industrialization. Thus, population growth must be restrained."[20]

Fourth, there was a contradiction between rapid population growth and improvement in livelihoods. In April 1957, in an interview with a reporter from the *Wenhui Daily*, Ma said "now, there is shortage in the supply of grain, pork, cloth and coal. All these conditions of scarcity are linked to the problems incurred by a large population."[21] After analysing the relationship between production and consumption, *New Population Theory* asserted that both accumulation and reproduction must be increased if people's living standards were to increase. Instead, increases in accumulation and production/reproduction, however, are constrained by a large population and massive spending on consumption. He concluded that population growth must be kept under control if people's lives were to be improved through promoting production.

Fifth, there was a contradiction between rapid population growth and the advancement of science and technology. Ma's book elaborated on the relationship between material production and these advances. Production is the basic driving force of scientific research, and material production is the foundation for advancement of science and technology. The book suggests that

> we must first advance technical equipment in the sectors of industry to promote productivity quickly and subsequently lay the material foundations for scientific research. At present, though conditions for scientific work in our country have been greatly improved, they still cannot fully meet the requirements of research because they are being constrained by current industrial levels and limited national financial resources. Only the acceleration of accumulation will help us out of this predicament. Efforts should be made to bring population under control so as to preclude population multiplication from holding back the progress of scientific research.[22]

Based on this analysis, Ma suggested that three detailed steps be taken to address the population problem in China.

In the first place, a new population census should be conducted to grasp the real picture of population growth within a span of five to ten years when a general survey was to be carried out, between 1958 and 1963. Population policy could only be made on the basis of a scrupulous and dynamic statistical analysis of real population growth. The estimated increase should be included in the second and third five-year plans, improving their accuracy.

In the second place, efforts should be made to propagate new customs, and to dispel traditional old-fashioned feudal beliefs like that stating that "may the family prosper five generations running", and "the worst form of unfilial conduct is to have no descendants. After the publicity worked to some extent, marriage laws should be amended to encourage late marriage. He considered it appropriate for men to marry at around 25 years old, while the proper age for women to marry would be around 23 years old. If such amendments did not effectively restrain population growth, then

---

[20] Ma (1979).

[21] Idem., p. 68.

[22] Ma (1979).

more drastic and effective administrative measures should be adopted. For example, those who have two children could be rewarded, those who have three be taxed and those who have four be heavily taxed. The revenues thus collected could fund bonuses for couples who have only two children. In the process, the national treasury would make no payment nor receive any revenue.

Last, contraception was recommended as a birth-control measure, while artificial abortion was rejected.

It is noteworthy that in a special section of the published version of *New Population Theory,* Ma detailed how "Malthus' population theory was reactionary". According to Ma, the Malthusian idea that population increases in a geometrical ratio and food increases in an arithmetical ratio was false, and that its intention was "to provide a theoretical base for the capitalist system and its government". Another section of the essay pointed out how his *New Population Theory* and Malthus' theory were based on different perspectives: "Malthus' starting point is to conceal the capitalist government's mistakes", while his own, entirely different, starting point was to improve people's material and cultural life via enhancing productivity. Ma also used examples from daily life to elaborate on the point that a large population does not help to realize mechanization and automation. Currently, he argued, it was inadvisable to develop a lot of large-sized industry. Instead, small- and medium-size industry in large quantities would provide more jobs for the labour force.[23]

There are two major reasons why Ma Yinchu would specially use two sections of his book to criticize Malthusian theory and specify the fundamental differences between it and his own. On the one hand, he was showcasing his understanding of and attitude towards Malthus' work. On the other hand, it reflected some trepidation that people might associate him with Malthus and see their theories as similar. So, Ma was eager to make a clear distinction between them, in order to avoid such a politically awkward situation. Unfortunately, fate did not favour him. Later, when *New Population Theory* was met with vicious criticism, his arguments in these two sections served, ironically, as evidence against him.

2.2.3 Two Rounds of Criticism Against Ma's New Population Theory

After the article *New Population Theory* was published in 1957, Ma Yinchu was excited for a while. Whenever he met his friends and acquaintances, he would say that the country had made such progress that the population problem could now be discussed openly, and he decided to conduct further research into the problem. Unfortunately, at the moment when Ma was passionate about serving his country, the Anti-Rightist Campaign began. At the beginning of the movement, the whole sociology school, which advocated restrained population growth, was wiped out and some of its chief representatives were labelled "Rightists". As the movement grew,

[23] Ma (1979).

Ma Yinchu was also enveloped by the storm. On May 4 1957, Chen Boda, an alternate member of the CPC Central Committee Political Bureau and a theorist in charge of ideological work, addressed the 60th anniversary of Peking University, remarking sarcastically that Ma Yinchu should make a self-criticism of his population theory. Meanwhile, his views that the rate of population increase had outstripped the pace of agricultural development and that treated human beings only as consumers, ignoring the fact that they were first producers, were criticized in the work report of the Second Session of the 8th National Congress of the CPC. While at Peking University, of which he was president, Ma Yinchu was singled out by name and criticized for his article on the population question. Subsequently, some essays critical of his theory, and others that criticised his other propositions such as "comprehensive balance in national economy" and "revolving-around theory" appeared in national newspapers and magazines. The main accusations against Ma Yinchu at that time were as follows.

Firstly, Ma was painted as a faithful advocate of Malthusian population theory, which denied the historical materialist principle that a large population is beneficial.

Secondly, he was accused of distorting Marx's theory of relative surplus population in capitalist countries and wrongly applying it in China's context. Ma's opponents believed that population in a socialist country should keep increasing rapidly and there was no such a problem as surplus population.

Thirdly, he was accused of directing his critical remarks about population quality at the people, and was thought to have no sympathy for the working class and was even considered hostile to them. So-called population quality insinuated that the workers and peasants had little education and therefore had poor knowledge and low cultural tastes.

Initially, Ma took those criticisms seriously and strove to locate the defects of his theory and to fix them. But he did not waver from his fundamental viewpoints. Nor did the criticism preclude him from his administrative work at Peking University. The 10th anniversary of the founding of the PRC approached after the first round of criticism, then criticism of Ma entered its second stage. At that time, I had just enrolled at Peking University, so I witnessed the criticisms of the university president. There was a rumour that Kang Sheng, a so-called "Party theorist", an alternate member of the Political Bureau of the CPC Central Committee, and deputy director of the Publicity Department of Central Committee had proclaimed, in a speech during the 37th anniversary celebration of the founding of CPC held at Peking University in 1958, that Ma Yinchu was a devoted follower of Malthus instead of Marx. There was another rumour that Kang Sheng had written letters to some leaders in the theory circle suggesting that Ma be censured, like Dean Gooderham Acheson, then US secretary of state. Big-character posters accusing Ma Yinchu of propagating a poisonous population theory appeared extensively on the Peking University campus and even the walls of his residence and his writing desk were covered with such posters. Though big and small gatherings intended to criticize Ma Yinchu were held repeatedly, he was seldom present at them. At one meeting, of more than 100 people, a professor fabricated a story about Ma's eating habits, using dirty words in his own dialect to slander Ma. Obviously, such a meeting went far beyond the academic realm and, instead, had become a vehicle for political ends.

Ma Yinchu had two choices in this situation: he could take the criticism and make self-criticisms accordingly, in which case everything would be all right for him; the other was to insist on his theory and refuse to self-criticize, in which case the consequences would be unpredictable. Several of Ma's good friends exhorted him to take the first path, among them, Zhou Enlai, the first premier of the PRC, who also urged him to make some self-criticism and talk about his problem in a sincere way in such a grave political situation. Ma understood that the critical essays against him were not well-intentioned, but he did not care much about his personal interests or his future. What worried him was the advice from Zhou Enlai, for Zhou was not only a leader above him, but also a very close friend who had rescued him many times from political danger and guided him in his participation in the revolution. Yet, as a righteous man, Ma chose to defend truth and resolutely took the second road despite the immense political and psychological pressure.

In an article in the 11th issue of the magazine *New Construction* in 1959, Ma clearly expressed his resolution and his complicated feelings, saying:

> I want to express my heart-felt gratitude to another friend and make an apology to him. He tried all means to rescue me when I was detained in Chongqing. In 1949, I went to Beijing to participate in the administration of state affairs as a response to his invitation via telegraph. I am deeply indebted to his past help and I will never forget his kindness and friendship. Yet in the current academic dispute, I have not taken his sincere advice, which has made me feel very uncomfortable. I have not followed his advice because I am very confident about my theory and I must insist on it for the sake of intellectual integrity ... I sincerely hope that my dear friend will remain broad-minded. I will be very fortunate if he does not see my decline of the proposal to engage in self-criticism as an offence.[24]

There is no turning back when you shoot an arrow from a bow. Ma Yinchu kept his promise and remained committed to his standpoint.

In the 1st issue of *New Construction* in 1960, Ma's article "A Restatement of My Requests" was published, reaffirming his decision to take up the challenge of opening a forum in the *Guangming Daily* to exchange opinions with his adversaries. He restated his remarks in the article published the previous year: "The challenge is reasonable enough, and I will take it with gratitude. Though an old man of nearly 80 years old, I will respond to the challenge single-handedly, fully aware of being outnumbered. I will defend my position until my last breath and will never surrender to the critics who are accustomed to applying coercion instead of persuasion."[25] In fact, many people approved of Ma's theory at the time, however, under the special historical circumstances of the time, no publication of any article that endorsed Ma's New Population Theory was permitted.

Besieged by opponents, Ma Yinchu was already 78 years old and this was not the first time he had taken action single-handedly. The first time, when he was a teenager, his father had wanted him to become an accountant but he refused, instead going alone to Peiyang University in Tianjin to continue his studies. Several years later, alone again, he travelled to the US to further his education at Yale and Columbia

[24] Ma (1979).

[25] Ma (1979).

universities. The second time, he expressed his opposition to the bureaucratic capitalism represented by the Big Four Family Clans,[26] when he was invited to deliver lectures on the economy to then-president Chiang Kai-shek (1887–1975).[27] Chiang was greatly offended by Ma's criticisms and challenges. Later, Ma became involved in the mass movement against civil war and widespread famine. He was consequently thrown into a concentration camp in Guizhou Province for his political offences.

The third of Ma's solo actions was very different from the first two. The one common point was his concern for the welfare of the people. Remarks he made in an interview for the *Wenhui Daily* reflected this: "Please think about the life of our people in the past. For most people, their grain lasted just six months and for the rest of the year, they had to survive on ground chaff and edible plants. Unfortunately, when famine occurred, even edible plants were not available. They had to subsist on the roots of grass and tree bark, and tens of thousands were forced to leave their hometown to beg for a living. Countless people even dropped dead while out begging."[28] After the founding of the PRC, productivity increased and people's lives improved though the improvement was seriously limited, mainly because of the high population growth. It was Ma's deep concern for the people's welfare, the development of productive forces, and the improvement of material and cultural life, that led him to publish his *New Population Theory* and to defend resolutely his academic integrity. We might assume that he never felt lonely when defending his stand single-handedly, thanks to this genuine concern for the general interest. Though confronted with more than 200 critical articles, and many organized meetings of criticism, he continued to defend his position, despite also being removed from his high government posts. His courage in adversity originated in his dedication to the people's well-being.

In the two waves of criticism of Ma Yinchu, the attitude of Mao Zedong, chairman of the PRC and General Secretary of the CPC at the time, was the source of much speculation. I too wanted to understand Mao's exact attitude towards Ma while I was editing *The Complete Works of Ma Yinchu*. Despite consulting a great deal of salient material, my efforts were futile, resulting in no direct evidence. Some believe that Mao approved of Ma's theory, as evidenced by the praise he gave him at the Supreme State Conference in 1957. Others believe that Mao was opposed to Ma's theory, since he cannot have been unaware of the severe criticisms of Ma and of his removal from his posts, and that Mao would not have turned a blind eye to Ma's suffering if he had supported his positions. Still others find it hard to say, presuming that there was both approval and disapproval, or that Mao had first approved but then turned to rejection of Ma's theory. Finally, though, it does not much matter whether there were any direct instructions or written comments on Ma's case from Mao. The

---

[26] The "Big Four Family Clans" dominated the Chinese national economy before 1949. Their chief political representative was Chiang Kai-shek.

[27] Chiang Kai-shek was president and highest military leader of China from 1926 to 1949. He retreated to Taiwan Island to continue his rule after his military forces were beaten by the People's Liberation Army.

[28] Ma (1979).

outcome for Ma and his theory was inevitable given the political, economic, and social circumstances and development in the middle and late 1950s. What matters is a practical and realistic understanding of Mao's thoughts on population and how they are incorporated into the system of Mao Zedong Thought. To this end, in the next section, I next share my limited view with readers for further discussion and investigation.

As is well known, on the eve of the founding of the PRC, US State Secretary Acheson alleged that China could neither solve its economic problems nor feed its people after the revolutionary victory. Chairman Mao responded sharply to Acheson's remarks: "It is an immensely good thing for China to have a large population. We can solve the consequent problems no matter how many times the population increases. The solution lies in production. Among all the things on earth, humans are the most precious. Under the leadership of the CPC, any miracle can be created, as long as there are humans." Mao also intended this assertion that revolution plus production can solve the problem of feeding the Chinese people as a maxim to refute all sorts of pessimistic views.[29] There have been many different interpretations of Mao's remarks. One is that they convey his idea of encouraging population growth, another is that they should be linked to his later comments on population to the effect that there are difficulties associated with a large population and thus birth control should be implemented. In my view, these words cannot represent Mao's core views of the population question for they were expressed to refute Acheson's negative assertions rather than to address the population problem specifically. In my estimation, Mao's basic views on population problem in China can be represented by two statements he uttered at other times.

The first of these important remarks was: "It is good to have a large population in our country, but there are also difficulties with it."[30] This view of population should be considered a mature idea because it was taken from Mao's speech "On the Correct Handling of Contradictions among the People", which is regarded as an exemplary piece of writing that elaborated on China's practical problems from the philosophical perspective of dialectical materialism and historical materialism. Mao was consistent in his view that it was good for China to have a large population.

On the eve of the founding of the PRC, Mao's article *On the Bankruptcy of the Idealist Conception of History*, refuted the fallacy that the revolution arose out of a large population. It also responded to Acheson's challenge by remarking that problems created by the large population can be solved, no matter how many times it increased, and that the solution lay in production. Because this article was written to respond to political needs it cannot be inferred that Mao advocated an ever-growing population. However, one point is quite clear: he was not afraid of a large population and he had held such a view during the revolution. He had led a revolution that overthrew Chiang Kai-shek's regime, defeated the Japanese invaders and liberated all of China through a strategy of encircling the cities from the rural areas and seizing power with military force. What was the decisive factor that contributed to

[29] Literature Research Office of the CPC Central Committee (1991).

[30] Literature Research Office of the CPC Central Committee (1999).

these victories? The people. Victory was achieved because the revolution relied on the mobilization and participation of the people.

After victory, capital was required to develop the national economy. The territory controlled by the PRC was the fourth largest in the world after the former USSR, Canada, and the US. Though China was endowed with abundant natural resources, the capacity to tap those resources was limited because it was a backward agricultural country, plagued by war for several decades. China's GNP was at a relatively low level. All this economic adversity was the natural result of invasion by the imperialist countries and the plundering of wealth by the Four Big Family Clans. As the PRC was being established then, its financial income fell short of its expenditure. There was also financial chaos and serious unemployment. In such tough circumstances, China's assets were only a large population and rich human resources, an incomparable advantage vis-à-vis other countries. If good use could be made of this relative advantage, then other advantages would follow, propelling economic and social development. Acheson had no such insight, so Mao criticized his idealist conception of history with the materialist conception of history. Mao considered it imperative to refute Acheson's negative proclamations.

Yet during the debate with Acheson, some of Mao's words went too far. For example, in saying that "it is an immensely good thing for China to have a large population", the adverbial modifier "immensely" is a step too far. Similarly, saying that "the consequent problems can be solved no matter how many times the population increases", was an exaggeration of the real possibilities. Mao arrived at the basic conclusions that a large population is not be feared, and that a large population "is a good thing" from the perspective of historical materialism. The view that a large population can also cause difficulties is based on a dialectical materialist analysis. To solve those problems, family planning should be promoted among the people. That is, a combination and consolidation of dialectical materialism and historical materialism is required to understand Mao Zedong's specific elaboration of a certain population problem and to apply it in different historical periods. Only with consolidation of the two philosophical methods can we correctly understand why Mao said "it is an immensely good thing for China to have a large population" in response to Acheson in 1949. If the two philosophical methods are divorced from each other or opposed to each other, it will be impossible to understand Mao's stand, methods, and views on how to solve China's population problems. Sometimes people have formed distorted views based on a limited understanding of Mao's specific sentence, taken out of the context.

Mao's second significant statement concerning China's population was "when the population in China reaches 700 or 800 million, it will be the time for Mr. Shao Lizi's proposals to be widely accepted."[31] Taken from a speech from Mao made to a Supreme State Conference, it was his spontaneous comment when Shao Lizi proposed the adoption of contraception and birth control programmes. Mao asked:

> Is it good to have a large population or a small population? In my opinion, it is good to have a large population at present. I am afraid the current population needs to continue to increase

[31] The National Population and Family Planning Commission (2007).

> a little. If we want people to practice birth control, there are two disadvantages before us. First, there are not enough contraceptives. Second, birth control methods are not propagated widely enough. Most peasants are illiterate and accustomed to marrying at an early age. We can't force them to practice birth control and they cannot control themselves. Human beings are still unable to control their own reproduction. Neither are several hundred million people in China able to keep their own actions under control. In my opinion, the situation will become tough when the population in China reaches 700 million. When it reaches 700 or 800 million, then will be the time for Mr. Shao Lizi's proposals to be widely accepted. When we talk about birth control now, people will turn a deaf ear to it. But that does not mean that we don't need to inform people of the importance of birth control. Or that we don't need to produce devices used for birth control. On the contrary, we not only need to spread relevant information of birth control, but also to try to do the job well. Personally, I endorse birth control and family planning as the Japanese people do. A large population can make the nation look more powerful. At the same time, we should also note the problems that ensue. Yet there is no necessity to fear the problems. Anyway, I don't fear them. If there is an increase of 200 million people, the population problem will be solved soon. Why? Because things will develop in the opposite direction when they become extreme. Now I think the population problem still hasn't reached its most serious point. China is a country with a vast territory and abundant resources. There is still some farmland that can sustain life. What can we do if there is not enough food for all the people? Eat less. To solve the problem, on the one hand, we should adopt birth control strategies. On the other hand, we should try to develop the habit of being thrifty.[32]

These words from Mao explicitly expressed his thoughts and schemes for implementing birth control. They contain several layers of meaning.

First, Mao endorsed birth control and family planning. He believed it was necessary to understand the significance of the population problem, but that there was no need to be too worried about by it. And at that moment in history, China needed a large population to look powerful.

Second, it was inappropriate to try to implement family planning on a large scale because the preparations were inadequate. Primarily, there were not enough contraception devices available, so family planning could not be practiced extensively. Nor were the ideas behind birth control sufficiently widely understood. People were not well-informed of the meaning and methods of contraception. And finally, the general public was still not mentally prepared for the change. Literacy among farmers was low, and they still followed tradition in marrying early. Most people still did not have the requirements for practicing birth control through contraceptive methods.

Third, Mao had conceived of his scheme for implementing birth control in the whole country. The general idea was to let the ideas follow their natural course of development and to change strategies when necessary. Sufficient preparations must be made before action was taken. For instance, patience is necessary for people to be properly informed. Patience is also needed to wait for people to improve their awareness of the consequences of a large population and the benefits of practicing family planning. At the same time, adherence to the philosophical principle that things will develop in the opposite direction when they become extreme must be sustained. If, say, there were an increase of 200 million people, then problems would

[32] The National Population and Family Planning Commission (2007).

emerge. But when conditions were ripe, the population problem would then be duly cracked.

Generally, the birth control actions taken coincided with Mao's predictions about the situation, following the scheme he had conceived. During the ten years following Mao's speech at the State Conference and another ten years of Cultural Revolution,[33] no major moves were made to control population increases or to encourage family planning practices. After the end of the Cultural Revolution, more measures began to be taken to control population growth and more strict policies were gradually enforced.

2.2.4 *The Value of the New Population Theory in Policy Drafting*

When Ma Yinchu presented his new population theory at a conference in 1955, it was immediately opposed by some representatives of the Zhejiang Panel of the National People's Congress. Later, though his ideas were appreciated for a short time, for a longer time Ma had to defend his theory single-handedly. Compared with the period of criticism, the period of being appreciated was merely a short episode for Ma. In 1960, the critiques of his theory had become so virulent that it was impossible for him to continue as president of Peking University so he submitted his resignation and it was accepted soon afterwards. Subsequently, he was removed from the Standing Committee of NPC, and his complete disappearance from academic and political circles as well as from social affairs followed. His new population theory and other economic theories have not been discussed publicly since then. However, that does not mean that they have been forgotten. Indeed, Ma's theory provides us with a clear theoretical basis on which to analyse the population question in China today and to formulate population policy. The value of his theory mainly lies in the following aspects.

First, Ma accurately identified and analysed the nature of the population problem in the PRC. The first demographic census conducted in 1953 showed that the total population had passed 600 million, and its release attracted the attention of far-sighted people in all sectors of Chinese society. In 1954, Shao Lizi (1882–1967), for

[33] In 1966, confronted with a grave and complex external environment at the time, the Communist Party of China was concerned to consolidate China's socialist state power, and made a wide range of efforts in this regard. However, Mao Zedong's theoretical and practical errors concerning class struggle in a socialist society became increasingly serious, and the CPC Central Committee failed to rectify these mistakes in good time. On the basis of an erroneous appraisal of the prevailing class relations and the political situation in the Party and the country, Mao Zedong launched and led the Cultural Revolution. Lin Biao and Jiang Qing took advantage of Mao's mistakes to commit many crimes that brought disaster to the country and the people, resulting in ten years of domestic turmoil which resulted in serious economic and social losses and setbacks. Finally, in October 1976, the Political Bureau of the Central Committee smashed the Gang of Four, putting an end to the catastrophic Cultural Revolution.

See https://language.chinadaily.com.cn/a/202111/18/WS6195aa45a310cdd39bc75fe0_2.html

example, a celebrated social activist and member of the Standing Committee of the NPC at the time, presented the first proposals to spread information about contraception, to instruct people in contraception methods, to provide contraceptives, to relax restrictions on birth control techniques, to increase the legal age of marriage and so on. His suggestions were intended to lighten household burdens and to promote maternal health but had scant concern for the size of the total population. Occasionally, Shao spoke to the effect that "it is beneficial to have a big population, yet there should be some restriction when faced with so many difficulties."[34] The views of the sociology school that had advocated restrained population growth reappeared in the 1950s. Their key ideas were clear, with the intention to convince that overpopulation was detrimental to efforts to raise household living standards and the development of the country. Essentially, they argued that only birth control can solve the problem of overpopulation. However, with the dread of being labelled followers of Malthus, their analysis of the nature of population problem in China was always evasive, and they avoided, for example, using the term "surplus population" in their research.

Top leaders of the CPC and the state as well as the central government had clear attitudes towards the population question. First, severe restrictions were to be lifted and new regulations concerning contraception and artificial abortion were to be formulated. Next, birth control was to be promoted, initially suggested by Deng Xiaoping. Thirdly, family planning was to be implemented, first proposed by Mao in 1957. Though these top leaders had expressed their views concerning population problems in some open speeches and some written instructions on the relevant documents, their chief concern was to improve women's reproductive health, to lighten people's heavy burdens, to provide better education opportunities for children, to offer greater job opportunities to young people and so forth, through birth control programmes. When difficulties experienced by the general population were discussed, they only remarked to the effect that "it is surely beneficial to have a large population" or "there are merits and drawbacks" at best.[35] Thus though the population problem was widely discussed by political leaders and scholars in China in the 1950s, the nature of the problem was not made explicit, out of apprehension or other concerns. Ma Yinchu was an exception.

Ma explicitly pointed out there was surplus population and that it was increasing too rapidly. His research made the following findings: As far as natural resources were concerned, there was little arable land per capita, and little petroleum and other minerals per capita, because of the large population. As far as industrialization was concerned, there were little capital and investment and few modern enterprises. As far as employment was concerned, there was a shortage of jobs despite a substantial workforce. As far as living standards were concerned, there was scarcity of grain, cloth, and other daily necessities. Meanwhile, production of such daily necessities increased slowly due to the large population and its rapid growth. As far as public services and facilities were concerned, though a large proportion of the population was teenagers and younger children, there was a lack of both primary and high

[34] The National Population and Family Planning Commission (2007).

[35] The National Population and Family Planning Commission (2007).

schools. Though numerous people were suffering from diseases, doctors and hospital beds were scarce. And scientific research was inadequate. As far as the developing state was concerned, capital accumulation and economic construction seemed not to keep up with population growth. The key point of Ma's new population theory was that capital accumulation and development of all sectors were held back by the large and growing population. What did he mean by "being held back by the population"? He meant that population oppresses productivity. Though the assertions "population represses productivity" and "surplus population" did not appear in Ma Yinchu's publications or speeches directly, his analysis had already identified China's population problem as one of overpopulation.

There are two types of population problem, under-population and overpopulation. The former is characterized by low birth rate, low and even negative growth, severe aging, and an under-supply of working-age people resulting in adverse effects on production, exchange, distribution, consumption, savings, investment, technological advancement, and social security. Under-population is a typical case of a society being constrained by productivity. It is typically a problem in developed countries such the US, Japan and some European countries. Overpopulation, on the other hand, is characterized by high birth rate, high growth rate, a young age structure, an oversupply of working-age people, and high pressure on education and employment offers. Overpopulation also affects production, consumption, accumulation, investment, technological advancement, the development of public services and so forth, and restricts economic and social development in a different way. It is a typical case of productivity being repressed by the population. Generally, the population problem in developing countries belongs to this category. In the 1950s, Ma Yinchu identified the population problem in China as overpopulation. His keen insights, solid academic training, and practical investigations of social realities all contributed to his distinctive research findings.

Second, Ma Yinchu was able to make proposals directly to top political leaders, exerting a special influence and achieving remarkable effects. In the 1950s, a considerable number of people had published articles and made speeches advocating birth control, but these articles and speeches mainly appeared in ordinary press outlets and academic conferences. Ma Yinchu's role was quite different. As a representative of the National People's Congress and member of its Standing Committee, he could carry out investigations in many places and hold meetings to listen to opinions from people of all walks of life, including workers, farmers, officials, and scholars in order to collect first-hand data. After that, he would pen reports based on his investigations and submit them to the NPC. His reports had to be taken seriously because they were among his official duties. On 5 July 1957, his article "New Population Theory" was published in the *People's Daily* as a written presentation by a NPC representative. And more noteworthy was the attention Ma was given by the then-chairman Mao Zedong following his confident speech, and his propositions for birth control and family planning at the Supreme State Conference. As already noted, we also know that other leaders of the CPC, including Liu Shaoqi, Zhou Enlai, Chen Yun, and Deng Xiaoping were all acquainted with Ma's population theory at that time, demonstrating his incomparable influence. No-one who held similar ideas about birth control had such an influence. Moreover, the long-lasting campaign of criticism of Ma Yinchu and his determined self-defence efforts served to spread his ideas further, while all the other scholars who advocated birth control were silenced in the Anti-Rightist Campaign. As a young university student at Peking University at that time, I had access to Ma Yinchu's population theory and the critiques written of him during the second round of criticism. When nearly 100 critical articles had flooded the newspapers and magazines, represented by the *Guangming Daily*, big-character posters covered nearly all walls in Peking University, and meetings of criticism were so visible and loud that it was impossible to turn a blind eye or a deaf ear. This experience led me to engage in research on demography and later to try to resuscitate Ma Yinchu's reputation.

Third, Ma had elaborated the basic principle for solving the population problem in a comprehensive way, giving due consideration to population size and quality. His "New Population Theory" mainly focused on explaining the contradictions between a large population and capital accumulation, industrialisation, technological advancement, ordinary people's lives, and science and education in social and economic development. To address these contradictions, he argued, population growth must be curbed. As an economist and a top expert in demography, he had to attend to the integrity of his theoretical system and the completeness of the guidelines and policies he was promoting. So when he concentrated on illustrating the importance of controlling population size, he also considered population quality. He had referred to this question in the *New Population Theory* article. Later, he published another article, titled *Why Emphasize Population Quality* and in yet another, *My Philosophical Thoughts and Economic Theories*, he explicitly stated that China should "enhance education levels" and "improve population quality". He had conducted research that associated population size with quality, showing that these aspects are not only interrelated but also reinforce each other. He commented: "In poverty-stricken China, where there is little capital and a large population, it is advantageous to organize the people and manipulate them as valuable resources, yet we should not ignore the disadvantages of a large population. A large population, of course, means enormous resources, but simultaneously it also means a heavy burden. My 'New Population Theory' recommends retaining the positive side of a large population and doing away with its negative side in order to preserve its advantages while eliminating its disadvantages. To achieve this goal, we should improve population quality and control population size"[36] Grasping the crux of the population problem in China, Ma positively illustrated the fundamental orientation required to solve it. His propositions not only went straight to the core of the problem at the time, but they also hold water today: simultaneously controlling population size and enhancing population quality represents the best strategy for solving the population problem in present-day China.

Fourth, Ma insisted on control of population size and on implementing birth control programmes and made detailed suggestions for doing so. At the end of February 1957, Chairman Mao presided over the Eleventh (Enlarged) Supreme State

[36] Ma (1959).

Conference, where he addressed unplanned reproduction, calling the situation anarchic for the lack of discipline and planning. At a conference held a few weeks later, he remarked: "Human beings should control themselves. I hope that people might have their children in a planned way. Sometimes, human reproduction might increase a little bit; sometimes, it might halt for a while. And some other times, we might reduce reproduction a bit, advancing in a wave-like manner."[37] It was at this conference that Mao complimented Ma Yinchu's positions. Two months later, on 9 May, Ma published his article "The Relationship between Population Problem in Our Country and the Development of Productivity" in the *Wenhui Daily*, stressing that "in short, the only effective way is to control population through practicing birth control." In other countries, the term *jihua shengyu* usually refers to family planning. In China, however, it refers to controlling population growth, controlling the number of births, to be exact. Ma Yinchu reasonably associated population control with having children in a planned way, an integration of ends and means. In terms of means and methods, he suggested the propagation of knowledge about family planning, the dispelling of traditional feudal beliefs, the advocating of later marriage, amendment of the Marriage Law, adoption of the use of contraception, rejection of artificial abortion, and the implementation of policies such as rewarding couples who have few children and taxing those who have more. These suggestions not only accorded with contemporary social realities but also showed a certain foresight whose influence is still felt in population policy today.

2.3 Twenty Years of Silence in Population Theory

Criticism of the Sociology School that had advocated restraining population growth, and especially the extensive and long-lasting criticism of Ma Yinchu's new population theory, affected Chinese society so adversely that in the following two decades, the population question was seldom discussed. As a freshman at Peking University, I was witness to the meetings of criticism of Ma Yinchu's theory held on campus. I also saw the two rounds of criticism of Ma's theories in newspapers and magazines. I was so baffled by all this that I frequented the university reading room to peruse all Ma's articles, including "New Population Theory" and other essays in *New Construction*. I also read his opponents' articles, including those published in the *Guangming Daily*. An image of Mr. Ma Yinchu, respectable President of Peking University, under siege kept appearing in my mind. I was immensely touched by his steadfast determination to stand up for his truth, even unto death. I was also impressed by his courage in accepting what he saw as reasonable elements of the criticisms and his willingness to change some details of his positions accordingly. The more I read, the more I felt that Ma was reasonable while his critics belonged to a different league. Few of the critical articles sounded convincingly scholarly, while the overwhelming majority of

[37] The National Population and Family Planning Commission (2007).

them took words from Ma's articles and deliberately misinterpreted them or decontextualized them. Through the misinterpretation of isolated words and phrases, these so-called critiques reduced Ma's views to "sheer Malthusianism" or attacked him politically with charges of "defaming socialism" or "serving the capitalists" and so on. Apart from such vicious tags, these critical articles also employed used quotations from Marx, Engels, Lenin and Mao in their bid to make Ma capitulate.

Such criticisms neither convinced Ma Yinchu nor had the intended effect on novice scholars like me. On the contrary, they aroused my indignation. The Chinese saying "he who laughs last, laughs best" prompted me to imagine that one day this history could be set right. Finally, my article *Defending Mr. Ma Yinchu against the Accusations about his New Population Theory* published in August 1979, in the *Guangming Daily*, after the Third Plenary Session of the 11th Central Committee of the CPC in December 1978, which had criticized the "excesses" of the Cultural Revolution and other political campaigns. Meanwhile, I compiled a collection of Ma's articles, speeches, and interviews with journalists about the population problem and the book was published by the Beijing Publishing House. The book was so popular that it was reprinted three times. Thus I felt quite relieved that my quest to do justice to Mr. Ma and his population theory was finally fulfilled after 20 years.

The two rounds of one-sided criticisms of Ma Yinchu's population theory had a dire impact on population research in China, especially since the study of demography and demographic theories were immature in China and consequently, partial and dogmatic theories were likely to be formed.

2.3.1 Treating the Population Question as a Political Issue

Because the population problem was treated as a political rather than a social issue, almost nobody dared to discuss it in the more than 10 years following Ma's academic obliteration. In his time, Ma's analysis of the population problem in China was denounced as "defaming socialism" (because it revealed socialism's weak points), and his New Population Theory was denounced as "a sharp sword" attacking both socialism and the CPC. Nearly all criticism of Ma Yinchu's proposals for controlling population linked it to Malthusian theory. Ma was condemned as a Chinese Malthus, as a devoted follower of Malthus instead of Marx. According to such critics, whoever discussed population-related problems, such as the lack of food, clothing, housing, jobs, education resources, and medical and public health services, was besmirching the image of socialism and defaming it. Whoever suggested a need to restrain population growth because of worries about overpopulation was labelled a Malthus-in-China. The resulting effect was to render the population question a forbidden political issue.

Demography emerged as an independent discipline in Europe in the seventeenth century. There had been discussions about population in more ancient times in both China and elsewhere. In ancient Greece, for example, in Plato's imagined "republic", there were proposals to keep the population size moderate—not too large and not too small. In ancient China, while Han Feizi discouraged population growth, Confucius advocated it. However, these discussions about population were not demography in any strict sense. Between the sixteenth and seventeenth centuries, some Europeans, including Giovanni Botero (1544–1617), John Grant (1620–1674), and William Petty (1623–1687) published research on population questions. Grant made the earliest statistical analysis of deaths, to discern the regular distribution of mortality, using parish registration records. Grant's work had inaugurated a new discipline-demography because of which Petty was called the father of "demography" though the term was not used in his time. At the same time, in *Political Arithmetick* and other publications, Petty analysed the relationship between population and wealth, especially between population and land. He also calculated labour forces as a fundamental national power, and designed the items and requirements of demographic statistics, contributing to the gradual establishment of demography.

In the eighteenth century, the development of capitalist economies raised many questions about population, encouraging further research into population and wealth, and the relationship between population growth and means of livelihood. Little progress was made in the field of demography except in a broad sense. With the industrial revolution in the middle of the eighteenth century, the capitalist economy took off, promoting the development of statistics. Statistics of population and general demography burgeoned. In 1790, the US government conducted the first census that comprised the whole country. Subsequently, some European countries including the UK, France, Norway, Belgium, and Germany also conducted censuses and many other specific surveys, gradually enriching data accumulation on population and promoting the development of demography as a field of study. In 1855, French scholar Achille Guilland adopted the term demography in his book *Éléments de statistique humaine ou démographie compare.* He too did systematic analysis of the factors that contribute to internal population changes, thereby marking what is considered the establishment of demography as a discipline. Two points can be made here about the formation of demography.

The first is that the birth of demography arose from the need for information about population changes and development. Demography developed special research objects and methods that cannot be replaced by other disciplines.

After its birth, demography developed along two paths: micro-demography, with its statistical focus on human reproduction itself, and empirical demographic research, which integrates social practices into the analysis, particularly cross-disciplinary research among demography and economy, demography and society, demography and resources, and demography and the environment. This research advanced, creating sub-disciplines and expanding the research scope of macro-demography along the way. But whether we refer to micro-demography or macro-demography, the field is a branch of science that studies population change and its laws of development. Demography does not belong to political science, still less is it a sub-branch of political science, though people may encounter different population problems in different periods, analyse them in different ways, hold different views about how to deal with them, using different methods in different societies with different degrees of economic development.

The various practical problems of population that Ma Yinchu discussed in his new population theory were broached for positive reasons. Ma wanted us to take the problems seriously, and to study and solve them; his intention was not to expose the weak points of socialism, and even less to defame socialism. To distinguish his thoughts from Malthusian thought, two sections specifically expressed his criticism of Malthus' thought; the first was headed "The Mistakes and Bankruptcy of Malthus' Population Theory", and the second was "The Differences between my Population Theory and that of Malthus".

First, Ma criticized the reactionary nature of Malthus' ideas: "Malthus' population theory was reactionary because he had written the book with the intention of providing a theoretical defence of the capitalist system and its government, by concealing the harmful measures taken by the British government. What Malthus' population theory told workers was tantamount to the lie that the poverty prevalent among workers was not the fault of government but was mainly created by the rapid increase of population and the slow increase of grain production. This stand was the starting point of his population theory and the point at which Malthus had committed a fundamental mistake."[38]

Next, Ma refuted Malthus' view that population increases in a geometrical ratio and food in an arithmetical ratio, using facts of economic growth and population increase in China and other countries and declared Malthus' ideas bankrupt.

However, the critics of Ma Yinchu' theory did not listen to his arguments. They firmly believed that Ma had attacked socialism and the CPC and that he was China's Malthus. The population problem was a political issue, so Ma must be condemned and repudiated politically. Afterwards, who dared to investigate the population question if that involved defying the general will of the people? In any case, it was impossible to have articles on overpopulation and related topics published. Consequently, the population question was a forbidden research area in the country with the largest population in the world for about two decades.

2.3.2 *Demographic Theory from "Eating Mouths" to "Working Hands"*

Hundreds of articles arguing against Ma's new population theory made a common criticism: that Ma Yinchu saw people merely as consumers, without first seeing them as producers. In other words, he only "noticed people's eating mouths while ignoring their working hands". A kind of dogma had gradually formed to the effect that there would be more labourers when there is a large population. With more labourers would come more accumulation. With more accumulation the economy would grow faster. Therefore, a larger population is preferable. In an article in the *People's Daily*, the author asked why people only noticed a person's mouth that eats without first noticing that he had a pair of hands that can almost do anything. He argued that the

[38] Ma (1979).

term "demography" ("人口论" in Chinese, literally "the theory of eating mouths") was problematic, and should be replaced by "theory of working hands" ("人手论"). Accordingly, there was, it was argued, an immutable rule that overpopulation exists only in capitalist societies and would never occur in a socialist society.[39]

This was regarded as authoritative demographic theory at the time. The dogma that a larger population meant more labourers and thus more production, more accumulation and faster development, was definitely out of tune with Ma's theory. Since it was formed in the process of criticizing Ma's population theory, his theory had, in fact, functioned as the counterpoint to the dogma. The following section presents some detailed analysis from three perspectives of demography, population economics, and population sociology. One key question is what is the relationship between human beings as producers and as consumers? Abstractly, the population is a unity of consumers and producers. Generally, a person creates more wealth in their lifetime than they consume. Hence there will be surplus, from which society can accumulate and develop. Seen from this angle, people are producers more than they are consumers. Similarly, on the individual level, we should first notice someone's working hands before their eating mouth. However, it should be noted that there are three preconditions for arriving at this conclusion, what I call, the three Don't Forgets:

In the first place, don't forget that there are objective time differences. Before people become producers, they are first and foremost consumers. We will again be mere consumers when we get old. We remain consumers from infancy to the end of our lives. Only during the working-age period (around 15 to 59 or 15 to 64, as identified by demography) do we become producers. Thus the time we spend as consumers is far longer than our time as producers. If fertility and birth rates are high, resulting in a large number of new-borns and a high percentage of young people, then the condition of insufficient production will be aggravated. In the 1950s and the 1960s, such circumstances prevailed in China.

Second, don't forget that certain preconditions are required before a person can become a real producer. Working-age people need certain means of production if they are to produce goods. If the means of production are insufficient, then it is futile to have so many working-age people. They will become only theoretical producers instead of real ones. In the 1950s, the main contradiction in China was a shortage of means of production rather than a shortage of labourers. A large population, poor economic foundations and low productivity were the salient socio-economic features. It was always a formidable task to create enough jobs for the working-age population. Some critics of Ma Yinchu concluded that the problem was a shortage of labourers rather than of overpopulation, for they adhered to the idea that "people are the most active element in productivity and are creators of inexhaustible social wealth" while turning a blind eye to the problem of insufficient means of production. Some even believed that the people can work miracles without material means. These opinions were more like hopes that served to comfort in the worrying situation of an ever-growing population.

---

39 Ruo (1959).

The third "don't forget" is that technological advances reduce requirements for human labour. With the continuous development of science and technology come rapid advancement in technological innovation and revolution in the chief material production areas, in both industry and agriculture. While the change in the number of workers in a certain sector or function is directly proportional to the increase in fixed assets, it is also inversely proportional to their technological equipment. The development of industrialization and modernization shows that labour will be transferred from material to non-material production, from agriculture to urban industry and commerce, and then from urban industry and commerce to tertiary industries, primarily the service industries. The view that the development of material production in agriculture and industry requires a large labour force was already out of date.

Another key question is what is the relationship between population change and accumulation? Assuming that people are mainly producers rather than consumers and that more people means more production, Ma's critics concluded that more people means more accumulation. And yet, as far as we know, when the rate of accumulation stays the same, the increase in a country's accumulation chiefly depends on increases in national income. The latter depends on the development of production and the conditions of labour and means of production. Only when there are sufficient means of production is it possible for the larger labour force to combine with the means of production to enter the stage of production and to increase production and accumulation. Conversely, if the means of production are insufficient then the increased labour force cannot enter the stage of production, no matter how numerous. Consequently, accumulation cannot be increased in the least bit. The greater number of labourers makes no contribution to increasing accumulation, yet they still need to be fed, clothed, housed and educated. With their consumption of the means of livelihood, social expenditures and the percentage of consumption in national income will rise. In these circumstances, if national income is fixed, then accumulation will surely decrease.

It should also be noted that there is a great difference between human reproduction and material production. The use value of a product made by a certain factory is fixed, and it does not have to change its form. The initial product of human reproduction, an infant, cannot realize its use value immediately after birth. Infants must be reared and educated before they can become real labourers and realize their use value. During the process of growing from baby to a child to a young person, we are merely consumers, decreasing rather than increasing accumulation. If there is an excessive rise in the juvenile population within a certain period, then an accordingly large percentage of expenditure will be spent on raising children and educating them. That means the decrease in accumulation and delay in industrialization which Ma Yinchu analysed in his new population theory.

The third key question for our purposes here is what is the relationship between population change and labour productivity? Ma Yinchu's critics usually did not deny that it was beneficial to improve labour productivity. However, they assumed not that population increases reduced labour productivity but rather that a bigger population means even bigger increases in labour productivity. Their idea was exemplified in the situation of a simultaneous increase in urban population and labour force, together with improvement of labour productivity. When agricultural labour productivity improved as the population increased, these critics seemed more convinced of the correctness of their theory. They used the examples that a lot of labour was needed to expand the amount of arable land, to exploit the mountainous areas and to develop a diversified economy during the 1958–1960 economic plan known as the Great Leap Forward. Furthermore, plenty of labour was also needed to construct water conservation projects, to raise per-unit yields through intensive farming methods, and the like. Hence more labourers guaranteed improved agricultural productivity. By producing evidence that there was a shortage of labour during the Great Leap Forward, Ma's critics also attempted to establish that labour productivity cannot be improved without continuous increases in the population.

The most basic relationship between labourers and production assets and labour productivity occurs when technological equipment remains unchanged, and the quantity of employed people is directly proportional to increases in fixed assets. When fixed assets increase, the number of employed people will rise. When fixed assets remain unchanged, the quantity of employed people is inversely proportional to their technological equipment. That is, when technological equipment improves, labour productivity increases and the quantity of employed people will decline accordingly. Therefore, changes in population and labour force are closely associated with employment and improvements in labour productivity.

In his "New Population Theory", Ma wrote:

> To promote labour productivity in industry, accumulation of capital should be enhanced. Moreover, each worker's technical equipment should be improved. Meanwhile, population should be controlled. The reason is that if population multiplies autonomously, it will be difficult to accumulate capital rapidly. The fastest way of accumulating capital is to promote labour productivity, and then workers' income will certainly be increased. If farmers' productivity cannot be improved proportionately, then the gap between workers' and peasants' incomes will widen, negatively affecting the alliance of workers and peasants. Therefore, it is urgent to control population growth. Otherwise, problems will become thornier in the future and it will be more difficult to solve them.

He elaborated on this point, saying that "with further development of socialism, mechanization and automation will surely be extended … Since our country has a large population, mechanization and automation should not be achieved within a short period of time. At present, we should not build many large factories. Instead, we should construct more medium and small ones, because these latter will offer more jobs … The large population has impeded high-speed industrialization in our country and makes it impossible for us to make enormous strides in economic development."[40]

However, the opponents of New Population Theory attempted to repudiate the good sense that it contained, in its propositions about the relationship between population change and labour productivity in population economics, for example. Faced with their theory that the faster the population and the labour force grow, the faster

---

[40] Ma (1979).

labour productivity would improve, people can do nothing but heave a sigh and wear a bitter smile.

2.3.3 Population Quality and Racial Eugenics

Ma's writing on the question of population quality was repudiated because it was said to "have originated in eugenics advocated by capitalist sociologists and racism propagated by fascists"[41] and to be "slander of the Chinese people and in the service of the imperialists."[42] As a matter of fact, "population quality" exists objectively; all matter contains quality and quantity, which are two inseparable aspects of its unity. Population is no exception. Whether Ma Yinchu's viewpoints on population quality were accurate and complete, we can decide at another time. But it is worth noting here that it was perceptive of him to discuss health quality, and to emphasize that the scientific, educational and cultural qualities constitute the core of the population's capacities. His research into population quality cannot be criticised as such; the charges of "slander", "service to imperialism" and the like were baseless. In the past, some negative influences were created by the criticisms of capitalist eugenics that made no distinction between the positive and negative aspects of eugenics and they hindered the improvement of population quality in China. Consequently, there was difficulty in finding employment for all those of working age. On the other hand, there was a severe shortage of the experts in some fields, engineers, and scientific and technological talents desperately required to develop the national economy. Thus there was an acute contradiction between population quantity and population quality. There are three aspects of population—quantity, quality, and structure. Though control of population quantity may represent the chief and dominant aspect of the population question in a given period, it is still indispensable to improve population quality and to adjust population structures. To resolve the population question in a comprehensive way, population quality and structure must be taken into consideration.

2.3.4 Rapid Population Growth Under Socialism

When criticizing Ma's view of restraining population growth, critics invariably stressed the rapid population growth and highlighted it. Some critical articles quoted from the *Textbook of Political Economy* published in the USSR: "The population will keep increasing quickly. People will enjoy a life with a high level of welfare, while the morbidity and mortality rates are very low. Meanwhile, people capable of working will be utilized fully and reasonably." Those critics believed that what was conveyed in this quotation was the essence of population development regulation

[41] People's Daily (1958).

[42] *Economic Studies* (1960).

for a socialist country. The rapid population increase in China demonstrated how it worked. Based on this understanding, they endorsed policies that encouraged population growth. According to them, "a socialist country will never fear rapid population growth. Instead, it will try its best to create all favourable conditions for rapid population increase, for it regards population growth as a great power that can accelerate socialist and communist construction and promote social development."[43] Though some critics also wrote approvingly of birth control to a certain extent, they did it not out of desire for population growth constraint but rather in order "to protect women's health, and to give children a better upbringing."[44] The notion that continuous rapid population growth is a rule of population development in a socialist country was taken from the *Textbook of Political Economy*. It was a dogma that met the need to criticize Ma's arguments for restraining population growth and it became an authoritative and dominant theory. Unfortunately, it dominated the field of demography in China for a long time, with far-reaching effects.

We can draw several important lessons from those years of confusion: First, we should understand the relationship between academic questions and political issues. Academic questions, especially those in the social sciences, are usually linked with politics. But since they belong to different categories, they should be treated using different methods. From the perspective of national political life, Ma Yinchu's presentation of his population theories at the Supreme State Conference, in his role as a representative of the NPC, became a political issue. The question of how to respond to his proposals and views was entirely determined by organs of state power. But the population problem he presented was an academic question in terms of content, and a theoretical question in nature. When dealing with academic questions, the principle "Let a hundred flowers blossom, and a hundred schools of thought contend" should be adopted. These questions should be solved through free and reasoned discussion and presentation of facts rather than by coercion by sheer weight of numbers. Where academic questions are concerned, truth does not always lie with the majority. However, during the period from the late 1950s to the early 1960s, this principle was violated in the so-called debates with Ma Yinchu. Especially with the vicious interventions by Chen Boda and Kang Sheng, the academic question was transformed into a purely political issue. Thus, criticism was blocked and views advocating restrained population growth vanished overnight. Moreover, the population question became a forbidden topic that no one dared to discuss any more. Consequently, almost no article on population problem was published in the following ten or more years.

Second, academic debate should be democratic. A strange phenomenon could be seen in the criticisms of and debates with Ma Yinchu: while there 100+ people on one side, there was only ever a single person on the other side. To quote from Ma Yinchu, "I fight for my standpoint single-handedly". Did it mean nobody endorsed his position? No. About this situation, Ma wrote: "Since the publication of my article in the 11th Issue of *New Construction*, not a few people have written to back me up. Some people dare not to sign their letters, only signing 'Respectfully Yours', although

[43] *New Construction* (1955).

[44] *New Construction* (1960).

they support my theory." Indeed, many people approved of Ma's population theory. His view of maintaining a comprehensive balance in the economy also won many supporters. But few people dared to stand up for him in public because of the political pressure. Such debates cannot lead to any truth.

Third, we should persist in adhering to the principle that practice is the only criterion with which to test truth. Before 1949, China had a high birth rate, but a high mortality and a low growth rate. After the founding of the PRC, there was a serious decline in the mortality rate, and the population grew rapidly, because the government under the leadership of CPC effectively resolved a series of problems, including unemployment, floods, famine, hunger, and epidemic diseases. Based on his own thorough investigations, Ma Yinchu's *New Population Theory* was written to address these new trends in population change. In sharp contrast, many people criticized him without engaging in field work themselves or analysing materials carefully to understand the new problems and contradictions emerging from these changes. Instead, their criticisms were dominated by dogmas, divorced from reality. The more severe the criticism of Ma, the further it was from reality, and the greater the harm to research into the population question. Thus the situation was averse to the formulation of scientific population policy and the solution of the problem.

As the criticisms of Ma's new population theory proceeded, one-sided theories were being formulated in the process. In theoretical circles and in public opinion the view that "the larger the population, the better it is" dominated. All these combined to make it impossible to confront the population question seriously. And thus explanations of prevailing population policies sounded feeble. In these circumstances, it became vital to have Ma Yinchu rehabilitated and have the false accusations against his theories reversed in order to resume studies of population theories. However, this wish remained a wish; the whole country was severely affected by the "Left" trend of thought at the time, and the entire society was reduced to political turmoil for ten years during the Cultural Revolution. Under immense political pressure, public opinion was strongly influenced by the view that "it is good to have a large population". So questions concerning population theory and population policy were handled strictly according to instructions from top Party and executive leaders. Consequently, there appeared the abnormal phenomenon that theory, policy, speech, and action remained separate from each other and incompatible with each other. This situation did not change until the guiding principle of the CPC—"to seek truth from facts"—was re-established at the Third Plenary Session of the 11th Central Committee of the CPC in 1978. Following the setting aside of the unjust accusations against Ma Yinchu's new population theory, scientific studies, including demography, began to flourish again in China and the "abnormal" separation of fundamentally connected elements began to change.

References

Chen, C. H. (1957, May 4). Discussion on the Population Problem of China during the Transitional Period. *Wenhui Daily*.

Chen, D. (1957). Birth Control, Late Marriage and the Population Problem in the People's Republic of China. *New Construction*, (5).

Economic Studies. 1960 (1).

Fei, X. T. (1957). What should Population Research Study, *New Construction*, (4).

Literature Research Office of the CPC Central Committee. (Ed.). (1991). *Selected Writings of Mao Zedong* (Vol. 4). People's Publishing House. pp. 1511–1512.

Literature Research Office of the CPC Central Committee. (Ed.). (1999). *Selected Writings of Mao Zedong* (Vol. 7). People's Publishing House. p. 228.

Ma, Y. C. (1959). My Philosophical Thoughts and Economic Theories, *Journal of Peking University* (Social Science Edition), (5).

Ma, Y. C. (1979). *New Population Theory*. Beijing Publishing House.

National Statistics Bureau of the PRC. (Ed.). (1986). *China Statistical Yearbook of 1986*. China Statistical Publishing House. p. 92.

New Construction. 1955 (12)

New Construction. 1960 (4).

People's Daily. (1958, June 6).

Ruo, S. (1959, April 15). Eating Mouths and Working Hands, *People's Daily*.

Sun, B. W. (1957, May 11). 800 Million is Probably the Optimal Population Size for China. *Wenhui Daily*.

The National Population and Family Planning Commission. (Ed.). (2007). *Population and Birth Control History in China*. China Population Publishing House.

Wu, J. C. (1957). A New Discussion on China's Population Problem, *New Construction*, (3).

Chapter 3
Growing Concerns About Population Policy

From the survey of population policies in the different dynasties of Chinese history as presented in Chapter 1, and the review of disputes about the population question in the 30 years following the founding of the PRC in Chapter 2, we can conclude that the role of policy and culture in population change, and their impact on it, should not be ignored or underestimated. However, in certain historic circumstances, neither policy nor culture can fully account for demographic change. Because the population question is essentially a development one, determined chiefly by economic, scientific and technological, political and social conditions, in order to investigate the causes and effects of population policy changes from 1949 to 1979, we must first examine those conditions in that period.

3.1 The Population Grows Despite Fluctuations

For 30 years after the founding of the PRC, China's population went through dramatic fluctuations, and even disputes over population policy caused population numbers to go up and down. The trends must be explained in relation to the prevailing political, economic, and social conditions, and special attention must be paid to economic development, since demographic changes are based on certain economic foundations. An analysis of population change between 1949 and 1978 is presented chronologically in the following section.

X. Tian, *Population Policies in the People's Republic of China*,
https://doi.org/10.1007/978-981-96-9658-1_3

3.1.1 Three Years of Recovery: A Quick Start to Demographic Transition (1950–1952)

Demographic transition theory is still regarded as one of the most mature theories of demography. There are three different stages of transition, namely the stage of high birth rate, high mortality rate, and low growth, the stage of high birth rate, low mortality and high growth, and finally the stage of low birth rate, low mortality and low growth. Countries with large populations have, almost without exception, gone through this transition pattern and followed similar trajectories. China has long been the country with the largest population in the world. And though it has experienced many population changes: increase—decrease—increase, it has never been a regular pattern of population transition. Since 1949, enormous changes have taken place in the fields of politics, economy, and society in China, providing the necessary conditions for the frequent demographic changes that took place. To explain why the first stage of population transition, characterized by a decline in mortality rates, occurred we must first discover what transformations took place in politics, economy and society.

Before the revolution, China had suffered from external aggression for more than 100 years and from incessant fighting among warlords for several decades. In addition, as Chiang Kai-shek and his forces retreated to Taiwan Island after their defeat by the revolutionary forces, they plundered what they could, leaving every sector of the economy wrecked and every industry in need of resurrection. The main task at the time was to carry through the revolution, defeating the Kuomintang (KMT) troops. The People's Liberation Army (PLA) led by the CPC marched down to the south of the country, clearing out the remnant KMT troops stationed in the south, south-west and north-west regions of China. After this series of successful military actions, the people's government was established and in 1951, Xizang was peacefully liberated, which marked the unification of the mainland of China, except Taiwan, Hong Kong and Macao. This liberation of the mainland of China and China's great victory in the War to Resist US Aggression and Aid Korea greatly consolidated national defence and inspired the whole nation to work for a better future. The Chinese people had been subjected to miserable suffering during the incessant wars of the previous 100 years, so when peace was installed, there began the task of creating the conditions necessary to the recovery and development of national production. The founding of the PRC and the reestablishment of a national economy enabled the beginning of peaceful construction.

A peaceful social order first made it possible for displaced people to return to their hometowns, and thereby created a peaceful environment for human reproduction. But if people are to enjoy their work, then economic growth must be promoted. Compared with the best economic year before Japan's invasion of China in 1937, agricultural output in 1949 had dropped by more than 20%, grain yield had dropped by 22.1%, and cotton output by 46%. There were more serious reductions in the primary and industrial sectors: light industry output declined by 30%, coal by 47.6%, steel production by 62.2% and the volume of freight transport declined by 57.4%.[1] With channels of commodity circulation blocked as well as domestic and foreign trade, no link in the production, exchange, distribution and consumption chain was stable. Merchants illegally hoarded goods, and banks and old-style private banks sought high profits through speculation. There was a desperate shortage of materials, with the result that prices spiralled up out of control. National revenues fell short of expenditure and people lived in extreme poverty, with vast numbers of urban dwellers out of work. In these circumstances, population reproduction was only simple reproduction, with the salient features of a high mortality rate, high infant death rate, short life expectancy, and a low increase rate.

Confronted by these difficulties, the nascent PRC government took a series of bold measures in an effort to improve the financial situation. These included the abolition of the imperialist countries' privileges, confiscation of bureaucratic capital, and land reform aimed at ensuring that peasants had their own fields to work. The prices of goods were stabilized, and public finances were centralized. Industry and commerce were reorganized and more jobs were offered. Movements were launched to fight against the "three evils" of corruption, waste, and bureaucracy and the "five evils" of bribery, tax evasion, theft of state property, cheating on government contracts, and stealing of economic information to increase production. Key projects were established to enhance the strength of the national economy.

After three years of recovery and construction, the war damage had been repaired. By the end of 1952, agricultural output had surpassed its highest previous level; transportation, domestic commerce, and foreign trade reached new heights, and people's lives were improved significantly. First, prices were stable, and there was a sufficient supply of daily necessities like grain and cloths, a clear contrast to the soaring prices before the founding of the PRC. Second, farmland was allocated to the peasants, who accounted for 90% of the total population, and who could now feed themselves and meet their basic needs. Third, the problem of the jobless four million urban dwellers was solved, with more than half of them finding employment, and most of the rest properly arranged. Industry and commerce in cities were restored and progressing. In 1950, labour insurance was introduced in medium- and large-size enterprises, and a medical insurance system covered public employees, including teachers, initiating the social security system in China. Fourth, there were rapid developments in education, medical services, and public cultural undertakings. All types of schools at all levels were re-opened and enlarged their enrolments. Hospital bed capacity doubled, and effective treatments were found for infectious diseases like tuberculosis. Cultural facilities like libraries and local culture centres were built. In other words, people's material and cultural lives were gradually improving. The progress made in all these fields indicated that the tasks set for the three years of recovery had been fulfilled and from 1953, the PRC entered the stage of large-scale economic development.

For most people, the achievements of these three years of economic recovery met the population's demand for material subsistence. The working-age population's demand for means of labour had also been met to some extent. Improvements in

[1] Liu and Wu (1985, p. 16)

Table 3.1 Natural population change in the PRC 1949–1952[2]

Rate change per 1,000 people per annum	1949	1950	1951	1952
Birth rate	36	37	37.8	37
Morality rate	20	18	17.8	17
Natural increase rate	16	19	20	20
Total fertility rate	5.80	5.81	5.70	6.05

medical care, health services and sports had reduced the mortality rate, and in particular the infant mortality rate, which plummeted. Social stability and unity, coupled with the drastically reduced risk of dying from war and pestilence, shifted population reproduction from a pattern of high birth rate–high mortality rate–low increase rate to one of high birth rate–low mortality rate–high increase rate. Table 3.1 demonstrates this.

Table 3.1 indicates that from 1949 to 1952, the birth rate in China remained at a high level, about 37 per 1,000 people, showing the continuity of high fertility rate. There was a relatively high mortality rate but this was gradually reduced, to 17 per 1,000 people by 1952. This change reflected the decline in the mortality rate. Conversely, the natural increase rate went up, reaching 20 per 1,000 in 1951 and 1952. This shows that the three-year national economic recovery efforts had resulted in a transformation to a pattern of high birth rate–low mortality rate–high increase rate, exhibiting new potential for population increases.

3.1.2 The First Five-Year Plan (1953–1957): The First Baby Boom

The three-year economic recovery saw a shift in the population reproduction from the high birth rate, high mortality rate, low increase rate pattern to the high birth rate, low mortality rate, high increase rate pattern, revealing a new trend of increase. Would this trend continue? Theoretically, the restrained growth position advocated by the sociology school had been criticized, public opinion was in favour of a population increase, and no policy on birth control had been formulated. Instead, local governments occasionally rewarded "hero mothers", who had had several children, indicating that official policy favoured a bigger population. More importantly, the central government had taken a series of significant measures to implement large-scale economic construction, exemplified by the launch of the first Five-Year Plan, creating a social environment favourable to fertility.

In 1952, the CPC Central Committee presented its general line for the transition: "there will be a transition period from the founding of the PRC to the time when socialist transformation will be basically completed. The general line and general

[2] National Statistics Bureau of the PRC (1986, p. 92).

mission lie in gradually realizing the country's socialist industrialization and socialist transformation of agriculture, handicrafts, and capitalist industry and commerce, over a fairly long period." This general line for the transition period was vividly called "the body and two wings": the body referred to socialist industrialization, and the two wings to the transformation of agriculture, the handicraft industry, and capitalist industry and commerce. By 1956, the transformation was almost complete; the industrial sector was either nationalized or functioned under collective or joint state-private ownership arrangements, and together accounted for more than 90% of national revenues, the percentage of self-employed and private businesses dropping dramatically. These conditions facilitated the planned economy, contributing to the formation of a highly centralized and unified economic system.

According to the requirements of this general line, the whole country participated in the first five-year economic development goals, with the focus on 156 large projects aided by the USSR, and the industrial construction carried out by 694 construction projects above the norm (in the First Five-Year Plan). By 1957, the First Five-Year Plan had more than fulfilled its quota. Fixed assets had risen by 46 billion RMB in capital construction investment. A group of industrial and mining enterprises, including large coal mines, iron and steel foundries, machine manufacturing factories, non-ferrous metal enterprises, and chemical fertilizer plants had been established. Many railways and highways had been built. The completion of large-scale irrigation and reservoir projects had improved capacity to resist disaster and mitigate disaster. With progress in social programs such as education, science, culture, health, and sports, people's living standards had improved markedly.

Between 1952 and 1957, the gross societal output value rose by 70.9%, from 101.5 billion RMB to 160.6 billion RMB. Specifically, total industrial output value grew from 34.9 billion RMB to 70.4 billion RMB, an increase of 128.6%, and its portion of the gross societal output value rose from 34.4% to 43.8%. Total agricultural output value went from 46.1 billion RMB to 53.7 billion RMB, an increase of 24.8%. Since total agricultural output value grew more slowly than that of industry, construction, transportation and commerce, its percentage of the gross societal output value declined from 45.4% to 33.5%. National revenue went up from 58.9 billion RMB to 90.8 billion RMB, with per capita income increasing 36.5%, from 104 to 142 RMB.[3] It is not difficult to see that construction plans for the First Five-Year Plan had been successfully implemented, fulfilling the plan's targets for both quality and quantity. In sum, in this period, the economy developed rapidly and steadily, and living standards were notably enhanced, creating new conditions for a population increase. These are the circumstances in which the PRC's first baby boom occurred (see Table 3.2).[4]

In 1952, the total population of China was 574.82 million. In 1957, it was 646.53 million—an increase of 71.71 million people in five years, an average of 14.34 million people annually, making an average increase rate of 23.79 per 1,000 people per year. This baby boom was characterized by on the one hand first, a continued high birth

[3] National Statistics Bureau of the PRC (1986, pp. 36, 37, 39, 52).

[4] Idem, p. 92.

Table 3.2 Natural population change 1953–1957

Rate change per 1,000 people per annum	1953	1954	1955	1956	1957
Birth rate	37.00	37.97	32.60	31.90	34.03
Mortality rate	14.00	13.18	12.28	11.40	10.80
Natural growth rate	23.00	24.79	20.32	20.50	23.23
Total fertility rate	6.05	6.28	6.26	5.85	6.41

rate followed by a slight drop. On the other hand, the mortality rate decreased by 6.2 per 1,000 people in five years, an annual average of 1.24 per 1,000. Thus the annual average population increase rate approached 24 per 1,000 people, creating the first rising peak in population increase.

3.1.3 The Great Leap Forward and Three Years of Low Fertility

The first baby boom was interrupted by the Great Leap Forward,[5] which began in 1958. Under the general banner of the general line of socialist construction and the slogan "Construct socialism faster and better with more energy and more effort in an economical way", across the country people participated in the movement in high spirits, with revolutionary enthusiasm. The movement to form rural people's communes had also been forging ahead and was later incorporated into the Great Leap Forward. However, the movement in the rural areas deviated from its original goals, misled by such extreme slogans as "the land will yield as much grain as people can boldly imagine". Numerous reports of extremely exaggerated agricultural output appeared in the newspapers. In the same year, at the supreme state conference, Chairman Mao remarked: "Is it good to have a large population or a small population? In my opinion, it is good to have a large population at present. I am afraid the current population needs to continue to increase a little … A large population can make the nation look more powerful. At the same time, we should also note the consequent problems. Yet there is no necessity to fear the problems. Anyway, I don't fear them. If there is an increase of 200 million people, the population problem will be solved soon."[6]

In May 1958, Liu Shaoqi, vice-chairman of the CPC, submitted a work report to the Second Session of the 8th National People's Congress on behalf of the Central Committee, in which he anonymously criticized the views of the sociology school that advocated restraining population growth, represented by Ma Yinchu. The report

[5] The Great Leap Forward was a mass movement that lasted three years, characterized by the setting of high targets for agricultural and industrial production across the country. The targets proved to be exaggerated and unrealistic and as a result, the movement turned into a major setback in the construction of Chinese socialism.

[6] The National Population and Family Planning Commission (2007, p. 44).

said: "Some scholars even assert that the progress in agriculture cannot keep up with the speed of population increase. They think when the population grows, consumption will increase but capital accumulation cannot increase simultaneously … They only see people as consumers: When there are more people, there will be more consumption. The problem is, they fail to realize that people are producers in the first place. When there are more people, it is more likely that they will increase production, thereby increasing capital accumulation. Obviously, the views of those scholars run counter to the principles of Marxism and Leninism."

On 10 December 1958, the *Resolution on some issues concerning People's Communes* was passed by the 6th plenary session of the 8th Central Committee of the CPC. The document set out the following prospects:

> In the past, people were often worried that there was not enough arable land for the large population. Yet the bumper harvest of 1958 disproves the worry. The experience of reaping a real bumper harvest from deep ploughing and careful cultivation, separate-layer fertilization and rational, close planting is to be recommended. If the experience is meticulously popularized, there will be enough arable land for the peasants. And then, the population is no longer in surplus. Instead, a shortage of labour force will be observed. This means a huge change. We should strive gradually to reduce, over a few years, the amount of arable land that is currently planted to crops to, say, one third of its original size, depending on local conditions. One part of the vacant land will be fallow, planted only with forage grass or weeds that can be used for fertilizer. Another part will be planted with trees or transformed into reservoirs. Ornamental plants of all colours are expected to grow on the plains, in the hills and on the water, creating a beautiful landscape in the country.[7]

However, instead, the Great Leap Forward ended in a massive devastation of productivity. Consequently, the commune movement gradually regressed to the primary level of agricultural cooperation and was later transformed into a system of contracted responsibility linking remuneration to output following the Reform and Opening-up policy. The destruction of productivity was a heavy blow to population production, and various evaluations have been made of its negative influence. Data provided by the National Bureau of Statistics clearly accounts for the influence if we compare the data before and after the Great Leap Forward. Between 1957 and 1961, the birth rate dropped from 34.03 per 1,000 people to 18.02 per 1,000, a decrease of 16.1 per 1,000 people. In contrast, the mortality rate increased from 10.8 per 1,000 people to 14.24 per 1,000, an increase of 3.4 per 1,000. More specifically, in 1960, the mortality rate rose to 25.43 per 1,000 people, an increase of 14.7 per 1,000 over the 1957 figure. As far the natural growth rate is concerned, it declined from 23.23 per 1,000 people in 1957 to 3.78 per 1,000 in 1961, a decrease of 19.6 per 1,000. It is noteworthy that the natural growth rate plummeted to -4.57 per 1,000 persons in 1960, a decrease of 27.57 per 1,000. See Table 3.3.[8]

---

[7] Peng (1997, p. 4).

[8] National Statistics Bureau of the PRC (1986, p. 92); Population and Employment Statistics Division of National Statistics Bureau (1995, p. 40).

Table 3.3 Natural population change 1958–1961

Rate change per 1,000 people per annum	1958	1959	1960	1961
Birth rate	29.22	24.78	20.86	18.02
Mortality rate	11.98	14.59	25.43	14.24
Natural growth rate	17.24	10.19	-4.57	3.78
Total fertility rate	5.68	4.30	4.02	3.29

3.1.4 *The Economy, the Cultural Revolution and the Second Baby Boom*

After the three years of severe national economic difficulties created by the Great Leap Forward, the CPC Central Committee adopted the Adjustment, Solidification, Enrichment and Enhancement policy, in September 1960, which achieved some preliminary success in 1962. Then in May 1962, the Central Committee held a conference that discussed and voted up a report on an economic adjustment plan drafted by the Panel Responsible for Financial and Economic Work. The conference also resolved to make further adjustments to the national economy by setting feasible economic goals. This "comprehensive adjustment of national economy" referred to the further adjustments to be made in the following three years, on the basis of the preliminary success achieved in in 1962. That is to say, the further adjustments would extend to 1965.

The economic problems induced by the Great Leap Forward had exerted more influence on birth control and population control than any other policy. Yet its effect was abnormal and unexpected. Unsurprisingly, when the economic difficulties were over and economic adjustments had been made, a new baby boom occurred, which could be seen as compensation or retaliation.

Between 1962 and 1973, the birth rate in China surpassed 30 per 1000 people. The mortality rate gradually decreased, from 10.02 per 1000 people to 7.04 per 1000 people and in those ten years, the natural increase rate averaged above 23.33 per 1000 people. In 1963, the birth rate even reached as high as 33.33 per 1000 people, the highest in Chinese history. During the second baby boom, 320 million people were born, and about 300 million of them are still alive at time of writing. These baby boomers constitute a disproportionately large proportion of the age structure of the total population, which will exert a deep and long-lasting influence on future population change and socioeconomic development that will be hard to change. Table 3.4 presents the statistical picture.[9]

Between 1973 and 1978, not only the birth rate but also the mortality and natural growth rates showed a slow downward trend. The birth rate dropped from 27.93 per 1,000 people to 18.25 per 1,000, a decrease of 9.7 per 1,000 people and the biggest fall of the three. The mortality rate decreased from 7.04 per 1,000 people to 6.25 per

[9] National Statistics Bureau of the PRC (1986, p. 92); Population and Employment Statistics Division of National Statistics Bureau (1995, p. 40).

Table 3.4 Natural population change 1962–1973

Rate change per 1,000 people per annum	1962	1963	1965	1968	1970	1973
Birth rate	37.01	43.37	37.88	35.59	33.43	27.93
Mortality rate	10.02	10.04	9.50	8.21	7.60	7.04
Natural growth rate	26.99	33.33	28.38	27.38	23.83	20.89
Total fertility rate	6.02	7.50	6.08	6.45	5.81	4.54

1,000, the smallest fall. The natural growth rate fell from 20.89 per 1,000 people to 12 per 1,000. The total fertility rate dropped from 4.54 per 1,000 people to 2.72 per 1,000, a decrease rate of 40%, which had shortened its distance from the replacement level and indicated the end of the second baby boom.

3.2 Growing Concerns About Population Policy

Each population change, whether an increase or a decrease, within the first 30 years of the PRC, was associated with population policy to a certain degree. As has been mentioned, this association was not close, because demographic change is also determined by changes to economic foundations and social structures. In particular, fertility policy is generally positively associated with population change. Fertility policy and population change complement each other, though in certain circumstances, they are negatively correlated. That is to say, fertility policy and population change contradict and yet also complement each other. Thus we must not only take note of the regular changes but also analyse the detail of changes occurring under special circumstances, taking into consideration the change of time, location, and conditions.

The following section presents the evolution of Chinese population policy through different periods.

During the three years of national economic recovery, the chief task of the new government was to stabilize the social order, carry out democratic reform, develop production so as to restore the economy to pre-civil war levels, and to improve people's lives. Regarding population size, the whole country was inspired by Chairman Mao's criticism of Acheson's negative predictions. Encouraged by Mao's analysis, the Chinese people dropped their pessimistic views of a large population. Within three years, the demographic characteristics of China had shifted from the pattern of high birth rate, high mortality rate, and low increase rate to one of high birth rate, low mortality rate and high increase rate. Policy in this period was mainly directed at protecting fertility and preventing spontaneous abortions.

On 20 April 1950, the central government health ministry and the Health Department of Chinese People's Revolutionary Military Committee released a document entitled *Restrictions on Abortions for Female Officials*. According to this declaration, an abortion could be performed only after a woman who worked as a public servant had gained her husband's agreement and his signature on the application for an abortion. The application would be approved only after a gynaecologist in a public hospital testified that the husband had signed the document. "Those who have abortions that were not approved, and those who performed the abortion will both be disciplined." Subsequently, in December 1952, Culture and Education Committee of the Government Administration Council replied to the ministry of health that it had approved the draft of *Tentative Measures for Restricting Birth Control and Artificial Abortion* and a draft of *Measures for Trial Implementation of Premarital Health Checks.* According to the first document, "those who violate the measures stipulated and perform a sterilization operation or an abortion without permission will be punished for performing an illegal abortion. Both those who have the operation and those who do the operation will be charged in a court of law … The doctors will be subject to disciplinary sanction." In addition, there was a regulation that required pharmacies to apply to the local public health service authorities for approval before being selling contraceptive devices. Stores and street sellers were also forbidden to sell contraceptives without permission.[10]

These regulations were mainly directed at women public servants who want to have abortions or to be sterilized, though the authorities also wanted to keep the illegal sellers of contraceptives under control. These restrictions seemed not to affect the general public or the great majority of women, for the common working women at the time didn't have birth control requirements. It should also be noted that there was a baby boom around the world after World War II, which was somewhat connected with the appeal of banning artificial abortion and the consequent formulation of such laws and regulations in many countries. The Chinese laws and regulations restricting birth control methods formulated in the early and mid-1950s can be seen as positive because they were directed at preventing artificial abortion.

There were, however, different views at the time. In August 1953, Deng Xiaoping, deputy-premier of the Government Administration Council at the time, instructed the ministry of health to rectify its restrictions on birth control and banning the import of contraceptives. Deng also urged the health ministry to formulate and issue *Measures of Birth Control and Artificial Abortion.* On 27 May 1954, Deng Yingchao, then vice-president of the Women's Federation, wrote to Deng Xiaoping to report the views of women public servants on the birth control methods. The next day, Deng Xiaoping made the following instructions, "I think birth control is completely necessary and beneficial. Yet the ministry of health does not seem to have a positive attitude towards it. The Cultural Commission can hold a discussion with the ministry of health to ask for their opinion on birth control. If the latter approves of birth control, then some effective measures should be taken." Premier Zhou Enlai once noted that "Comrade Deng Xiaoping was the first to advocate birth control."[11]

In February 1955, the leading Party group in the health ministry submitted a report on birth control to the Central Committee, in which they made a self-criticism of their hasty formulation against birth control policy for their failure to take the

[10] The National Population and Family Planning Commission (2007, pp. 17–18).

[11] The National Population and Family Planning Commission (2007, p. 20).

issue seriously and to examine it carefully. The ministry had neither examined the issue carefully nor reported the situation to the Central Committee in timely fashion and asked for instructions. Therefore, they should take their negligence of the Party's leadership in the issue of birth control as well as their serious mistakes of decentralism as a warning. They also said that "according to the spirit of the Central Committee of the Party, in the historical conditions of China today, we think birth control policy should be properly applied. In the future, people's voluntary birth control actions should not be opposed." They also proposed guidelines and detailed measures for birth control. On 1 March, the Central Committee issued the following instructions on the above report: "The Party approves of implementing birth control appropriately. The Party committee in different regions should properly publicize the policy among officials and common people (not including people in the areas inhabited by the minority ethnic groups) so as to enable people to have a correct understanding of birth control."[12] Then in 1956, in the Advisory Report on the Second Five-Year Plan for Developing the National Economy, Premier Zhou Enlai reiterated that appropriate controls should be practiced concerning fertility. The amended draft of *Guidelines for National Agricultural Development from 1956 to 1967* advocated having children in a planned way.[13] On 1 March 1957, Chairman Mao delivered a speech at the 11th enlarged Supreme State Conference during which he explicitly advocated birth control, and family planning.[14] This indicated that the Central Committee (represented by Mao) had decided on a new understanding of birth control, shifting from restricting it to approving of it. Moreover, the new approach included the first expressions on family planning or having children "in a planned way."

However, the birth control and family planning proposals made during the first five-year plan period were still preliminary and restricted. In the report submitted to the Central Committee by the leading Party group in the health ministry, the word "appropriately" was used more than once. First, they stated that, at present, birth control should be practiced "appropriately", and next that birth control should never be restricted and should be advised "appropriately". Then, promotional and pedagogical material on birth control should be done "appropriately" so that the public would have a correct understanding of birth control. Yet the publicity should not be developed into a campaign. The same word "appropriately" also appeared twice in the instructions made by the Central Committee to the report: "The Party approves of practicing birth control appropriately", and "The Party policy (on birth control) should be publicized appropriately."[15]

This series of "appropriately" appeared in the aforementioned documents from top to bottom and vice versa, indicating that although restrictions on birth control had been lifted, an emphasis remained on an appropriate *scope* for applying birth control, and appropriateness in carrying out the task of the advocating and publicizing it. Why such emphasis on "appropriateness"? The root of the explanation lies in the

---

[12] Peng (1997, p. 1).

[13] Idem, p. 3.

[14] The National Population and Family Planning Commission (2007, pp. 24–25).

[15] Peng (1997, pp. 1–3).

purposes of birth control measures. According to the report made by the Party group in the health ministry, "there are currently over 600 million people in China, with an annual increase of about 12 million to 15 million people ... This rapid population growth will induce difficulties both for the country and for individual households for the time being. For the nation, there is not sufficient time to provide nurseries, schools and living materials for all the newborns. For individual households, more children will mean heavier burdens on the parents, which will affect the latter's work, study, and daily life. What's more, that will affect the upbringing of the children."[16] The instructions appended by the Central Committee to the report said that birth control should be implemented for the sake of the country, individual households, and the new-born generation. This new instruction was a broadening of the earlier purposes of applying birth control, which were to safeguard the health of mothers and children, and so on. However, the nature of population problem—overpopulation—was still not named. Neither was it possible to do so. As a result, the purpose of restricting population growth was evaded. Out of a fear of being labelled followers of Malthus, the health ministry's report specified that the birth control advocated in China's context had nothing in common with Malthus' population theory or neo-Malthusianism.

The view that "birth control should be practiced appropriately" put forward in the mid-1950s was still under discussion by the top leaders. It was, at best, trialled chiefly in big cities or some heavily populated rural areas. Though top Party leaders were vacillating between favouring a large population and a not-so-large population, the total population in China reached 672,950,000 in 1962, a net increase of 131,280,000 people, since 1949. At that time, China had recently transcended its immense economic difficulties and was about to experience a new baby boom. The fact of rapid population increase was impossible to ignore. From December 1962 to 1965, a series of reports on the application of birth control was submitted to the Central Committee. Meanwhile, Central Committee of CPC and the State Council issued a set of instructions on the implementation of family planning.[17] A new spirit was evident in the instructions and reports, enriching the policies concerning population control and family planning. They are summarized in the following section.

First, family planning was an established policy in China's socialist construction. The *Instruction on Advocating Family Planning Conscientiously by the CPC Central Committee and the State Council* (the *Instruction* for short) was issued on 18 December 1962. This document, in which the term "family planning" was adopted for the first time, functioned as a guideline for population work in this period. According to the *Instruction*, "implementing family planning is an established policy in our country's socialist construction. Carrying out this policy conscientiously over a long time will be conducive to preserving the health of mothers and children, to nurturing and cultivating the next generation, to tapping the potential of workers of both genders in production, in work and in study, and to the health and prosperity of our nation.

[16] Peng (1997, p. 2).

[17] Idem, pp. 4–6

Therefore, advocating birth control and family planning meets not only the needs of the public but also the needs of developing socialist construction in a planned way in our country. Advocacy of family planning in our country should never be mistakenly equated with Malthus' reactionary population theory. In view of the slackening of work on birth control and family planning in recent years, the Central Committee and the State Council see it as imperative to reiterate that the Party Committee and government at all levels should attach importance to and reinforce their leadership in carrying out this job".[18]

These words at the beginning of the *Instruction* were of great significance. In the first place, they determined the orientation of family planning: it was "an established policy" and one that must be "carried out conscientiously over a long time". Second, they explicated the reason, aim, and significance of the family planning project. The expression "be conducive to", used four times in the document, did more than merely draw attention to the benefits for mothers, children and family, it also attached importance to "being conducive to the health and prosperity of our nation". Then in 1963, the *Minutes of the Second Conference on Work in the City*, approved and issued by the Central Committee and State Council, pointed out that family planning would bring benefits in six areas: it would be beneficial "to carrying out socialist construction in a planned way; to controlling population increases in cities, alleviating the burden on agriculture, mitigating the supply of the basic necessities for people in the cities, and easing housing shortages, as well as the shortage of schools and other public facilities; to national financial accumulation and improving living standards; to workers' production levels, routine work, daily life and study; to preserving mothers' and children's health; and to nurturing and educating the next generation."[19] This document raised the implementation of family planning to the matter of concern for the whole population and for the country's national interests, and explained its relationship with the planned economy, financial accumulation, consumption, population control and so on. This was an elevated understanding, one with theoretical and practical significance. Finally, the document specified that in recent years, work on birth control and family planning had slackened, which was a weakness. Thus the Party Committee and government at all levels were required to "attach importance to and reinforce the leadership in implementing birth control and family planning. They should also put the task on their agenda. These requirements provided organizational guarantees for implementation of the policy.

Second, the primary aim of family planning was to lower the birth rate and the natural increase rate. The 1962 *Instruction* explicitly specified that "the natural increase rate of population should be appropriately controlled". The 1963 *Minutes of the Second Conference on Work in the City* also specified that "all cities should exert efforts to lower the birth rate." Specific goals were set in the document: "efforts should be made to lower the natural increase rate in city to below 20 per 1000 people within three years of adjustment. During the period of the third five-year plan, the natural increase rate should be lowered to below 15 per 1000 people. During the period of

---

[18] Peng (1997, p. 4).

[19] Idem, pp. 5–6

the fourth five-year plan, it should be lowered to below 10 per 1000 people."[20] This was the first time that the CPC Central Committee had specified a goal for the birth rate, and it was the first of its kind in management by objectives in the work of family planning.

Third, family planning was mainly carried out in cities and heavily populated rural areas. China is a country with a vast territory and diverse natural conditions, therefore, there are imbalances in the development of its economy, science and technology, culture, and society. There have also been great differences in population density, quality, structure, and ethnic composition. In the early 1960s, many parts of China were still isolated, afflicted by difficulties in transportation and communication. Thus the implementation of family planning policies had to be flexible, a single approach could not apply to all cases. In view of this, the policy had to be explicitly differentiated. The 1962 *Instruction* and several aforementioned documents issued by the Central Committee specified that birth control measures were to be advocated and family planning be promoted in cities and populous rural areas. Thus family planning was implemented in some regions only, rather than all over the country at the same time. Family planning was on trial and to be extended only step by step.

Fourth, the promotion of family planning had to be competently done, and technical instructions about contraception were to be highlighted. The *Instruction* contained specific rules about publicity and education for family planning. Its content should focus on explaining the benefits of family planning, including for mothers' and children's health, and the benefits for individuals and their families. Promotional material was mainly directed at young married workers in factories and mines, and city residents of both genders, as well as young married men and women in the heavily populated rural areas. Among young unmarried workers, apprentices, tertiary students, and peasants, there should also be publicity about restricting early marriage and appropriate advocacy for delaying marriage. In institutions of higher education, the moral education of students should be enhanced. Young people should be advised to focus on their studies instead of falling in love and marrying at an early age, since this would hinder their studies and aggravate their financial burden. The principle of the publicity was to be realistic, to adopt different methods when targeting different groups of people, and to avoid distorting family planning notions. There should be an awareness of people's customs and traditions, and their ways of thinking as well. Advocates should avoid inventing slogans, setting unrealistic goals, starting competitions, and so on. There should be no extensive publicity in areas inhabited by minority ethnic groups, or in sparsely populated areas, or those prone to natural disasters. But technical instruction in the use of contraceptive devices should be provided to those individual workers and peasants who had had too many children in too short a time and those clearly in need of family planning advice. Publicity should be oral and small scale. Where necessary, booklets about birth control and family planning, which included information on contraception, should be printed and circulated within a limited readership. Feeding information about birth control in public magazines and newspapers was to be avoided.

[20] Peng (1997, p. 6).

Fifth, the focus of family planning should be on contraception, and the production and provision of contraceptives should be guaranteed. The *Instruction* explicitly specified that the most effective form of birth control was contraception and that people must be made aware of the harm to women of induced abortions. At the time, contraception information was not easily accessible, and the quantity and quality of contraceptives was insufficient to meet needs. To preclude people from having private abortions that might result in life-long disease or injury or even death for want of access to professional practitioners, health departments should formulate specific regulations and create conditions to help people to have medical abortions or sterilization operations. The *Instruction* also stressed that the production and provision of contraceptives should be assured. The quality of contraceptive products should be promoted. There should be more product varieties and product sizes. Supply systems should be improved to make the product available in more stores to facilitate purchase. Furthermore, the *Instruction* specified that the price of contraceptives should not rise in any shop.

From the five aspects analysed here, we can conclude that population policy—mainly family planning—during the period of national economic adjustments from 1962 to 1965 was gradually becoming more comprehensive. The series of documents discussed affirmed that family planning was an established policy that must be carried out over the long term. They identified the aims, significance and contents of the policy, the methods for carrying it out, and the basic principles to be observed during its execution. They also stressed that publicity and education should be done conscientiously, that priority should be given to contraception, and that there should be technical instructions about birth control methods. Though research into population theory was still "forbidden" and no relevant essays had been published during this period, the *Instruction* and other documents transmitted by the Central Committee demonstrated that understanding of population problem in China had taken a big theoretical step forward. Understanding of the benefits of population control had been deepened; whereas previously it was seen as benefitting only women and individual families, there was now a general recognition that family planning would facilitate the development of the planned economy, and thus the nation's health and prosperity. Progress in the understanding of the population question was also reflected in the gradual establishment of offices and organizations charged with the work of promoting family planning.

Before 1962, the health ministry oversaw family planning, but in 1962 the Bureau of Family Planning within the Department of Women and Children was created. In 1963, the *Minutes of the Second Conference on Work in the City* demanded that a family planning commission be created in both central and local governments, to take up this work. Accordingly, the Family Planning Commission (FPC) under the State Council was created in January 1964, with Zhou Rongxin, then secretary of the State Council, as director. Subsequently, FPCs were set up in 25 provinces, autonomous regions, and municipalities. Organizations called Family Planning Technical Instruction Commissions were established to reinforce technical teaching about family planning methods in provinces and autonomous regions where there were no FPCs. With the improved understanding, a sharpened policy, and the establishment

Table 3.5 Baby quotas set by local governments 1962–1965

Location	Year	The baby quota
Tianjin City	1962	Reasonable to have two children, three if parents' personal circumstances permitted it
Wendeng County, Shandong Province	1963	Requirement for later childbearing, longer birth spacing, and fewer children
Shanghai City	1963	Advocacy of fewer children, longer birth spacing and later childbearing in urban areas
Liaoning Province	1963	Restrictions on having a third child
Anhui Province	1964	Advocacy of late marriage among public servants, and white collar workers in enterprises and public institutions
Guangdong Province	1964	Advocacy of later childbearing, longer birth spacing, and fewer children
Yunnan Province	1963	Proposition that one child is not deficient, two are adequate, and three are too many
Shan'an Xi	1962	Ideally two children, and at most three

of organizations in charge of family planning affairs, more specific policies were formulated and proposed by governments in many provinces, autonomous regions and municipalities, forming the embryo of the subsequent population and family planning policies. See Table 3.5.[21]

Unfortunately, from 1966, the Cultural Revolution disrupted the national economy and brought it to the brink of collapse, reduced the normal social order to the brink of anarchy, and had a negative effect on every social undertaking. However, different undertakings were impacted to different degrees, and even the same project was differently affected at different times. During the ten years of turmoil, population policy and family planning was more seriously affected in the earlier stage of the Cultural Revolution than in its later stage. For this reason, the following analysis of population policy in this special period is divided into two parts. According to the requirements of policy and effects of its implementation, the first part discusses the earlier stage of Cultural Revolution, from 1966 to 1970, and the second part covers the later years, leading up to the "reform and opening-up" policies from 1971 to 1978. Although the Cultural Revolution was ended in 1976, its after-effects lasted until the opening of the Third Plenary Session of the 11th Central Committee of the CPC in 1978. That is why the second part of this analysis covers the two years after the official end of the Cultural Revolution.

Generally, in the early years of the Cultural Revolution (1966–1970), the population and fertility policy of the previous era remained essentially the same. This is verified by the Central Committee's instructions on the report entitled *Several Issues*

[21] The National Population and Family Planning Commission (2007, pp. 58–59).

concerning Family Planning presented by Qian Xinzhong, health minister at the time, which said:

> It is extremely important to implement family planning. In 1962, the Central Committee instructed Party committees at all levels to reinforce leadership in family planning policy implementation. The experience in recent years demonstrates that successful family planning meets the needs not only of the general public, but also of planned socialist development in all sectors in our country. In the meantime, as long as the working methods are suitable, and people are willing to practice it, there will be encouraging results in the work of family planning. The report on the issues of family planning prepared by Qian Xinzhong is commendable. When the third five-year plan begins, local Party committees should summarize their experiences and promote family planning step by step in cities and heavily populated rural areas, according to the instructions made by the Central Committee and State Council in December 1962 and the instructions on Qian's report. Through their work, we hope the rate of population growth will continue to decline so that the needs of our developing economy will be satisfied, and the people's lives improved.[22]

However, because of the chaos created by the Cultural Revolution, especially in its early stage, the leading organs of government and the Party were so seriously disrupted that they were semi or fully paralysed most of the time, unable to maintain a normal working order. Because of the chaos, much work, including family planning, was neglected. To be precise, the work of carrying out family planning at that time was frequently nobody's business, which meant ineffective implementation of population policy.

In 1970, the world population reached 3,696,990,000, an increase of 1,177,520,000 people or 46.7% in 20 years, at an average annual rate of 1.94%. This excessive growth caught the attention of thoughtful people in every country. Two highly influential events took place in 1972. One was the publication of *The Limits to Growth*, authored by scholars from the Club of Rome, (D.H. Meadows, D.L. Meadows, J. Randers, and W.W. Behrens III). The report shocked the world. The team fed data about population increases, agricultural production, non-renewable resource depletion, industrial output, and pollution generation in the past 70 years into a computer model and predicted that "the limits to growth on this planet will be reached sometime within the next one hundred years."[23] They wrote that "the most probable result will be a rather sudden and uncontrollable decline in both population and industrial capacity."[24] As soon as the report was published, it was translated into many different languages and it elicited strong reactions. It made assumptions and novel arguments that the authors believed and responded to the pressing questions about the future of humankind, the limited world resources and unlimited human needs, advising, as the naturalist John Muir had in the previous century, "not blind opposition to progress, but opposition to blind progress."

Most economists opposed the Club of Rome's report while most ecologists approved of it. Other economists and ecologists approved of some of the views expressed, and disapproved of others. Among demographers, too, the report had both

---

[22] Peng (1997, p. 8).

[23] Meadows et al. (1972, p. 23).

[24] Idem, p. 23.

supporters and opponents. Most agreed with some of the arguments but objected to the rest. It can be acknowledged in any case that the report proposed a novel perception of population growth, and its analysis of the relationship between humans and natural resources, food production, industrialization, and environmental pollution furthered our understanding of the consequences of population growth.

The second of the highly influential events of 1972 was the June U.N. Conference on Human Environment, held in Stockholm, which adopted the *Declaration of the United Nations Conference on Human Environment*, known also as the Stockholm Declaration. It stressed that humans were not only the products of our environment, but that they also shape it. When making plans of action, humans must examine their effects on the environment. The 1972 Declaration proposed a series of concepts like "development should meet the requirements of environment", "non-destructive development", and "continuous and sustainable development". This was the first conference at which governments of all countries had discussed environmental problems, how to preserve the global environment and linking population change to environmental transformation. The question of how to achieve sustainable development of population and environment was presented to the world.

In the early 1970s, China was still a closed country, so it is not easy to calculate the influence of these two events on population policy in the PRC. But there were some historical coincidences. As mentioned earlier, Chairman Mao once remarked, "when the population in China reaches 700 or 800 million, it will be the time for Mr. Shao Lizi's proposals to be widely accepted". The Fig. 800 million caused great concern, because it was a critical juncture in the history of Chinese population development. The total population of China exceeded 800 million at the turn of the 1970s. From then, population problems became increasingly prominent, even severe, coinciding with the publication of the Club of Rome report and the UN Stockholm Declaration. Yet at that time, the social turmoil of the Cultural Revolution was still going on in China. While officials were transferred to work in farms in remote rural areas to have their ideological thinking remoulded, urban young people who had received secondary education were sent to live and work in production teams with local peasants in the countryside, which relieved the population pressures in the cities to some extent. These policies worked against the urbanization trend, so it was unlikely that they could persist. Instead of solving long-standing population problems, they could only provide a temporary buffer to their negative effects. When the population passed 800 million in the early 1970s, everyone in China grew somewhat concerned. In these circumstances, it was pressing to formulate policy that could bring the population growth under control. Hence, the confusion in family planning and other population policies during the early stage of the Cultural Revolution was brought under control and policies restricting on the number of children families had were created, sending new signals to Chinese society.

First, the new population policies specified targeted birth numbers. Population and fertility policy in the 1950s and 1960s usually made little more than some general appeals, but also stressed that Party committees and government at all levels should reinforce their leadership in family planning and tackle the work seriously. Sometimes, the population index was mentioned in the development plans for the national economy. For instance, in the report made by Qian Xinzhong in 1966, the draft for the third five-year plan proposed that the total population of China should be kept below 800 million by 1970. Some provinces and municipalities, including Shanghai, Tianjin, Hebei, Shandong, and Sichuan also set population targets. In the 1970s, specific targets were set. For example, in June 1973, the *Report on Issues of National Economic Plans* by the State Planning Commission stated: "we should strive to reduce the net population increase rate in cities to 10 per thousand people and to reduce that in rural areas to below 15 per thousand people in 1975." In February 1978, the State Council's *Report on National Family Planning Work Meeting* proposed that the natural population increase rate in three years be brought down to below 10 per thousand people. The plan was publicly released in the Work Report presented to the first session of the Fifth National People's Congress. This control target was reiterated in the *Report on the First Meeting Held by the State Council Leading Group for Family Planning,* commissioned by the CPC Central Committee. The report further proposed that the natural population increase rate be brought down to 9 per thousand people by 1985.[25] We can see that targets for fertility and population increase rates were not only set, but were also gradually tightened in this period.

Target numbers for births, too, were gradually becoming more specific. As mentioned, local government had made various statements about the desirable number of children per family, such as "ideally two children, and at most three", "later childbearing, longer birth spacing, and fewer children", "one child was is not deficient, two are adequate, and three are too many" and so on. These requirements were not the same across the country. Then in the 1970s, there emerged national standards about the number of births. According to the *Report on the First Meeting Held by the State Council Leading Group for Family Planning,* transmitted by the Central Committee in October 1978, "each couple is advised to have one child at best, or two children at most. The birth spacing should be at least three years. The local government can issue detailed statements of the goal of later childbearing and longer birth spacing according to their population planning objectives." Regarding the age of marriage, the report proposed that "in the rural areas men and women be encouraged to marry at 25 and 23 respectively, while this age will be slightly higher in the cities.[26] This was the first time that the Central Committee had specified in a document that one child per couple would be best and they should have two at the most. It was then just a short step to calling upon couples to have only one child. As is well known, the setting of targets was a typical characteristic of social and economic development during the period of the planned economy. In the 1970s, China's population policy, which had a focus on birth control, finally began to specify targets for family planning.

Second, the birth target was incorporated into national economic plans. During these years of the planned economy, the economic and social development benchmarks issued by the state were mandatory and must be fulfilled. To guarantee the achievement of the birth target it was incorporated into the system of benchmarks

---

[25] The National Population and Family Planning Commission (2007, pp. 86–87).

[26] Peng (1997, p. 14).

of national economic development from 1973 onwards. As mentioned, the June 1973 *Report on Issues of National Economic Plans,* the State Planning Commission specified that the net population increase rate in the cities should be kept to 10 per thousand people and reduced in the rural areas to below 15 per thousand people by 1975. This was the first time a population increase goal was incorporated into national economic plans and hence into the planned economy management system. The 7th clause in *Report on National Economic Plans in 1975,* made by the State Planning Commission, approved by the Central Committee, prescribed that "family planning is advocated by Chairman Mao, and it is imperative to control the population. To abide by the instructions of Central Committee of the Party, Party Committees at all levels should put family planning on the agenda and an official should be made responsible for this task. All resources should be mobilized to promote family planning and people's awareness of later marriage and family planning." Under the conditions of the planned economy, it was very important to include such population benchmarks in the national economic development plans because once they were, they would be put on the agenda of Party Committees and governments at all levels and be implemented conscientiously.

Third, guidelines for implementing the population policy were proposed. When the population increase index was incorporated into plans of national economy in 1973, the Leading Group and Office of Family Planning was set up by the State Council again to take charge of family planning work, with Hua Guofeng appointed as the head of this office. In 1976, Wu Guixian, then vice-premier succeeded Hua and in June 1978, the post was occupied by Chen Muhua, who concurrently served as vice-premier. After taking office, Chen held meetings to summarize the experiences of family planning work, proposing a guideline for carrying it out: "the Party Secretary should lead the work of family planning while the whole Party should engage in it. Education about family planning should be carried out effectively. Some typical work experiences should be applied to guide the future work. Research on family planning should be promoted, techniques to manufacture effective and safe contraceptive devices should be improved, and family planning measures should be put into effect. The public should be mobilized to practice family planning and the policy should be implemented persistently."

This guideline highlights something of the nature of Chinese society: "the Party Secretary should lead the work of family planning" points to the Party's leadership in family planning through the incorporation of population and family planning into the Party's work schedule. How did China succeed in bringing its population growth under control? There is a common perception outside China that it is a highly organized society. The subtext is that there is a bottom-up leadership system in the Party, which serves as an organizational guarantee for implementing policies such as family planning. In other countries, too, efforts have been made to curb population growth, but without success. Successive Indian governments strove to impose population control policies, including compulsory sterilization under Indira Gandhi's prime ministership, but these efforts were often futile.

Chen Muhua's guideline also attaches importance to advertising family planning goals. The publicity was to clarify the reasons for controlling population size and the ensuing benefits to the country, the family, and individuals. Some typical successful cases of the policy should be identified, she suggested, to serve as convincing examples in the promotion of family planning. A fine example has great appeal. When the public, especially the numerous peasant population sees those successful examples, their awareness of family planning practices will be heightened.

The third significant aspect of the guideline was the importance it attached to technical instructions about contraception. Safe, convenient, and effective contraceptives should be produced and made available to the people, which would also encourage their acceptance of the policy.

Finally, the guideline stressed that the ideological work required to make a success of the population policy should be carried out carefully and conscientiously. The public should be mobilized and organized to carry out family planning policy consistently. Under the historic conditions of the time, the implementation and application of the guideline demonstrated the specific characteristics of family planning work in China and it succeeded in controlling population growth to a certain extent.

The fourth signal of the tightening of population policy was that the family planning policy was written into the Constitution of the PRC. On 5 March 1978, the Constitution of People's Republic of China was approved by the first session of the 5th National People's Congress. Article 53 stipulates that "women are entitled to the same rights as men in the fields of politics, economy, culture, society, and family life in all aspects. Men and women should receive equal pay for equal work. They are both free to choose their spouse. Marriage, family, mothers, and children are under the protection of the state. And the state advocates and implements family planning." This was the first time that the state had put family planning policy into the Constitution, thereby providing the supreme legal guarantee for controlling population growth and implementing family planning policies.

3.3 The Testing of Theory in Practice

There were both encouraging achievements and discouraging setbacks in population theory studies in the first 30 years of the PRC. Specifically, there were both innovative findings and the expression of metaphysical views. Both the sociology school that advocated curbing population growth and Ma Yinchu's new population theory adopted theories from sociology, economics, and statistics to investigate and analyse the newly emerging population problems. They presented their understandings of the problems and put forward theories and propositions for population policy. Chapter 2 made a preliminary presentation of their views. Since there had been no narrowly demographic population theory developed in China since 1949, the theories and approaches of this group of sociologists and Ma Yinchu were ground-breaking. In the following section, their main research findings are first recognized and presented systematically.

The sociologists initially developed their population theories under the influence of Malthus' ideas. But after criticising Malthus' population theory to varying degrees, they accused Malthus of defending capitalism and then proceeded to develop their own independent theory and policy propositions for curbing population growth. They developed a "moderate population density" position, emphasizing that the Chinese population would grow beyond the "moderate" limit of 800 million if no intervention were made. Such a situation would slow the pace of industrialization and the modernization of agriculture; the transition to socialism would be greatly lengthened if there were rapid population growth. The latter would result in a disproportion between living labour and materialized labour. They recommended that a policy of population control be implemented conscientiously. Some of these scholars pointed out the gravity of the population problem by contrasting the fertility, mortality, and population increase rates before and after the founding of the PRC. In doing so, they suggested solutions to the problem from economic, political, and social perspectives. They believed that none of those solutions would tackle the problem from the root; only birth control measures could resolve the matter fundamentally. They further explained that birth control was positive rather than negative, because "birth control will not only greatly benefit personal family life, but also the country's reconstruction, and the nation's health and prosperity". For specific methods of birth control, they unanimously favoured late marriage, and they were mostly against sterilization and artificial abortion.

In contrast, Ma Yinchu's new population theory mainly discussed existing population problems from the perspective of economics, especially financial economics. He pointed out that the problems were created by contradictions between rapid population growth and capital accumulation, improvement of productivity, supply of raw materials for industry, promotion of living standards and the development of science and education. He used statistical data and applied statistical approaches to construct an explanation that was convincing on the macro level. He suggested that another census be conducted, that policy be based on statistical analysis, and that new concepts of fertility (eliminating old feudal fertility beliefs) be popularized. He also suggested that the marriage law be amended to advocate later marriage; and to reward the couples who had two children but tax those who had three or more. The taxes collected would be used for the bonuses paid to those who had fewer children.

Thus it can be seen that discussions of the population problem question in the 1950s, together with the many essays written and statements given to the press on the subject, can be categorized as research in the specific fields of demographic economics and demographic sociology, *avant la lettre*. They represented core academic views and methods as well as the direction of demographic research within a certain period and as such, it is a pity that these publications were unfairly criticized. Yet ironically, they were preserved as objects of criticism. In contrast, hundreds of articles were churned out to criticize Ma Yinchu's ideas. These texts, heavily popular and political, represented the leading trend in politics, the economy, and society at the time. It is absurd that such metaphysical views, forms of argument and conclusions so dominated Chinese political and academic circles for two decades. Indeed, it is difficult for those who did not live through those times to understand. Fortunately, though these abstract theories dominated public opinion and slogans like "the more people, the better" were shouted loudly, practice was the criterion of truth after all. To people's disappointment, their personal experiences told them that a large population did not bring benefits but instead trapped them in poverty. Almost 20 years later, demographic studies in China began to undergo changes. A new book on demography appeared.

In December 1977, *Demographic Theories* compiled by the Demography Research Centre at the Beijing Economics Institute (presently known as the Demography Research Institute of the Remin University of China) was published by the Commercial Press. Despite being a slim volume, of only 108,000 characters, it had a certain impact, since there had been no publication on demography for a long time.

More than a year later, the editorial board of the *Economic Research Journal,* sponsored by the Research Institute of Economics at the Chinese Academy of Social Sciences, received two contributions concerning the little book, which I was asked to review. The authors approved of some parts of the book but raised objections to its main points. For example, they disapproved of its criticisms of the restrained population growth position taken by Ma Yinchu and other scholars, and "the law of diminishing fertility of soil". They also raised doubts about the assertion that the capitalist system created the absolute impoverishment of the proletariat, and the argument that "humans are the most precious among all things in the world". They maintained that such views were aligned with the "left dogmatism" that was used to criticize Ma Yinchu's theory. Hence, the argument went, the book should clarify its standpoint and eradicate the negative impact of this "left dogmatism". Having reviewed the commentaries, I advised that the criticisms of the book were at least partially correct but that they should not be published in the *Economic Research Journal* for the following reasons.

First, *Demographic Theories* was the first monograph on demography and population in 17 years. Instant criticism of such work would be detrimental to the development of the science of demography and population studies. In the past, such scholarship had suffered from too much criticism. So many theories had been repudiated and very few established. Consequently, people were baffled about what are true theories. It would be better to show some tolerance and not to criticize the book in a hurry, despite some defects in its theorising. On the other hand, waiting would do no harm, for time is the most impartial judge.

Second, I argued, we should see the achievements in demographic research historically. *Demographic Theories* was published at the end of 1977, before the third plenary session of the 11th Central Committee of the CPC at which Ma Yinchu's new population theory was rehabilitated. Though the Cultural Revolution had ended and the so-called Gang of Four (the most influential faction in the CPC, that had attempted to seize leadership during the Cultural Revolution) had withdrawn from the political arena, it took time to eliminate their "left dogmatism", to reverse the false accusations against Ma Yinchu's work, and to set things right in the field of demographic theories. It would be preferable not to criticize articles and books published before the rectification of earlier mistakes.

Third, though some theoretical perspectives in *Demographic Theories* were questionable, its basic views deserved to be affirmed. For instance, the starting point of its explication of population problems in China was that it is a good thing to have a large population in China despite the difficulties it creates. It maintained that family planning should be implemented to control the pace of population growth methodically. Long passages in the book were devoted to discussing the rationale and the potential for implementing family planning programmes, as well as the principles and methods of formulating population development plans. Finally, the book was enlightening in its presentation of some novel views. The editorial board decided to adopt my recommendation not to publish.

Population studies cannot be divorced from prevailing scholarly conditions or the general social environment. On 10 May 1978, *Practice as the Only Criterion of Testing Truth* was published in the 60th issue of *Theoretical Trends*, sparking a massive discussion on the criteria for truth. The mistake of blindly adhering to all instructions and decisions made by former Chairman Mao Zedong was criticized. The National Symposium on Theoretical Demographic Research, held in November 1978, reflected the positive change in demographic studies. More than 30 essays were submitted to the meeting, on topics ranging from the law of population development of socialist systems to the relationship between population and the "four modernizations".

Regarding the first of these topics, most scholars approved of the view, expressed in *Demographic Theories*, that "population increase in a planned way is an objective law of socialist population reproduction", though some argued about the nature, content, wording, and mechanisms of the law of socialist population development. The former dogma of population development in a socialist country—that "The population will keep increasing quickly", which originated from the *Textbook of Political Economy* published in the USSR—was broken to some extent. The law of population development was revised as "population should increase in a planned manner". However, the conference papers were based on similar theoretical foundations that assumed that since socialist national economies develop proportionally and according to plans, human reproduction must develop in a planned manner too. Yet though they were partially obscured by familiar quotations from the works of Marx, Engels, Lenin, Stalin and Mao Zedong, the authors presented some sound arguments.

The symposium discussions on the second topic were more thorough. Starting from the fact that the Chinese population was growing at a fast pace, the scholars shared the view that the population growth rate should be kept at a low level in a planned way, because it would be beneficial to an efficient use of the work force, to financial accumulation, to the whole country's scientific and cultural qualities as well as education levels, to labour productivity, to material and cultural life, to women's emancipation, and so on. Thus, control of the population growth rate was a prerequisite for realizing the "four modernizations".

This symposium was held with the approval of Chen Muhua, alternate member of the Political Bureau of the Central Committee, vice-premier of the State Council and leader of the Leading Group of Family Planning Affairs under the State Council. She attended some sessions, listened to some of the discussions, and delivered a long speech on the relationship between theory and practice, the current questions that required investigation and the promulgation of family planning theories.

According to Chen, theory comes from practice and in turn guides practice. When studying theory, we should primarily associate theory with practice, be it practice in China or in foreign countries. Statistical data indicated that the population growth rate was not high in the US. In France, the total population had remained around 50 million for many years. The population in both the German Democratic Republic and Federal Republic of Germany was declining. Meanwhile, in some other countries, the growth rate was very high, and there were other countries in which the birth rate was high but the population increase rate was not. Research was needed into the following questions: Why were there such differing situations in the various countries? What are the characteristics of their social policies? What are the features of their economy and what economic policies were in place? Why was population size affected in these ways? In the 1970s in Czechoslovakia, I observed that women were unwilling to have children despite government encouragement to do so, in the form of a beautiful pram for every new-born baby, and long maternity leave and family allowance payments. People were still reluctant to have children in Czechoslovakia, while in China, some people were unwilling to have fewer children despite the ubiquity of family planning programmes. Why was that? Such population research required international comparisons.

When we criticize Malthus' population theory, we need to find out how popular it was in capitalist countries. Did those societies adhere to Malthus' theory or others? Which theory was functioning? We said that humans were the most precious treasure, for humans can work and create wealth. All wealth was created by humans. There would be no wealth if there were no humans. No matter how efficient automation might be in the future, a human will still be required to press the machine buttons. As we also know, people are both producers and consumers. When production levels are not high or labour productivity is very low, our consumer aspect will be prominent. A very big population will impose a heavy burden on society thus it is reasonable that different countries will adopt different population policies according to their divergent situations.

Theoretical research must be integrated with practice: theory comes from practice and in turn guides it. Theory enables us to act consciously. Though family planning had been carried out in China for more than two decades, our demographic studies were still limited. Theoretical studies in this field had to be reinforced to inform more conscious family planning work. Practice would put our ideas to the test and decide whether they are truthful. We should then adhere to the correct theories while rectifying the wrong ones. We should raise practice to the level of theory and apply the theory in practice, then again elevate the new practice to the level of theory. The process must be repeated several times before a theory is established. This requires emancipated minds brave enough to investigate research subjects that used to be "forbidden". In academic research, the principle of "no highlighting someone's shortcomings, no labelling, and engaging in no personal abuse" suggested by Deng Xiaoping should be adopted. In the past, Chinese scholars dared not analyse population problems from an economic perspective lest they be rebuked as Malthusians. In the late 1970s, we started to analyse the problem from an economic perspective and no one was labelled a Malthus follower. Since economic problems created by a large population exist objectively, it was foolhardy to turn a blind eye to them.

Taking the situation of the time into consideration, Chen Muhua suggested a range of research questions. The following are worthy of special attention.

First, she asked what was the optimum tendency for population development in China? Was it good for the population to level off, or to increase slowly, or to increase at a lower speed? Or was it good for population to level off in one period, to drop a bit in another period, and then to increase a little in still another period?

Second, with the improvement of living standards, and better medical services, life spans had extended, Chen noted. In the past, folk wisdom said that a man seldom lived to be 70 years old, but by now, it was not rare to see people in their 70 s. There were predictions that this would result in an aging society and some warned that in that case, the quality of the work force would deteriorate. These questions require investigation.

Third, Chen noted that in the late 1970s, about 18 million babies were being born and about six million people were dying every year in China—a net increase of about 12 million people per annum. If population growth levelled off and remained the same, would this situation be advantageous or disadvantageous to production? After the population has levelled off for several years, should it rise again? If the population should increase, what would be the optimum growth rate –5 per 1,000 people, 7 per 1,000, 8 per 1,000 or 10 per 1,000? Efficacious population policy must be informed by reliable theoretical study based on a consideration of a broad set of possible circumstances.

Fourth, from 1973, the population index began to appear in the national plan. The current index was determined according to the proposal put forward in the government work report made to the fifth National Congress. Chen asked if this were reasonable or unreasonable? That too can be up for discussion. Of course, the index can be changed. It does not matter whether all opinions are mature or not, the question can be discussed by the meeting and suggestions made to the relevant authorities.

Chen Muhua also said we should not only conduct theoretical studies, but also offer ideas for the promotion of family planning policy, especially to help the public change their traditional mindsets. For example, there was a strongly prevalent idea that a married couple must have a son or sons to take care them when they are old. Why was this the case? In the capitalist countries, people were not raising children in order to receive care in return when they grew old. When children become adults, Chen argued, they should fend for themselves: adult children should move out of their parents' homes and establish their own homes. They should make a financial contribution if they dine at their parents' house and vice versa. Both parents and grown-up children should live independently. Why did such practices exist in other countries? Why was there a different story about raising children in China? Perhaps because of the influence of traditional thoughts of Confucius and Mencius, or perhaps it was related to certain social policies. The popular practices in the Western countries should be explored, she suggested, and the results made known to the public.

Another example she cited was that in China it was difficult for married daughters to look after their parents because sons-in-law who live with their parents-in-law were usually looked down and mocked by their neighbours. The usual practice in the countryside when production teams allocated tasks was to assign a son-in-law some dirty, tiresome, and less-paid jobs. Such practices and discrimination against sons-in-law were both wrong, she stated, and scholarly articles should criticize this phenomenon. But they should deal with specific phenomena in specific ways. Situations vary from case to case. Was it always the case that a young married woman lived with her husband's parents? Did all parents prefer to have sons? City dwellers usually preferred to have a daughter instead of a son, for a daughter was considered warm, like a cotton-padded coat in winter to her parents. Mothers were willing to live with their married daughters. The situation in the urban areas differed from that in the countryside. All these situations and problems required theoretical investigation that in turn could serve the real tasks at hand. Articles were required about how to address the problems incurred by traditional thinking and customs in order to remove barriers to effective family planning.

To carry out family planning effectively, people from many sectors including medical services, women and children's health services, and education, should be involved. If we do a good job educating children, they will become people with amazing capacity. Since modern production systems demand skilled workers, scientific and cultural education must be made available to all Chinese citizens. As science and technology advances, it will in turn facilitate the implementation of family planning. The fact that people continued to have children in an unplanned way reflected the low education level of the country to some extent. If family planning were effectively carried out, health and living standards would improve. And if people's considerable worries disappeared, they would be more motivated to practice family planning.

Finally, Chen Muhua told the scholars present that we should continue the previous correct practices in publicity work on family planning, and supplement it where necessary. In the past, people were hesitant to popularize family planning and felt embarrassed to do so in public. But family planning is an important project, so we should do the job confidently. Research should be conducted and books should be written on family planning so that people will be well-informed.[27]

Chen's speech called on participants of the symposium to open their minds and to contribute suggestions to solving the population problems in China by conducting research that integrated theory and practice. Her encouraging speech launched a new phase of population research, implying that the invisible "restrictions" imposed by earlier political pressures had been lifted. In a word, her talk actively promoted demographic studies. The truth is that during the 17 years when the circle of researchers in demographic studies was "asleep", population problems continued to be referred to in speeches by Central Committee members and in the Party's documents, and understanding of the problems had grown gradually. Marked progress can be detected in the deepened comprehension of the benefits of controlling population growth

[27] Collection of Papers Presented at the National Symposium on Researches of Demographic Theories (internally circulated). (1978, pp. 2–4).

and implementing family planning. Previously, only the commonplace benefits of improving mothers' and children's health and alleviating family burdens had been recognized. Now, it also was recognized that family planning was a significant project that concerned the development of the national economy, of culture and science, and the project of "four modernizations". In sum, it was a project vital to the future and fortune of China. "Overpopulation" had already been mentioned in some documents, and this paved the way for the tightening of population and fertility policy.

References

Collection of Papers Presented at the National Symposium on Researches of Demographic Theories (internally circulated). (1978). pp. 2–4

Liu, S. N., & Wu, Q. G. (Eds.). (1985). *A Brief History (1949–1983) of Socialist Economy in China.*, Heilongjiang People's Publishing House. p. 16

Meadows, D.H., Meadows, D. L., Randers, J. & Behrens III, W. W. (1972). *The Limits to Growth—A Report for the Club of Rome's Project on the Predicament of Mankind*. Universe Books. p. 23

National Statistics Bureau of the PRC. (Ed.). (1986). *China Statistical Yearbook of 1986*. China Statistical Publishing House.

Peng, P.Y. (Ed.). (1997). *A Complete Record of Family Planning in China*. China Population Publishing House.

Population and Employment Statistics Division of National Statistics Bureau. (Ed.). *A Handbook of the Key Data of Chinese Population (1995)*. p. 40

The National Population and Family Planning Commission. (Ed.). (2007). *Population and Birth Control History in China*. China Population Publishing House.

Part II
The Whole Story of the One Child Policy

Chapter 4
Setting Things Right in Demographic Theory

In the late 1970s, the People's Republic of China finally made a historic breakthrough in its development with the reform and opening-up policy. This breakthrough began with the serious desire to "set things right" and clarify matters in the field of theory. Demographic research was no exception to the ensuing investigations and revisions.

4.1 Rehabilitating Ma Yinchu's New Population Theory

As discussed in Chapter 2 above, the theoretical dogma "the larger the population, the better" was formed when the sociology school that advocated restrained population growth and Ma Yinchu's new population theory, especially the latter, were criticized. When the principle of "seeking truth from facts" was restored, a process of rethinking was set off in all fields of society. It became imperative to loosen various restrictions in thought and to break away from fixed theoretical ideas and codes of conduct. Since theoretical revisions are crucial to the development of scientific research, they become crucial to the implementation of policy as well. Had Ma's population theory not been rehabilitated, it would have been impossible to revise certain confused interpretations and to bring a different rationality to the field of demographic research. It would have been difficult, for example, to re-examine cases of unjust charges in prior political movements and to reverse wrongful verdicts declared against certain intellectuals. It would also have been difficult to dispel the fear that people felt during such campaigns of criticism. The situation should be understood in the light of the demographic, economic, political and social conditions at the time.

The two waves of criticism of Ma's population theory involved many facets of the population question, and the development of a plethora of metaphysical views. Since the 1970s, the state had enforced a family planning policy to constrain population growth, and had made remarkable progress in this. Scholars in demographic studies had also endeavored to broach some issues that were hitherto regarded as "forbidden".

X. Tian, *Population Policies in the People's Republic of China*,
https://doi.org/10.1007/978-981-96-9658-1_4

Yet, the fear remained and very few articles were published in the field, in which the authors still did not forget to make criticism from time to time. Though they criticized Malthusian ideas about population, Ma also become a target of criticism. This situation did not change until the third plenary session of the 11th Central Committee of the CPC, which restored the principle of seeking truth from facts. How was this principle to be restored in demographic research? The rehabilitation of Ma's population theory became crucial to such questions.

As seen in Chapter 2, Ma analysed the existing population problems in China in practical terms, through his field work, and proposed strategies to overcome them. Disappointingly, he was unjustly criticized and removed from office, following which no one dared to conduct research on demography for more than ten years. Scholars were afraid of being subjected to such severe criticism. Scholarly thought in the field could not be freed and personal anxiety dispelled until Ma's theory was rehabilitated. The nature of China's population problem could not be faced squarely, nor could the principle "let a hundred flowers blossom and a hundred schools of thought contend" be followed.

My article "In Defence of Ma Yinchu's New Population Theory," published in the *Guangming Daily* in 1979 must be mentioned in this context. An academic colleague wondered publicly why I had written the article and edited the book *New Population Theory*. He surmised, in an internet essay that the Central Committee of the Party wanted to have Ma Yinchu rehabilitated, and had assigned the task to the Economics Research Institute of the Chinese Academy of Social Sciences (where I was working). Perhaps because I formerly worked at the ministry of education and was an alumnus of Peking University, or perhaps because I had some unique understanding of the issue, the task of Ma's rehabilitation was allocated to me. However, his guess about my motives was not correct.

As described in Chapter 2, as a freshman at Peking University in 1959, I had witnessed the second wave of criticism of Ma' theory. Ignorant as I was at the time, I began to seek out Ma's and his opponents' writings. I found Ma's views reasonable, whereas the criticisms of him were all similar and full of popular political slogans. As Ma increasingly disappeared from public view, my indignation about his treatment grew. His resolute self-defence pushed me to start collecting relevant materials on the whole case. Over the years, I continued to think occasionally about his case, expecting that the day would come when justice would be done for Ma.

In 1976, with the political power of the Gang of Four crushed, I wrote an article defending Ma. Given that the slogan "we should resolutely adhere to all the instructions and decisions made by Chairman Mao" was still very politically influential, I reckoned that my essay would not be published. When the third plenary session of the 11th CPC Central Committee restored the principle of seeking truth from facts, I revised the article several times and finally entitled it "In Defence of Ma Yinchu's New Population Theory," and submitted it to the *Guangming Daily*. I chose this newspaper first because it was suited to such theoretical articles, being a major publisher of essays on education and science. Second, this was the very newspaper that had spearheaded criticisms of Ma, so it had an obligation to express its current attitude to that past movement against him. Apparently, the newspaper took my piece seriously and said it would be published as a key article. Before it was published, two short essays expressing indignation about the injustice meted out to Ma appeared in the paper. An editor told me these essays were a sort of "appetiser", while mine would serve as a general apology and settlement of the error it had committed in the criticism of Ma Yinchu. When my article was finally published, an editor's note expressed this intention, stating that I had not written the article at anyone's behest or order but that instead, it came about of a long-cherished wish of my own. It should be added that though it had been in preparation for two decades, my article was not the first, as some colleagues declared, "that aimed to rehabilitate Ma Yinchu". My article exerted more influence because of its length and comprehensive analysis.

The rehabilitation of Ma's theory not only began a new chapter in demographic research, but also provided theoretical and public support for population policy in the new period. From my perspective, the following theoretical questions were clarified:

First, the relationship between population and the "four modernizations" was explained. Modernization in the sectors of industry, agriculture, science and technology, and national defence has been the dream of all ethnic groups in China since the first Opium War (1839–42). But what is the relationship between the "four modernizations" and population change and development? In his new population theory, Ma had pointed out that realization of "four modernizations" mainly depended on improvements in labour productivity, science and technology, and education and health. Thus, we should first control population growth, then improve productive capacities through developing science and education, which would be achieved chiefly through controlling population size and enhancing population "quality". Based on the presumption of a labour shortage in the Great Leap Forward movement,Ma's critics believed that the population (and the labour force) should continue to grow in order to accelerate industrial and agricultural modernization. In their opinion, rapid population growth would be the fundamental driving force behind the four modernizations. Having Ma's theory rehabilitated meant reversing this mistaken assertion. We must recognize that during different historic periods, there have been divergent, even completely different, population requirements due to differing levels of productive forces. To achieve the goal of socialist modernization, we had better adopt advanced technology, upgrade labourers' technical equipment, and improve the organic components of production so as to achieve mechanization, automation and electrification of production and to significantly improve labour productivity. Socio-economic development in China would certainly shift from a primary dependence on increased numbers of labourers to a dependence on their improved quality. This would be built on the foundation of improved intelligence, scientific knowledge, and education attainments; an improvement in the degree of civility is both prerequisite and guarantee of these.

Different population requirements are determined by the varying requirements for capital in different historic periods. Economic development in any society requires certain resources and capital. The division of social and economic development stages is not marked by the products being produced, but by how production proceeds and the kind of tools used in production. Thus, economic development of human society can be divided into three economic stages, and within each the dominant means of production, core capital, and chief driving force are highly interrelated. See Table 4.1.

The different economic and social forms require very different populations and labour forces. Primitive, slavery and feudal societies, which together have lasted the longest in human history, are collectively known as the agricultural and pre-agricultural societies. In those societies, though means of production were continuously evolving, they basically comprised hand tools, and human hands and arms were the source of power. In the face of nature, people's capacities were limited. Social and economic development was determined by natural capital in these primeval times, and climate, rainfall, land, and forests, together with the quantity and quality of animals and plants, were decisive. The function of population and labour forces was mainly related to quantity. Ancient civilizations were usually first created in regions with favourable climates, fertile land, and flourishing animal and plant life, and countries with ancient civilizations would burgeon thereafter. In the mid-eighteenth century, steam engines replaced the water-mills and the spinning jenny replaced the spinning wheel. Productive forces made an unprecedented leap and industrial societies replaced agricultural ones, bringing new population requirements, including changes to working age. The shift from dependence on population size to dependence on population quality took place.

At the beginning of industrial society, development was chiefly promoted by output capital. When the primary goal of a traditional industrial period was to increase the quantity of goods, requirements for population and productive forces were not yet high. There was a slight difference between the production of relative surplus value and absolute surplus value. The above-mentioned shift gradually took place in the later phase of industrial society. After World War II, the new technological revolution driven by microelectronic techniques occurred. New materials, resources and technologies emerged one after the other, and development cycles were constantly shortened. We currently live in post-industrial times, which are led by the biological sciences and marked by the development of intelligent tools. Population quantity and quality requirements have also undergone revolutionary change. The main signs of post-industrial society are the replacement of machine tools by intelligent ones and the replacement of output capital by human capital. Human capital refers to the

Table 4.1 The dominant means of production, core capital, and driving forces in different economic and social forms

Economic and social forms	Dominant means of production	Core capital	Chief driving force
Agricultural and pre-agricultural societies	Hand tools	Natural capital	Labour power, animal power, wind power, wood
Traditional industrial society	Machine tools	Output capital	Coal, oil, electricity
Post-industrial society	Intelligent tools	Human capital	Electricity and nuclear energy

value of knowledge, skills, experience, and health that people possess. A well-known Chinese entrepreneur once said, "a good enterprise design will never work out well without talented workers". The most important competition in a modern society is the competition for talented personnel and for human capital. Thus, today, social and economic development primarily depends on human capital—the capacities of population and labour forces rather than population size. There is demand neither for continuous population growth nor for rapid population growth. Any economic development still relying on externally expanded reproduction that mainly depends on a bigger labour supply will be as backward as a spinning wheel. The pressing need now is for improvement in a population's capacities, that is to say, the comprehensive improvement of its health, intelligence, scientific knowledge, education, and civility.

Second, the confusion between the theories of "working hands and eating mouths" was cleared up. As discussed in Chapter 2, in the critiques of Ma's new population theory and others that advocated restrained population growth, the argument was often made that Ma only noticed people's "eating mouths" while ignoring their "working hands", meaning that he failed to see that people can work with their capable hands. This theory of working hands was little known in demography. But it is unscientific to talk about a person's mouth and hands separately. The critics' intention was to emphasize that people's hands represent production, which is highly valuable and important, while mouths symbolize consumption, which is of little value or importance. The logic behind the argument is that since a person has a pair of capable hands, he or she can create infinite wealth. The more people, the more wealth will be created. Thus, the more people, the better.

The rehabilitation of Ma's theory challenged this overemphasis on production and neglect of consumption. It also eradicated the influence of such metaphysical views and restored the notion that a person is a unity of producer and consumer. Theoretically, if population changes occur alongside economic and social development, and resources and environment, then a stable population size is optimum, no matter whether people are seen as producers or consumers. If there is a labour shortage, then it is beneficial to increase the population. If there is surplus labour, then population growth will be detrimental to the development of economy and society and to a healthy balance between population, resources, and the environment. From this perspective, at the time when Ma's theory was being criticized, China was overpopulated. While acknowledging that people are producers and the most active element in productive forces, we must also note the underemployment that occurs with an increase in population and the labour force, along with an intensified shortage of consumer goods. Population reproduction takes a long time and is characterised by a steady increase, and it takes about 20 years before a child is old enough to join the workforce. So, population growth first means an increase in consumption and the aggravation of the contradiction between the shortage economy and over-population. The working-age population needs adequate means of production and jobs, and they cannot become real producers without both. Because there was surplus population and labour in China at the time, it was illogical to emphasize people as producers while neglecting their functions and effects as consumers.

Third, the theory that there are two types of production, namely production of population and production of material resources, was developed when some confusion in demographic theory was clarified. In contrast to the "working hands" theory, the creation of a "two kinds of production" classification was related to a specific historic background. Since Ma's theory had only just been rehabilitated, population theorists still resorted to the elaboration of themes in classical Marxist texts. Demographers began to seek theoretical bases in works by Marx, including *The German Ideology*, and Engels, including *The Origin of the Family, Private Property and the State*. The following remarks from Engels were often quoted, "According to the materialistic conception, the determining factor in history is, in the final instance, the production and reproduction of the immediate essentials of life. This, again, is of a twofold character. On the one side, the production of the means of existence, of articles of food and clothing, dwellings, and of the tools necessary for that production; on the other side, the production of human beings themselves, the propagation of the species."[1] Following this argument of Engels, the demographers now argued that the production of material means is the foundation and population reproduction must accommodate it. Meanwhile, population production is not passive, but rather can promote or slow down the production of material means. Over time, this view was summarised as "two types of production"—of population and of material resources.

In reality, in China, it was pointed out, there are several situations in which these "two types of production" cannot accommodate each other. Specifically, the total population cannot accommodate the means of livelihood, the working-age population cannot accommodate the means of production, people's scientific and cultural capacities and education levels cannot accommodate existing economic progress and scientific advancement, and the urban and rural population distribution cannot accommodate the industrial structures. Therefore, it was suggested, when striving to develop the production of material means, we should try to control population size, improve the population's capacities, and adjust population structures so that the production of material means can accommodate the rate of human reproduction. In some specific historic circumstances, this notion played a clarifying role in demography and economic theory.

Fourth, research began in China into the notion of an "optimum population", first proposed by British economist Edwin Cannan in the late nineteenth century. Cannan argued that under certain conditions of production, there is an optimum population size, which will ensure the maximization of income distributed. If the population is above or below the optimum level, then distributed income per head of population will decrease. Based on this, Hugh Dalton proposed a mathematical formula:

$$M = (A\text{–}O) \div O$$

where O represents the optimum population, A the actual population size, and M the degree of maladjustment between the two. When M is positive, there is overpopulation and when M is negative, there is population shortage. When M equals zero, it means optimum population has been achieved. In the 1970s, French demographer Alfred Sauvy further developed the theory of optimum population, proposing two

[1] Compilation and Translation Bureau of the CPC Central Committee (1972).

concepts: the economic optimum population and the power optimum population. The first concept refers to the population size that will maximize economic benefits and the second refers to the population size that enables a country to obtain maximum national power. The latter is more desirable than the former. Despite much dispute about the theory, it remains important among demographic theories. However, in China it was long criticised, dismissed as one of the theories that whitewash the capitalist system, a judgement that impeded demographic research in China. The rehabilitation of Ma's new population theory and the views of the sociology school confirmed the significance of optimum population as a demographic theory and encouraged research related to it. In the early 1980s, an upsurge in work on optimum population resulted in interesting findings in several disciplines, and innovations in research methods appeared.

Fifth, the law of socialist population development was revised. Following the criticisms of the idea of restrained population growth came the doctrine that "the law of population development in a socialist country is that population will keep increasing quickly", taken from the *Textbook of Political Economy* published in the USSR, which was incorporated into demography, economics and sociology curricula. Later, this became a dogma that seriously impeded the development of demographic theory in China. After the "Gang of Four" was demolished, this was modified to some extent, becoming "the law of population development in a socialist country is that population should increase in a planned manner". Population "should increase in a planned manner", in alignment with the planned economy. In short, socialism was bound up with "population growth". In other words, the development of socialism was bound to be accompanied by population growth.

The rehabilitation of Ma's theory allowed a new perspective on the demographic transition. Through investigating the inevitability of the evolution of the reproduction pattern from high birth, high mortality, and low increase to high birth, low mortality, and high increase, and then to low birth, low mortality and low increase, the dogma that the population was sure to keep increasing in a socialist country could be cast off. It was suggested that the law of population change and development was not only constrained by production relations and the nature of social system, but also by biological or natural laws, the level of social productive forces, resources, and food, and the environmental conditions. Henceforth, the essence of socialist population development law is that population change and development should be commensurate with socio-economic development, and should be coordinated with changes in the environment and available resources, and should promote social progress. The details of this accommodation are reflected chiefly in the following aspects: the total population should match the means of livelihood, the working-age population should match the means of production, and the degree of aging should match social security provision. As well, the population's skills should keep pace with scientific and technological advances, urbanization should respond to changes in industrial structures, and regional population distribution should match the allocation of productive forces. Accordingly, when promoting production development and social progress, it is imperative to adjust the size, skills, and distributions of the population to improve the degree of accommodation continuously. Yet there might be different focuses in different historic periods. In one period, adjustments might focus on population size while in another they might focus on quality and in still another, on population structures.

From a certain perspective, perhaps, other discussions about population problems in China do not belong to the category of demography. Yet, this is a history of a period when demographic research in China had just been restored, in the late 1970s and early 1980s. In a departure from existing dogmas, the shift in how major population and development issues were seen showed that the demographic problem in China was that over-population was stifling productive forces. Based on this recognition, strategic decisions such as the intensified control of population growth and specifically the one-child policy were proposed. The recognition also provided a foothold for subsequent population policy adjustments. There was henceforth greater awareness of the contradiction between over-population and low capacities, leading to a recognition that while curbing population growth, we must also improve the health, education level and civic awareness of the population. The rehabilitation of Ma Yinchu's ideas ushered in an era of investigating and analysing China's population problems in a realistic way, thus preparing the way for the formulation and implementation of population policies from the 1980s.

4.2 Clarification of the Nature of Population Problems

Opinions about population in China can differ dramatically. Those who advocate strict control of growth, for example, argue that Canada has a territory of 9,985,000 square kilometres, almost the same size as China but population density in Canada is 4 people per square kilometre, only 2.7% of that in China. The US is also a similar size to China, but its population density is 36 people per square kilometre, less than one quarter of China's. The Canadian and US economies are highly developed, which is directly related to their small population size and large amount of arable land. In contrast, those who advocate encouraging fertility and continuous population growth argue that Japan has a territory of just 378,000 square kilometres, less than 4% of that of China, yet its population density is 348 people per square kilometre, nearly two and a half times that in China. India's territory is 2,980,000 square kilometres, less than one third the size of China, yet its population density is 450 people per square kilometre, three times that in China.[2] Compared with all these countries, is China's population size too large or too small? Should the state continue to control population growth? Or should it encourage growth and implement policies conducive to that? Simple contrasts cannot lead to a sensible answer.

[2] National Statistics Bureau of the PRC (2018).

Where is the fundamental divergence in these opposing views on population? Demographers have investigated population development history, demographic transitions and the stages within it, population mobility and migration, regional distributions of population, and the relationships between population and resources, environment, economy, social development, as well as sustainable development. All their research shows that population change and development are determined by socio-economic development and constrained by the level and character of productive forces. Different levels of social productive forces and their different natures, which are chiefly determined by the tools of production, determine the population reproduction type, indicating the diversity of population problem. In the contemporary world, there are more than 200 countries and a wide assortment of ethnic groups and races. The nature of population problems falls to two categories.

The first category is when population limits productive forces. This situation mainly exists in developing countries, especially low-income countries where social productive forces are not developed. Most of them are in a transition from traditional agricultural society to industrial society. Here, manual tools still dominate and the mechanized big industry is not advanced, thus there is low labour productivity. In such circumstances, the general population and the labour force usually do not have sound health, or high levels of intelligence, scientific capacity or education levels. In such societies, science, education, cultural programmes and public health services are underdeveloped. Generally, the urbanization rate is low, and there is a wide urban–rural gap. In most of these societies, the population is still in a stage of high birth, low mortality, and high increase. In a few of them, population production is still at the stage of high birth, high mortality, and low increase, where the high fertility rate is a prominent characteristic. Using UN data, Table 4.2 below presents contrasts between total fertility rate, gross birth rate, gross mortality rate, gross increase rate, infant mortality rate and life expectancy estimates for both sexes combined in developed countries and developing ones (excluding China), from 1980 to 1985 and from 2010 to 2015.[3]

Table 4.2 shows that from 1980 to 1985, the total fertility rate in the developed countries fell to 1.9, below the replacement level of 2.1. The birth rate dropped to 14.9 per 1,000 people, mortality dropped to 9.5 per 1,000 people, and the natural increase rate fell to 5.4 per 1,000 people. The population reproduction type in these countries had shifted from a high birth, low mortality, high increase pattern to one of low birth, low mortality, and low increase. In contrast, in the developing countries (excluding China), population reproduction was typically high birth, low mortality, and high increase; the total fertility rate was 4.9, more than double the replacement level. Birth and natural increase rates remained high. Three decades later, in these same developed countries, total fertility rate had fallen to 1.7 (a drop of 10.5%), the birth rate declined from 14.9 per 1,000 to 11 per 1,000 (a decrease of 26.2%), and the natural increase rate declined from 5.4 per 1,000 people to 0.8 per 1,000 (a massive 85.2% drop). These changes indicate that with the birth, mortality and natural increase rates continuously at such low levels, population reproduction in the

[3] U.N. Population Division (2009).

Table 4.2 Population reproduction types in the world, the developed and the developing countries

1980–1985	The world	Developed countries	Developing countries (excluding China)
Birth rate (per 1,000 people per annum)	27.9	14.9	36.3
Mortality rate	10.3	9.5	11.7
Natural increase rate	17.6	5.4	24.6
Total fertility rate	3.6	1.9	4.9
Infant mortality rate	74.0	18.0	92.0
Life expectancy estimates for both sexes combined (years)	61.7	72.1	56.7
2010–2015	The world	Developed countries	Developing countries (excluding China)
Birth rate	19.4	11.0	23.5
Mortality rate	8.3	10.2	6.1
Natural increase rate	11.1	0.8	17.4
Total fertility rate	2.5	1.7	2.9
Infant mortality rate	43.0	6.0	52.0
Life expectancy estimates for both sexes combined (years)	68.9	78.0	65.2

developed countries has entered the middle and late period of the post-demographic transition. In the developing ones, though the total fertility rate had dropped to 2.9, it was still above replacement level. It needed to drop by a further 27.6% before the population reproduction pattern shifted to the low birth, low mortality, low natural increase pattern. Similarly, the birth and natural increase rates also needed to decrease by a large degree to be on par with the situation in the developed countries.

If we compare infant mortality rates and life expectancy estimates for both sexes combined at birth from 2010 and 2015, we find significant differences between the developed country data and those of the developing countries. In the former, the infant mortality rate was six per 1,000 people, while in the developing ones it was 52 per 1,000 people, a wide gap. Life expectancy in the developed countries was 78 years, while in the developing ones, it was 65.2, also a large gap. From the perspective of socio-economic development, it was found that from 2010 to 2015, most developing countries were at the stage of development trap, with high fertility rate, low productivity, and low yields. As a result, significant migrant outflows of both the general population and the labour force there have become normal. Between 2010 and 2015, the average population outflow amounted to 2,298,000 people on average in each of the developing countries annually. Why this movement of people and workers? In general terms we can note the universality of push and pull factors in migration. Population outflows are the function of both internal push and external pull factors. Specifically, net emigration is an autonomous solution to the contradiction between a shortage economy and over-population. We can thus define the current population problem in the developing countries, especially those with low incomes, as a contradiction between shortage economies and over-population.

The second category of population problems is where productive forces constrain the population. Seen from the perspective of productive forces, the history of socio-economic development is one in which production tools have been improved, updated, even revolutionized. Primitive peoples used stones and sticks, and they struck stones to make fire. Then, people invented and used copper and iron tools and bows and arrows. In the agricultural field, production tools have shifted from traditional tools such as shovels, pickaxes, and ploughs, to tractors, planters, and harvesters. In the industrial field, there has been a shift from traditional technology such as lathing, riveting, forging, and welding to the use of new energy, new materials and "intelligent" tools. Each major technology has improved productivity hundreds of times.

With the current development of new technologies such as gene engineering, aerospace and clone technology, and self-driving vehicles, intelligent tools will liberate humans from the front lines of work and production. The replacement of human labour by such tools will reduce rapidly the quantity of workers directly engaged in the production of material goods. With the eruption of intelligent software management tools, the number of people indirectly engaged in production and in public health, public works and public utilities will decrease even further and at a faster pace. In the period of intelligent tools, the pressure on populations and labour forces is immense. Intelligent tools relieve people not only of previously strenuous simple manual labour, but also from complex and mental work. They release people not only from the production of material goods, but also from the production of cultural products. They release both blue-collar and white-collar workers from their work. They benefit both cities and rural areas. And they benefit both those who work together in close proximity and those who work while in communication at a distance. The power of the new intelligent tools to change the population and labour forces is immense and far-reaching. What is worthy of our attention is that after World War II, the new technological revolution was dominated by the discipline of the life sciences. Development increasingly depends on the advancement of science and technology and on the comprehensive improvement of human capacities. It is increasingly obvious that today, national competition mainly lies in the competition for talented personnel and the accumulation of human capital. Therefore, socio-economic development no longer depends on increases in the general population and labour forces, but increasingly on the improvement of human capacities.

From this standpoint and the perspective of strategic development, China's intellectuals need to analyse the national population—its size, quality, and structures—against the background of its contemporary forms. To which category does the population problem in China belong? How is such a problem to be solved? The key question is the nature of the problem, which must be investigated in the context of contemporary national socio-economic development. The nature of the population problem is usually clear within a specific period. Since socio-economic development advances continuously, the nature of the population problem changes accordingly. Sometimes, its nature might be in transition. The argument in this book is that China's original population problem was over-population, where the size of the population constrained productive forces. But after 40 years of reform and opening-up policies, the surplus population is gradually disappearing and China is in a transition from over-population to shortage, where productive forces are constraining the population. Now, the task is to complete the transition and to guard against the risks that under-population might incur.

There are three layers of meaning in such a recognition, which are discussed in some detail in the next and other chapters, so they need be only briefly summarised here.

First, the population problem in China was originally one of over-population, wherein it held back productive forces. Population density was well above the world average, being 2.5 times greater than it in 2017. Natural resources per capita in China were at a very low level. There was then, both relative and absolute over-population in China.

Second, the nature of population problems is relatively stable and transitions last a long time. Whether transitions occur and how long the transition periods last are determined by the population's natural changes, and the new situation and socio-economic conditions. A transition is a cumulative process, from quantitative to qualitative change. The nature of a population problem does not change until accumulation has reached a certain level and quantitative change reaches the turning point where qualitative change ensues. When it happens, the situation will shift from over-population to population shortage.

Third, China is currently situated in such a transition. There are divergent interpretations of the degree of transition and whether it has reached the point of qualitative change. My view is that China remains in the transition process because it is still overpopulated and its population still holds back productive forces. Therefore, the over-population problem must be overcome first if China is to get out of the development trap associated with it. Moreover, the risks that will be incurred by a population shortage once the transition is completed must be foreseen. That is, the country should be prepared to transcend another type of development trap.

References

Compilation and Translation Bureau of the CPC Central Committee. (Ed.). (1972). *Selected Works of Karl Marx and Frederick Engels* (Vol.4). p. 2

National Statistics Bureau of the PRC. (Ed.). (2018). *China Statistical Yearbook of 2018*. China Statistical Publishing House. p. 935

U.N. Population Division. (Ed.). (2009). *World Population Prospects the 2008 Revision* (Vol. 1). New York.

Chapter 5
Context for the One-Child Policy

No discussion of the formulation of the one-child policy can bypass the 1980 symposium on the population question convened by the General Office of the CPC Central Committee (henceforth, General Office) and commissioned by the Secretariat of the Central Committee (the Secretariat), and the report submitted to the latter following it. Both the fertility policy that was promulgated after the meetings, and the *Open Letter to All Members of the Party and Youth League Members on the Question of Population Growth in Our Country from the CPC Central Committee* (henceforth, the *Open Letter*), reflected the spirit of the symposium and the final report. I explained this in my book *China's Population Policy over 60 Years*. Here, I elaborate on some of the arguments made in that book, using new documentary evidence that was not available when it was written.

5.1 Symposia on Population Problems

Between late March and early May 1980, the Secretariat commissioned the General Office to convene five symposia on the population question. These meetings were very important for several reasons. First, since they were convened by the General Office and commissioned by the Secretariat, they can reasonably be called Central Committee symposia. Second, government and Party leaders associated with population affairs, experts from the fields of the natural and social sciences were invited to attend, at which all participants' opinions were heard. Third, the report that was submitted to the Secretariat was a comprehensive analysis of the population situation at the time, the existing problems and the strategies for tackling them, and also the contemporary policy, possible changes to it, and suggestions for such adjustments. In an atmosphere in which total support was expected for the one-child policy in the past, it was inappropriate to reveal the context of that policy at the time. So the detailed situation was rarely known among the public. More than four decades has elapsed

X. Tian, *Population Policies in the People's Republic of China*,
https://doi.org/10.1007/978-981-96-9658-1_5

since then: the economy has undergone transition, society has been completely transformed, and the population has experienced significant changes. These shifts have been conducive to the revealing of some documents that were originally not known to the public, which can clarify the facts and assist our understanding of the policy through explanation of the context in which it was formulated.

5.1.1 Background to the Symposia

Since I was entrusted to draft the report to be submitted to the Secretariat, Feng Wenbin, the host of these symposia and the first deputy director of the General Office at the time, explained to me the reasons for holding the meetings. He talked about the population situation, the socio-economic situation, the relaxation of dogma,and Party leaders' instructions about population problems in particular. In order to gather information and anticipate possible problems before making a new population policy, this series of symposia was convened to hear opinions from all sectors. Since a report on the meetings was required to be submitted to the Central Committee, it was necessary to explain the socio-economic development context of the symposia. To be precise, there was an obvious problem of over-population in the late 1970s and population reproduction was still at the high increase stage. There was a prominent shortage economy at the beginning of the period of reform and opening-up, thus it seemed imperative to liberate and develop the productive forces. Following the restoration of the principle that we should seek truth from facts, reform in all aspects, including population policy, had to be vigorously promoted. In the wake of these changes, there were many appeals for an integration of theory and practice.

5.1.1.1 The Population Situation

As mentioned, the first 30 years after 1949 witnessed the transition of the population reproduction pattern, the first baby-boom, the first period with a low fertility rate, and the second baby-boom. The second baby-boom occurred mainly during the Cultural Revolution and the economic, political and social situations all contributed to its appearance. Whatever the specific reasons, over-population was becoming so serious that it was affecting socio-economic development and the improvement of material and cultural life.

The population situation at the time was mainly reflected in the following aspects: First, China's population was large and increasing at great speed. Already by the end of the 1960s, it had reached 806.71 million and by 1979, 975.42 million. In 1949, the Chinese population was 541.67 million, and it grew to 646.53 million in 1957; nine years later, there were 745.42 million people in China. By 1971, there were 852.29 million and six years later, 949.74 million. These data show that it took respectively eight, nine, five, and then six years to grow by 100 million. The time taken to increase by 100 million people was shorter by about three years in the late 1960s and the 1970s

Table 5.1 The three categories in a population age structure

	Young	Adult	Elderly
Persons 0~14 years old	above 40%	30%~40%	below 30%
Persons aged over 65	below 4%	4%~7%	above 7%
The ratio of those over 65 to those under 15	below 0.15	0.15~0.30	above 0.30
Median age	under 20 years	20~30 years old	above 30 years old

than in the 1950s and the early 1960s. Calculated by arithmetic average, the birth rate was 33.93 per 1,000 people in the 1950s, 33.3 per 1,000 in the 1960s and 24.45 per 1,000 in the 1970s, being slightly in decline. The mortality rate also continued to drop. In the 1950s, the increase rate reached 21.8 per 1,000 people, 18.43 per 1,000 in the 1960s and 19.17 per 1,000 people in the 1970s. The average annual increase rate between 1949 and 1979 reached 19.8 per 1,000 people, a strong growth trend.[1]

Second, the Chinese population was young, and with a strong increasing tendency. Demographic categories usually divide a population into young, adult and elderly. Up to now, there has been a consensus within academic circles in this division into three categories in a population's age structure. Yet there is slight divergence in the standards of division. The Table 5.1 below shows the generally accepted divisions.

So what was the population age structure in China at the time of the symposia? Such data was limited in China at the time, but the demographic statistics and forecasts released by the UN Department of Economic and Social Affairs are complete and reliable. A comparison of data from 1950 and 1980 shows no obvious change in population age structures in China. It was mainly a young population with a strong growth trend. It was also in transition from a young to an adult age structure.

Comparing Tables 5.2 and 5.3[2] above, we find no great change in the age structure in China between 1950 and 1980. The percentage of children and young adolescents increased by 1.9%, and the proportion of people over 65 years increased by only 0.2%. The ratio of those over 65 to those aged 0~14 years remained unchanged, and the median age fell by 1.8%. Generally, the age structure had begun to transform into an adult structure but it still contained characteristics of the young structure, with a strong growth trend.

A comparison of the Chinese data with those in other countries shows that China's population age structure fell somewhere between that of the developing countries and the world average, and there was a highly significant difference between China and its counterparts in the developed world. In 1980, the percentage of children and adolescents aged 0~14 years was lower than in the developing countries by 4%, and slightly higher than the world average (by 0.3%). The proportion of those aged over 65

---

[1] National Statistics Bureau of the PRC. (Ed.). (1990).
National Statistics Bureau of the PRC. (Ed.). (2018).

[2] U.N. Population Division. (Ed.). (2009). World Population Prospects the 2008 Revision (Vol. 1). New York.
pp. 48, 50, 52, 184.

Table 5.2 Contrast in population age structure between China and the world in 1950

	The world	Developed countries	Developing countries	China
Persons 0~14 years old	34.1%	27.4%	37.3%	33.6%
Persons over 65	5.2%	7.9%	3.9%	4.5%
Ratio of over 65 to under 15	0.15	0.29	0.10	0.13
Median age	24.0 years	29.0 years	21.6 years	23.9 years

Table 5.3 Contrast in population age structure between China and the world in 1980

	The world	The developed countries	Developing countries	China
Persons 0~14 years old	35.2%	22.4%	39.5%	35.5%
Persons over 65	5.9%	11.7%	4.0%	4.7%
Ratio of over 65 to under 15	0.17	0.52	0.10	0.13
Median age	23.1 years	32 years	20.1 years	22.1 years

in China was higher than in the developing countries by 0.7%, and lower by 1.2% than the world average. The ratio was similar of those over 65 to those aged 0~14 years and the median age: it was higher in China than in the developing countries but slightly lower than the world average. There were remarkable differences between the data in China and those in the developed countries that were already in the middle of an elderly age structure.

The third aspect of the population situation at the time was that the large rural population became an important factor in precluding a decline in the fertility rate. Urbanization is a tendency in population change and development throughout the whole world, including both developed and developing countries. It is a broad concept but in demography, urbanization refers to rural migration to urban areas, describing the urban–rural distribution of the population. From the perspective of labour economics, it is the distribution and structure of employment in the primary, secondary, and tertiary industries. From the perspective of economics, it describes the changes in industrial structures in agriculture, manufacturing, and the service industry. From a sociological perspective, it refers to the replacement of agricultural methods and ways of living by the methods and ways of industrial society.

The process of turning a rural population into an urban one is a process of social change in population and employment structures, the upgrading of industrial structure and the substitution of traditional civilizations by modernity. Thus from a certain perspective, a history of human civilization, especially of modern civilization, is one of continuing urbanization. According to UN data, the world's urban population accounted for 29.8% in 1950 and this increased to 39.6% by 1980. On 23 September

Table 5.4 Comparison of birth, mortality and natural increase rates in urban and rural areas in China (per 1,000 people)

	In cities			In the countryside		
	Birth rate	Mortality rate	Natural increase rate	Birth rate	Mortality rate	Natural increase rate
1955	40.67	9.30	31.37	31.74	12.60	19.14
1959	29.43	10.92	18.51	23.78	14.61	9.17
1963	44.50	7.13	37.37	43.19	10.49	32.70
1964	32.17	7.27	24.90	40.27	12.17	28.10
1971	21.30	5.35	15.95	31.86	7.57	24.29
1975	14.71	5.39	9.32	24.17	7.59	16.58
1980	14.17	5.48	8.69	18.82	6.47	12.35

2008, the UN declared that the global urban population had now surpassed the rural population and that the rural population was no longer the majority around the world.

Both the birth rate and the natural increase rate in China's urban areas and rural areas experienced historic turns in 1964. Previously, both rates were higher in the urban areas than in the countryside. Since 1964, both rates in the rural areas surpassed their counterparts in the urban areas and the gap tended to widen. This transformation largely determined population and population policy changes into the future. As an ancient civilisation, China had created an extraordinary farming culture and civilization. However, the more advanced a farming culture and civilization, the bigger the resistance to industrial revolution. Under Chinese conditions, not only was it difficult for the industrial revolution to develop, but also was there strong opposition to it.

When the PRC was founded, the urban population numbered 57.65 million, 10.6% of the total population. China was then a typical agricultural country. After 70 years of industrialization and urbanization, the urban population has grown fast, and the urban–rural distribution has changed greatly. But this process was far from smooth. There were advances, retractions, and fluctuations. From 1950 to 1960 there was rapid urbanization, the portion of urban population growing from 11.2% to 19.7%, an increase rate of 0.85% per annum. The period from 1961 to 1980 saw fluctuation. From 1961, the urban population declined for 12 years, and in 1972, it fell to 17.1%. Then there was some recovery, with the portion increasing back up to 19.4% in 1980. These 20 years of fluctuation aggravated the serious problem of the growing population and in 1964 both the birth rate and natural increase rate in the rural areas surpassed those in urban areas and in the following years, both rates in urban areas were lower than their counterparts in the countryside, and the gap tended to become wider, as shown in Table 5.4 below.[3]

Since 1964, the gap between the natural increase rate in cities and the countryside grew. In 1980, the difference in the birth rate was 4.65 per 1,000 people and the difference in the natural increase rate was 3.66 per 1,000, for the mortality rate in

[3] National Statistics Bureau of the PRC. (Ed.). (1986).

the rural areas was higher than that in the urban areas by 0.99 per 1,000 people. That is to say, the birth rate in the countryside was 32.8% higher than in the urban areas, and the natural increase rate in the countryside was higher by 42.1%. At the time, the rural population made up 80% of the total population in China. This situation was a considerable obstacle to decreasing the birth rate and controlling population growth. Even since the reform and opening-up policy, there are still differences in fertility behaviour in rural and urban areas, which limits the efficacy of the fertility policy in certain ways.

The fourth aspect of China's population situation at the time was there was steady improvement in population capacities, but the general level was not high. Demographers hold varied opinions about the elements of population capacities. One view is that population "quality" contains three elements: physical health, education levels, and political or moral awareness. Another is that population capacities comprise just the first two elements: physical health and education achievements. I am in favour of the first view but we would do better not to name the third element because such qualities should be measurable and comparable, using indexes. Since political awareness and moral awareness are conceptual terms, they are hard to quantify. I suggest therefore that the third element of population capacities be referred as civic awareness. It refers to the state and degree of social progress that a population in a country or a region embodies. It encompasses the total population's value system, moral outlook, legal awareness, and cultural awareness. It is represented by the degree of civilization and social progress, so it is measurable. For example, benevolence and generosity within a society can be measured in their acts of donation, charity activities, caring for those in need, and the like. People's civic and environmental awareness is reflected in behaviour such as whether they queue to board public transport, or toss their litter to the ground. Whether it is a harmonious society, and people's understanding of the rule of law are reflected in changes in the crime rate and the incidence of vicious crimes in the society. Civic awareness reflects the nature of a population's capacities in one aspect, and plays a crucial role in promoting social progress and harmony. Therefore, detailed analysis of these three elements of population capacities described are presented in the following sections.

First, physical fitness. In the first 30 years of the PRC, physical health among the Chinese had clearly improved. The infant mortality rate dropped from about 150 per 1,000 people in the 1940s to 34.68 per 1,000 people in 1981. In the 1940s, people's life expectancy at birth was 35 in rural areas and about 40 in the towns and cities. This rose to 66.77 by 1981.[4] These improvements were very encouraging. The rate of birth defects and maternal deaths also dropped significantly. Yet, the physical health of Chinese population was still not good. According to UN data, from 1980 to 1985, the global infant mortality rate was 78 per 1,000 people, 87 per 1,000 people in the developing countries, and 15 per 1,000 people in the developed ones. At the same time, average life expectancy in the world was 61.4, 58.6 in the developing countries,

[4] The Development and Planning Department of the National Population and Family Planning Commission, & China Population and Development Research Center. (Eds). (2004).

and 73 in the developed ones.[5] Thus by the end of 1970s and the beginning of 1980s, the physical health of the Chinese had surpassed the world average. Yet it was still a long way from that found in the developed countries. Since physical health is the foundation of all population capacities, it should be promoted significantly.

Second, scientific knowledge and education achievements. China's development relied on the transformation of a big country with a large population and rich human resources into a strong country with strong human capital. The key step was to improve scientific knowledge and education achievements. Current knowledge is the sum of predecessors' practical experiences and the improvement of human capital cannot be achieved without education. Nowadays, science and technology advance at a tremendous pace. Education is the only means by which an increasing number of people can gain a good command of the latest knowledge in science and technology and enable a country to develop a leading edge in scientific and technological development. Since the founding of the PRC, with the development of education and science, the education level of the Chinese population improved remarkably. This could be seen from the rise in the portion of the total population enrolled in school. In 1949, students at various levels in all kinds of schools accounted for just 4.76% of the total population. By 1979, the portion had risen significantly, to 21.41%. In the same period, the portion of college students rose from 2.2 per 10,000 people to 10.5 per 10,000, while the portion of middle-school pupils rose from 23 per 10,000 people to 621 per 10,000. Primary school pupils went from 450 per 10,000 people to 1,510 per 10,000 people.[6] A comprehensive index that reflects the population's education level is the average schooling of all citizens in a country. No such data is collected through the census in China, so we must make general calculations via certain methods.

When I headed a key national project "Investigation into the Elderly Population and Reform of the Old Age Security System in China" more than 20 years ago, I conducted research into the education level of some population groups. The findings showed that the average schooling of those who had received college education was nearly 16 years. Among those who had received high school or technical secondary school education, it was 11 years. The average schooling of those who had received junior middle school education was eight years, among those who had received elementary school education it was four years, and among those who were illiterate or semi-literate (those who know only a few characters), it was just 0.25 years. I then made a calculation according to the following formula:

$$\frac{\mathrm{U} \cdot Y_1 + Q \cdot Y_2 + H \cdot Y_3 + M \cdot Y_4 + L \cdot Y_5 + I \cdot Y_6}{U + Q + H + M + L + I}$$

U represents the number of people who have received college education. Y_1 represents the average years of education that a college student receives. Q represents the number of people who have received technical secondary school education. Y_2

[5] U.N. Population Division. (Ed.). (2005).

[6] National Statistics Bureau of the PRC. (Ed.). (1986).

represents the average years of education a student receives in technical secondary school. H represents the number of people who have received high school education. Y_3 represents the average years of education a high school student receives. M represents the number of people who have received junior middle school education. Y_4 represents the average years of education that a junior middle school student receives. L represents the number of people who have received elementary school education. Y_5 represents the average years of education that an elementary school student receives. I represents the number of illiterate or semi-literate people. Y_6 represents the average years of education that those people have received. According to the survey, Y_1 is equal to 16; Y_2 is equal to 11.5; Y_3 is equal to 11.5; Y_4 is equal to 8.9; Y_5 is equal to 5.5; Y_6 is equal to 0.25. When the data is put into the formula, we arrive at the average years of education the general population has received.

Why had people who have received certain level of education not received the prescribed years of education? Mainly because a minority of people dropped out of school before graduating. In the initial period of reform and opening-up there was a relatively high incidence of pupils dropping out of school. Why did a group of people have just three months of education? Mainly because many literacy classes at different levels were opened all around the country, where people were taught to recognise scores or even hundreds of Chinese characters through training lasting between one and six months. Using our formula, we worked out that the average schooling of people above six years old across the country was 2.25 years in 1964 and 4.21 years in 1982. The general education level of the Chinese population was still very low despite the obvious improvement indicated by the data. It was not only much lower than that in the developed countries but also lower than the world average level, and even lower than that in many developing countries. This was a serious obstacle to developing China into a strong modernized country.

Similarly, scientific and technological capacities were not high. In 1980, in the state-owned work units there were 657.9 scientific and technical personnel per 10,000 workers on average, accounting for 6.6% of all the workers. Among them, there were 232.2 engineering technicians per 10,000 workers on average, accounting for just 2.3% of all the workers, a very low proportion. There were 38.8 agricultural technicians per 10,000 workers on average, accounting for 0.4% of all the workers, this too was a rather low percentage. There were 190.8 health workers per 10,000 workers on average, accounting for 1.9% of all the workers. There were 40.3 scientific researchers per 10,000 workers on average, accounting for 0.4% of all the workers. Its portion was on par with that of the agricultural technicians. There were 155.8 teaching staff per 10,000 workers on average, accounting for 1.6% of all the workers. By 1989, there were 4,652 research and development institutions affiliated to the departments of local (county and above) and national governments. There were 450,369 staff in those institutions, including 161,875 scientists and engineers (making up 35.9% of all staff) and 107,585 other technical personnel (comprising a further 26.9% of all staff). In a word, the scientific and technical capacities of the whole population were not high.[7]

[7] National Statistics Bureau of the PRC. (Ed.). (1990).

Finally, we analysed people's cultural and moral fitness at the time. Physical fitness and scientific and education levels are easily observed, and hence easily attract public attention. Relatively speaking, cultural and moral fitness is not easy to evaluate directly. We can hardly expect people to establish a harmonious relationship between themselves and nature or among themselves in a country where people lack independence and a desire to innovate, where they have little understanding of democracy or the rule of law, and where they are dishonest and unfriendly with each other, lacking a sense of civic responsibility. Yet such cultural and moral fitness is indispensable to the harmonious development of a society and guarantees social progress.

This so-called cultural and moral fitness mainly refers to a core value system, moral codes, and an awareness of the law. The core value system refers to the dominant values of a certain period in a society and which are accepted by the great majority of the population. The moral codes refer to citizens' awareness of their rights and duties, and their attitudes to social engagement and national solidarity. They also include the characteristics of honesty and amiability. Legal awareness refers to people's understandings of democracy and the rule of law, including the principles that all citizens are equal before the law, that their rights and freedom are protected by law, that they have the responsibility to maintain and supervise legal justice and to assure that government officials serve with clean hands. China was long celebrated as a land of civility that has valued civility and etiquette since ancient times. And there can be little doubt that the Chinese people's cultural and moral fitness improved a lot after 1949. However, with the development of commodity production and the adoption of a market system, as commodity production and exchange values have grown so have traditional human characteristics such as family affection and friendship weakened. Residual traditional uncivilized behaviours together with the new ones have made it harder to promote cultural and moral fitness. Indeed, this seems a daunting task in contemporary China.

The fifth characteristic of the population situation at the time was the regional distribution structure of the Chinese population is unbalanced, and more population problems have been seen in the central and western regions of the country. On this subject, the Aihui-Tengchong Line (or Hu Line), drawn by renowned Chinese human geographer Hu Huanyong in 1935 to compare population density in the southeastern and northwestern areas of China is salient. Since its creation, this line shows slight changes to the population geographic distribution pattern because it not only shows population distribution but also differentiates between different terrain and climates.

China's land total is 9,600,000 square kilometres. This includes about 3,200,000 square kilometres of mountainous terrain (33% of the total territory), plateaux covering about 2,500,000 square kilometres (accounting for 26% of the territory), basins covering an area about 1,800,000 square kilometres (19%), plains totalling about 1,150,000 square kilometres (12% of the territory) and hills covering about 950,000 square kilometres (10%). Land that is less than 500 m above sea level covers an area of 2,417,000 square kilometres, constituting 25.2% of the territory, land 500–1,000 m above sea level covers an area of 1,625,000 square kilometres, constituting 16.9% of the total, land 1,000–2,000 m above sea level covers about 2,399,000 square kilometres (25%), terrain that is 2,000–3,000 m above sea level totals about 676,000 square kilometres (7% of the territory), land that is more than 3,000 m above sea level covers an area of 2,483,000 square kilometres, constituting 25.9% of the territory.[8] The Aihui-Tengchong Line is an oblique line that runs from the northeast to the southwest of the country. The major mountains, including the Altai Mountains, Tianshan Mountain, Qilian Mountain, Kunlun Mountains, Tanggula Mountains, and the Himalayas are located mainly on the northwestern side of this Line. In the same area, land that is higher than 2,000 m above sea level covers a vast area of China's territory, and most of the plateaux consist of land that is higher than 3,000 m above sea level. According to the population contour distribution principle, most plateaux are unsuitable for human habitation. Though there is also a vast area of relatively low-lying land in the northwestern part of China, it is mostly covered by desert; the main ones, including the Gobi, the Taklimakan and Tengger Deserts are located on the western and northern side of the Line.

In terms of climate, 32% of China's territory is in the humid climate zone. The semi-humid zone covers 15% of the territory, arid zones make up 31%, and semi-arid areas 22%.[9] Most parts of northwestern part of China fall in the arid and semi-arid zones, apart from a few places that have relatively plentiful rainfall. The arid and semi-arid regions are unsuitable for human habitation.

Thus there are vast differences in population density across China. The general tendency is that population density increases as we cross from the northwest to the southeast of China. The reason that population density in the northwestern region cannot increase substantially is its aridity. It is reasonable, then, to believe that the Aihui-Tengchong Line mainly follows the population distribution contour and the climate humidity line. Yet if we divide the densely populated river basin regions, such as the Yangtze River basin and Yellow River basin, into two sections, the division is different from that described by the Line. I return to this point in the conclusion of this book.

Here, we should have a clear understanding of the population situation in the western region of China against the backdrop of economic and social development in the region. In a paper entitled "The Emphasis of Western Region Development should be on the Accumulation of Human Capital", submitted to a tribunal on the development of China's western region in 2000, I argued that among natural capital, productive capital, human capital and social capital, there is more abundant natural capital in the 11 provinces, autonomous regions and municipality of China's southwestern and northwestern regions, but its distribution is very uneven. In this region, there was a shortage of productive capital, but human capital was the scarcest. The same was so with the social capital that is associated with human capital. Lack of the latter created a bottleneck that restricted development in the western region, so emphasis should be put on the accumulation of human capital. The strategy for developing the western region must include reinforcement of productive investment. Yet more important was the reinforcement of investment in human capital. The key

[8] National Statistics Bureau of the PRC. (Ed.). (1986).

[9] Idem, p. 5.

to western region development lay in whether we could continually strengthen the accumulation of human capital.

Apart from the differences in population density, there are also structural differences in fertility rates, age structure, degrees of urbanization, physical fitness, and education achievements and so on among the populations in western, central, and eastern China. Generally, when the total population reached 900 million in the mid-1970s, fertility rates in the western and central parts of China were higher than in the east. Specifically, in 1980, among the 11 provinces and autonomous regions in western China, in terms of total fertility rate (TFR), only those in Shanxi and Sichuan provinces were lower than the national average (2.24). In Inner Mongolia Autonomous Region and Gansu province the rate was slightly higher than the national average, in Guizhou and Yunnan provinces, the TFR sat between 3 to 4 and in contrast, in Qinghai, Ningxia, Xinjiang, Guangxi and Xizang, it was above 4, the highest in China. The TFR in most of the eight provinces in central China was slightly higher than the average national rate, and in the rest of the provinces of this region it was slightly lower than the national average. Except for a handful of provinces, the TFR in eastern China was lower than the national average.[10]

In terms of age structure, the populations in western and central China at that time were respectively in the middle and late stages of the adult age structure. In eastern China, it was approaching the elderly structure. The degree of urbanization in eastern China was far higher than that in the central and western regions. While urbanization in some eastern provinces reached the upper middle of the S-curve, some western provinces and autonomous regions remained in the initial stage, near the bottom of the curve. In terms of physical fitness and education levels, the differences were more marked among the different regions. In 1981, national life expectancy at birth was 67.88 years but it fell below this level in western China. The lowest life expectancy (only about 60 years) was found in Yunnan and Qinghai provinces, and the Xinjiang Uighur Autonomous Region. In central China, it was roughly the same as the national average. And in eastern China it was significantly higher than the national level. My own research showed that the education index corresponded to the average schooling rates and according to the formula described above, in 1987 the national average number of years spent in school in China was 4.645. Besides Shanxi province and Inner Mongolia and Xinjiang Autonomous Regions, the average years in school in the other provinces of western China were all below the national average: in Xizang, the average was 1.472 years, the lowest in China, in Qinghai, it was 3.132 years, in Yunnan, 3.265 years, and in Guizhou, 3.51 years. The education level in central China was on par with the national average. In particular, in Jilin and Heilongjiang provinces, it was slightly higher than the national average. Education levels in eastern China were markedly higher than the national level. Noteworthy is that the average schooling in Beijing was slightly over 7.121 years, the highest in the country. In Shanghai it was 6.932 years, and in Tianjin 6.341 years, both much higher than the

[10] Department of Population and Employment Statistics, National Bureau of Statistics of China. (Ed.). (1995).

national average.[11] In sum, there were many differences between western, central and eastern China in terms of population size, population quality and population structure. The differences indicate that there were more, bigger, and more complex population problems in western and central China than elsewhere.

5.1.1.2 The Economic Situation

The economic situation in China around 1980 is known, so here, I expand briefly on just two points: the macro-economic situation and living conditions, and the employment situation. Both are closely linked to population production, people's daily lives, and population policy.

Table 5.5 below shows some representative output, output values, and per capita levels of the macro-economy at the time.[12]

According to Table 5.5, socio-economic development in China had progressed remarkably since 1949. Although industrialization had been proceeding in China for some time and the economy had grown, general economic development, per capita production level, in particular, was still low. What we see are only the gross outputs of products in the field of manufacturing, agriculture, and the service industries. The real average per capita consumption would have been much lower because the output would go through distribution and redistribution and not all of it would enter the consumption phase. Take grain production and consumption for example: the grain produced is unprocessed and grain loses some weight after being processed. Some will be used as raw material for industry, some will be saved as seeds, and some will be reserved for other purposes. Thus the volume of grain distributed per capital will decrease to varying significant degrees. The real distribution occurs after a portion of the products have been deducted for various purposes, and according to specific plans. The public could buy them with coupons allocated to them. There was widespread use of coupons in those decades.

To illustrate with stories of personal experiences. In Beijing, I would receive a grain coupon (entitling me to 16.5 kg grain.) In 1959, the grain ration had been 18 kg for a college student in Peking University. But during the period of the "three-year difficulties", this ration fell to 16.5 kg. In the 1960s, when I was working, the grain ration stayed the same for a student. For a common resident in Beijing at the time, the ration was 14.25 kg. The vegetable oil ration was 250 g, as it was for meat and pastry (this weight was deducted from the grain ration). Vegetable and cloth coupons were dispensed annually. People who wanted to buy durable consumer goods such as bicycles, sewing machines, and watches had to buy the item by producing a specific ticket for it, or an industrial ticket. Even to buy snacks like peanuts or sunflower seeds during the Spring Festival, a specific ticket was required. And nor were such goods acquired simply by producing an appropriate ticket; they were received only after payment had been made and the specific ticket had been produced. In the

[11] Population Research Institute of Chinese Academy of Social Sciences. (Ed.). (1989).

[12] National Statistics Bureau of the PRC. (Ed.). (1986).

Table 5.5 Major Chinese economic indexes in 1980

Item	Absolute number	Per capita
Population	987,050,000	
All workers in society	418,960,000	
Total product of society	853.1 billion yuan RMB	864.3 yuan RMB
Grain output	320,560,000 tons	325 kilos
Cotton output	2,707,000 tons	2.7 kilos
Oil plants output	7,691,000 tons	7.8 kilos
Pork, beef and mutton output	12,054,000 tons	12.2 kilos
Aquatic products output	4,500,000 tons	4.6 kilos
Bicycles	13,024,000	0.01
Sewing machines	7,678,000	0.008
Watches	22,155,000	0.02
Household refrigerators	49,000	5 refrigerators per 100,000 people
TV sets	2,492,000	0.003
Washing machines	245,000	2.5 washing machines per 10,000 people
Raw coal output	620,000,000 tons	0.6 ton
Raw oil output	105,950,000 tons	0.1 ton
Steel output	37,120,000 tons	0.04 ton
Overall labour productivity of state-owned industries	12,080 yuan RMB per person per year	
Annual average salary of employees	762 yuan RMB	
Higher education enrolments	1,144,000 persons	1.2 per 1000 people
Hospital bed capacity	1,982,000 beds	2.0 per 1000 people
Number of doctors	1,153,000 persons	1.1 per 1000 people

cities, people's income was similar at the time. Since low wages were commonplace, there was usually little money left after the purchase of rationed food and other daily necessities. There were also people on even lower incomes who could not afford even to buy rationed goods. They had to apply for social relief. A 1980 survey showed that the per capita net income of rural residents was 191 yuan RMB while the annual average salary of a city worker was 762 yuan RMB. A worker can spend as much as 439 yuan RMB on daily necessities on average annually.[13]

When I began middle school in October 1953, the school authorities read aloud "The Resolution on Planned State Purchase and Planned Supply of Grain" to us. Later, the scope of planned state purchase and supply was enlarged to include cotton, clothes, and cooking oil. The school authorities explained that the reason those goods

[13] National Statistics Bureau of the PRC. (Ed.). (1986).

were purchased and supplied according to state plans was not because our country did not have those things nor because we did not have enough such things, but because there was a planned economy in China so goods were supplied according to economic plans. Subsequent practices established that the policy of planned state purchase and supply played a significant role in securing adequate supplies and stabilizing prices in China. Nearly 30 years has elapsed since the introduction of that policy, so we can contrast the economic situation in the 1980s with the situation when the planned state purchase and supply policy was introduced. People would wonder why the policy of planned state purchase and planned supply was still carried out in 1980. Did it meet the needs of planned economy or of the shortage economy? Despite divergent opinions on this question, one point is unmistakable: there was still a shortage economy in 1980. Not only statistics but also survey data showed this and personal experiences confirmed it.

There was also growing employment pressures. The shortage economy was not reflected only in the shortage of consumption goods; even more pressing was that labourers lacked technical equipment and the employment situation was worrying. In a highly centralized planned economy (though not in the collective economy in the rural areas), state-owned enterprises had to offer jobs to all working-age city dwellers. Indeed, it was the only way to solve the urban employment problem. Together with the economic rises and falls and the unmet demand for technical equipment, the number of jobless people and the unemployment rate both stayed stubbornly high. In 1980 alone, nine million people entered the job market in the cities and towns. Added to those already unemployed (5,415,000) this made for an unemployment rate of 4.9%.[14] Since there were problems with the methods for calculating the unemployment statistics, the real rate might have been higher.

In the rural areas, there was the production team. Those who could work automatically joined the collective agricultural production so, technically, nobody was unemployed in the countryside. Despite this kind of automatic employment, the number of potentially unemployed people kept growing. The rural areas absorbed not only the local workforce but also the workers sent there from the cities. For example, in the 1950s, a counter-urbanization trend transferred officials from cities to the rural areas to resolve the problem of "three breakthroughs"—the growth in urban populations, in the numbers that consumed commodity grain, and in gross payrolls. In the 1960s, under Mao's 1966 May Seventh Directive, hundreds of thousands of cadres, "persons in power within the Party who took the capitalist road", "reactionary" academic authorities, and "the stinking ninth category"[15] were sent from their original work units to the May 7th "cadre school", where they worked on farms to "remould" their minds. In addition, more than 160 million students left their urban homes to settle and work in remote rural areas, mountainous areas, and border regions for ten-plus years. Though these movements were not intended to solve the

[14] National Statistics Bureau of the PRC. (Ed.). (1990).

[15] A label given to the intellectuals during the Cultural Revolution, who came ninth on the list of undesirables after landlords, rich peasants, counter-revolutionaries, bad elements, rightists, spies, traitors and capitalist roaders.

urban employment problem, they in fact helped to alleviate it. However, while these movements began well, they ended with individuals returning to their native places spontaneously and silently so that in the end, the unemployment pressure returned to the cities.

5.1.1.3 The Political Background

The most influential political event was the Third Plenary Session of the 11th Central Committee of the CPC, at which the ideological guideline of seeking truth from facts was restored. Henceforth a series of policies was formulated. This was a historical meeting that determined the future of the Party and the state because it reversed the blind adherence to all instructions and decisions made by Mao Zedong, and sparked a process of clarification. Soon afterwards, the reform and opening-up policy was implemented. The meeting also had a positive influence on population research and population policies because it allowed some practical and scientific demographic theories and population guidelines and policies to be established. Chapter 4 has explained this in detail, but it is useful to add here that since the population policies were created by the CPC Central Committee, and implemented by the Party and the state, the attitudes and actions of the leadership were crucial. Given their influence on policy formation, some further understanding of the leaders' attitudes is in order.

In January 1975, a then deputy-premier Deng Xiaoping was assisting Premier Zhou Enlai with the work of the State Council. As part of his duties, Deng wrote instructions on a report by the Health Ministry. The instructions said, "Special attention should be paid to the issue of birth control." In April the same year, in a meeting with Tayebi Rabbi, member of the Algerian Revolutionary Committee and the Minister of Ministry of Agriculture and Land Reform, he said,

> It is troublesome to have a large population and unlimited population growth is a serious problem. The current Chinese population is nearly 900 million. Since 1949, the annual population increase in China has been 2%. At present, we advocate birth control, and the policy has worked in some places. We will try our best to keep the increase rate below 1.5% by 1980 and below 1% by 1985. To achieve this goal, we must do a lot of educational and organizational work.

In August, when Deng met an Austrian delegation from the Chinese Studies Association, he said, "It would be marvellous if the population would stop growing. However, few people want to have fewer children though that is linked to higher living standards and education attainments. Rapid population growth does not indicate that the economy is advanced; instead, it indicates it is backward. From this viewpoint, China is a developing country." In 1979, Deng addressed a Central Committee theoretical discussion "A big population has both advantages and disadvantages. When production is not developed enough, feeding the people, educating them and supplying them with jobs become serious problems. We will intensify our efforts to promote family planning work. Yet even if the population stops growing in a few years, the problems induced by a large population will continue to exist for a certain period … It is not an easy job to change a situation of limited arable land and yet a large population, especially when peasants make up a substantial portion of that population." Deng persisted in using per capita index as the yardstick to measure economic development within a certain period. Meeting the Japanese prime minister Masayoshi Ohira in 1979, he emphasized that per capita income in China would still be low by the end of the twentieth century. Even when it had become a relatively rich country in the Third World, say the per capita income reached USD 1,000, it would still be backward compared with the Western countries. The use of per capita indexes made it imperative to engage in exerting great efforts to enhance economic development on one hand and to keep population growth under control on the other hand.[16]

Chen Yun was a deputy premier by this time, but he did not speak much about population policy. What remarks he did make on the question were specific and concrete. In March 1979, when making inspections in Jiangsu, Zhejiang and other provinces, he made five suggestions for carrying out family planning work. First, family planning publicity should be vigorously promoted; second, family planning legislation should be adopted and implemented, and the policy should be that a couple is permitted to have only one child. Third, there should be rewards for the families that have only one child, fourth, prenatal care and birth control techniques should be improved, and fifth, the social security system should also be improved.[17]

Li Xiannian, another deputy premier at the time, delivered a speech at the First Meeting of the Leading Group of Family Planning Work under the Leadership of the State Council (henceforth the Leading Group) in 1978 in which he said, "Family planning is a strategic issue, so we should consider it from the overall situation and from a strategic point of view. Family planning is an issue that concerns the realization of 'four modernizations' and the prosperity of our nation. The Party Committee at all levels should put family planning work on their work agenda, discuss it often and work on it consistently." In April 1979, at a work conference held by the Central Committee, Li said, "We should do the ideological education work conscientiously. We should work out practical and effective methods to promote the idea that one couple had better have only one child."[18]

In this series of speeches and instructions made by Central Committee leaders, the goals and requirements of family planning were specified. In October 1978, the *Report on the First Meeting of the Leading Group of Family Planning under the Leadership of the State Council* proposed that one couple should have only one child, two at the most, and there should be at least three years between the two children. In September 1979, the Provincial Party Committee of Liaoning formulated some regulations about family planning, according to which, "it is generally advocated that each couple have only one child. There will be strict restrictions on permitting a couple to have a second child. Couples will be prohibited from having more than

[16] The National Population and Family Planning Commission. (Ed.). (2007). *Population and Birth Control History in China*. China Population Publishing House. pp. 90-92.

[17] The National Population and Family Planning Commission. (Ed.). (2007). *Population and Birth Control History in China*. China Population Publishing House. p. 93.

[18] Idem, p.92.

two children." Likewise, in Beijing, Tianjin, Shanghai, and Jiangsu, Jilin, and Shanxi provinces, the documents of 1979 also included the requirement that each couple was encouraged to have only one child. In Shifang county in Sichuan province, a pilot program had begun in January 1979, to implement a one couple, one child policy. The Leading Group promoted their experiences in other places in China.

Encouraged by such speeches, and Party and government documents, and the experiences of the pilot program, 136 couples from Rongcheng County, Yantan district, Shandong province, with Jü Hongze and Jü Rongfen as their representatives, started a petition called *To Have Only One Child for the Sake of the Revolution.* The petition said: "We, the 136 couples who have signed the petition, following the instructions of the Central Committee, have decided to have only one child, thereby giving up having a second child, because our personal well-being is linked to the welfare of the whole nation. When we give up having a second child, we will make a greater contribution to realizing the four modernizations. Only when we achieve these four modernizations can our life in old age be guaranteed." Similarly, 44 faculty members at Tianjin Medical College signed a petition that they were willing to have only one child per couple.[19] After these pioneering petitions were broadcast by the news media, public opinion rapidly swung to the idea that each couple should have only one child, and became widespread throughout the society.

5.1.1.4 The Theoretical Background

In 1980, the debate about practice being the only criterion for testing truth was basically over: it was concluded that all theories, guidelines, principles, and policies should be tested against practice. This criterion was to be used to test the truth of all kinds of theories, including demographic theories. In demographic studies, the doctrine "the more people, the better" came into being when the views that advocated restraining population growth and Ma Yinchu's new population theory, were criticized. As has been discussed, in the 1980s Ma's new population theory was rehabilitated and the view "the more people, the better" was ridiculed. The real consequences of an ever-growing population were that it was more difficult to feed, house, educate people, or to offer them jobs and medical services. The population problems turned out to be more acute even than that. Rehabilitating Ma's theory enabled a better understanding of the nature of the population problem in China, and of the relationship between population growth and economic development, between total population and consumption, between the working-age population and employment, and between population quality and modernization. It also allowed a better understanding of the relationship between population and education, and science and technology, and medical care and health service. Doing justice to Ma's new population theory provided the theoretical basis for controlling the size of the population and improving its capacities.

[19] The National Population and Family Planning Commission. (Ed.). (2007). *Population and Birth Control History in China.* China Population Publishing House. p. 96.

5.1.2 A Summary of the Five Symposia on Population Issues

From late March to early May of 1980, under the instructions of the Secretariat, five symposia were convened by the General Office. The first and second were held in the West Building of Zhongnanhai (the offices of the CPC Central Committee and the State Council) in late March and early April of 1980. Feng Wenbin, then first deputy director of the General Office, presided over both meetings.

According to *The Report on the Symposium on Population Problems* (first draft for discussion) and the minutes of the meeting of the Leading Group at the Fourth Meeting Room of Zhongnanhai on 25 April 1980, 63 people attended the first meeting. Among them were 25 leaders from various departments of the State Council. The others included 19 experts in demographic research, and 19 medical workers, specialists in neurology, gynaecology and obstetrics, urology, genetics, and contraception. The 25 leaders were from the Central Committee of the Communist Youth League of China, the All-China Women's Federation, the Ministry of Civil Affairs, the National Ethnic Affairs Commission, the State Planning Commission, the State Labour Bureau and so on. Fewer demographers were present at the second symposium.

At the beginning of the meeting, Feng Wenbin explained its aim: the Central Committee had decided to study the population issues, therefore, the General Office had been commissioned by the Secretariat to convene a meeting to discuss population policy for the next 20 years and beyond and to solicit suggestions from those present, representing different sectors of society. They included some heads of ministries and commissions under the State Council involved in population issues, and experts in social and natural sciences. Feng summed up population and family planning work in China, then raised several questions, including what population policy should be adopted, whether the one couple, one child policy was feasible, what issues might arise after the policy was adopted, and how should they be dealt with. He asked the participants to voice their opinions freely. Initially, they were not enthusiastic, so Feng asked workers from the Leading Group to present the family planning policy. Then participants were asked to discuss it, which they did for more than two hours. The meeting was adjourned at five o'clock in the afternoon, and participants were asked to prepare for a second meeting.

Several days elapsed before the second meeting was held, in the same meeting room, and in a lively atmosphere. Many a question was raised and some solutions volunteered. For example, one participant suggested that when there was only one child in a family, the child would probably be arrogant and spoiled. How could we prevent such a tendency and bring the children up well? The questions that attracted most attention were those about whether Chinese society would become an aging society, and whether there would be labour and military personnel shortages. Many participants volunteered solutions to these questions though they were generally not very convincing and further investigations were needed to evaluate their feasibility. Thus the second meeting was also adjourned while investigations were made into possible answers to the questions.

On 9 April 1980, the third meeting was held, and it was split into two panels. One panel met at the Guangxi Hall of the Great Hall of the People and the other at the Shanxi Hall. I participated in the first. I recall that an expert in the medical field named Chen Xuesi took the lead to express his opinion on the population questions. He mainly addressed the point that an only child might have some problems of character, therefore family education would be very crucial. He also said that population size and quality were two aspects of the population question, which both deserved attention. To improve population quality, we must begin with the care given to the infants. If a couple is encouraged to have only one child, then special attention should be given to issues such as genetic research, the incidence of birth defects, education about prenatal and postnatal care, and the availability of the various services. Another speaker emphasized that the population question was a world problem. From the 1830s, the global population took 100 years to grow to one billion people. Then the time it took to increase by the same amount shortened to 30 years, then 15 years. The population was growing exponentially, so we must find ways to control it. Several other participants suggested that a national population commission be created, a reliable social security system be introduced, and a long-term population plan be developed. First, population growth should be strictly controlled within 25 years—one population reproduction cycle; next, restrictions on population growth should be lifted a little to prevent a serious aging problem from developing. My records of the meeting are incomplete but covered the main issues discussed and the basic views of the participants.

On 25 April 1980, the fourth meeting was held at the Hall of Diligent Administration in Zhongnanhai, also chaired by Feng. This meeting mainly discussed the first draft that was to be submitted to the Secretariat; I was assigned to draw up the report. The participants confirmed that the report reflected the consensus that we should not only strictly control population growth but also address the potential threats to social stability. This consensus had been reached at the previous meeting. Feng remarked that the population guidelines, plans, and policies should be broadly inclusive. Improving population quality in its aspects of physical fitness and intelligence was not the only matter to be addressed relating to the one-child policy. When a couple was permitted to have only one child, some pregnant women might run away from home to hide. Inappropriate measures taken to deal with such situations could threaten social stability. The constitution forbade people to drown infants and if that phenomenon still occurred, it would prompt criticism from the international community. Effective measures should be taken both to apply the family planning policy and to address various doubts about it at the same time. We did not want to create antagonism towards the policy. General principles of population policy should be established. Finally, Feng summarized the following points. First, participants shared a unified understanding of population problems. Were the population to reach 1 or 1.2 billion in the future, it would not necessarily mean that we had not taken measures to control its growth. There should also be research into the population increase rate, for example, what was the optimum increase rate and adjustments should be made accordingly to achieve the target. In a word, the goal of population policies was positive, for it was concerned with the fundamental interests of the entire country. The policies were concerned with citizens' whole lives, from the moment they were born until they became old and needed to be provided for. Second, we should think about solving problems through keeping control of the processes. Strong public antagonism would have a negative influence, threatening social stability and unity. This possibility had not been taken into consideration before. How could we avoid it? Proposals for dealing with such eventualities should be made to the Secretariat. Third, some steps must be taken to encourage people to have fewer children. The most effective measure was to promote the use of contraceptive devices. Couples who had only one child should be rewarded. The aging problem was not a fundamental problem, but we should take appropriate measures to build good old people's homes. He also said that there should be practical and realistic replies to the important questions raised at the meeting. Documents could be attached to the report and on 15 May, there would be discussions at the Secretariat.

Several other participants also spoke at the meeting. For example, the deputy minister of the health ministry, Hu Zhaoheng, stressed that we should associate population questions with socio-economic development. Population and economic development forecasts should be made for the ten or twenty years. Li Chang from the Chinese Academy of Sciences underscored that population policy should be inclusive, of size, quality, genetic research, and child welfare. It was an admirable goal to control a generation, and a series of complete population policies was needed. The speech by Wu Jieping, a medical professor, highlighted the importance of considering population quality alongside population size. Everyone in China should be mobilized to improve population quality through altering old customs and breaking bad habits. Chen Dao from the Chinese Academy of Social Sciences and several other participants mentioned that the old generation of revolutionaries such as Mao Zedong and Zhou Enlai had affixed much value to the population questions, and we should learn from their instructions and apply them.

The fifth meeting took place on 30 April 1980, again chaired by Feng Wenbin. Attendance was small and the meeting focused on further discussion of and amendments to the report. Feng reemphasized the points highlighted at the previous meeting and amplified a couple of them. He said:

> Participants have reached consensus about the need to control population size. Some people have dispelled their reservations about "a single child" policy. We cannot achieve the goal of birth control by merely talking about it, neither can we achieve it by extending the marriage age in the Marriage Law. The solution to birth control mainly lies in contraception and the supply of contraceptives. Social stability and unity are crucial and if the birth control question cannot be settled then they will be disrupted and new elements of instability will arise. As a result, the achievement of the 'four modernizations' will be affected.

Wu Min and several other participants from the medical field remarked that contraceptive problems should be resolved and that devices must be safe and effective. With reinforced family planning work would come increased costs. As well, inadequate childcare should be dealt with. Hard work was required to eliminate feudal beliefs, and all people around China should act together. Wang Xinfa and some legal experts stressed that public security organs should be responsible for the collection of population statistics and other accurate data.

Feng asked me to revise the report that would be submitted to the Secretariat, on the basis of such discussions. After I did so, the revised report was subject again to further discussion and revision. Finally, the *Report on Symposia on Population Problems* (draft version for discussion) came into being. I was told to keep a copy of the report together with its five appendices, since I had drafted it, and to prepare for consultations when leaders inquired about the symposia or the Central Committee held meetings to discuss family planning work. What I said was reported in the minutes of the meeting of the Leading Group in the fourth meeting room of Zhongnanhai on 25 April 1980: "When Li (Xiuzhen) made a presentation about the symposia (on population problems), she reported the issues discussed at the meeting and the people present". Li stated in her report that 'at the second symposium, Feng (Wenbin) suggested that there should be no big meeting and we should prepare a document first. Tian Xueyuan,[20] from the Chinese Academy of Social Sciences can take charge of drafting the report. Several discussions were held during these meetings, and there are five documents attached as appendices to the main report…"[21] I was also asked by the Leading Group to draft the *Outline of the Presentation on Population Problems*, of which I also kept a copy. The relevant documents and meeting minutes are provided here for readers' reference. See the pictures 5.1, 5.2, 5.3, 5.4, 5.5, 5.6, 5.7, 5.8, 5.9 and 5.10).

The Outline of the Presentation on Population Problems (henceforth *the Outline*) was linked to the series of symposia on population issues and population policy, so some information about it is in order. During the series of symposia, I was invited by the Leading Group to draw up the draft *Outline*. I was a little confused about why Chen Muhua, head of the Leading Group and alternate member of the Political Bureau of the CPC Central Committee, had been absent from the symposia. Judging from the minutes of the meeting of the Leading Group on 25 April, Chen was not in Beijing at the time. When asked to draft the outline, I carried out the task based on the fundamental spirit of the first draft of the report on the symposia and my own understanding. I submitted the finished assignment to then-deputy Premier Chen Yun's office and later received a printed copy. On the whole, the document sent back to me had adopted my version of the Outline, but contained some revisions too. The first paragraph of the revised document said: "The Secretariat will deal with the population issue directly, allowing it to be properly settled under the Secretariat's strong leadership." This sentence was illuminating: the *Outline* that the Leading Group needed and the report on the symposia held by the General Office were not the same things. The reason why I was asked to draft both documents was probably because I could keep the two consistent with each other. Yet, how the *Outline* was used later was unknown to me. In July and August of the same year, I was also invited by the Leading Group to draft a report (for discussion) that was intended to be submitted to the third session of the Fifth National People's Congress. This

[20] Tian Xueyuan is referring to the author of this book.

[21] Readers can refer to the minutes of the meeting held by the Leading Group of Family Planning Work under the Leadership of the State Council on 25 April 1980 in the fourth meeting room of Zhongnanhai.

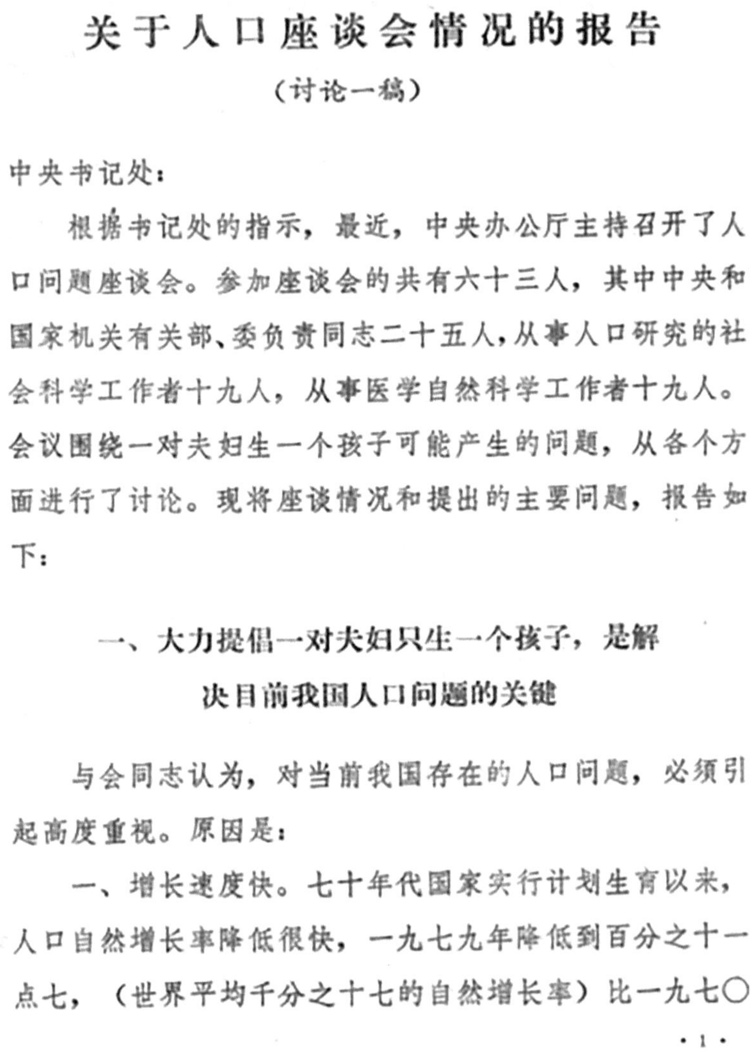

关于人口座谈会情况的报告

（讨论一稿）

中央书记处：

根据书记处的指示，最近，中央办公厅主持召开了人口问题座谈会。参加座谈会的共有六十三人，其中中央和国家机关有关部、委负责同志二十五人，从事人口研究的社会科学工作者十九人，从事医学自然科学工作者十九人。会议围绕一对夫妇生一个孩子可能产生的问题，从各个方面进行了讨论。现将座谈情况和提出的主要问题，报告如下：

一、大力提倡一对夫妇只生一个孩子，是解决目前我国人口问题的关键

与会同志认为，对当前我国存在的人口问题，必须引起高度重视。原因是：

一、增长速度快。七十年代国家实行计划生育以来，人口自然增长率降低很快，一九七九年降低到百分之十一点七，（世界平均千分之十七的自然增长率）比一九七〇

·1·

Picture 5.1 The report on symposia on population problems (draft version for discussion) (first page)

draft was consistent with the report on the symposia on population problems. The Leading Group is the leading organ that implements population policy formulated by the Central Committee. Thus the *Outline* was an important document. A brief introduction of the *Outline* might be useful here. It contained five parts, whose main content is presented below.

First, the *Outline* noted that implementation of the family planning policy was going well, especially in controlling population growth. Three points are worthy

〔附件一〕

目前我国存在的人口问题

一、总人口同生活资料的增长不相适应。

据统计，一九五三至一九七八年，全国居民消费总额增长二点八倍①，由于同期人口增长百分之六十六，按人口平均的消费额只增长一点三倍。每年新增加的消费额中，百分之五十八左右被新增加的人口消费掉了，用于提高原有居民消费额部分只占百分之四十二，人民生活水平提高受到限制。从一九五七至一九七七年，粮食和人口的增长率持平，每人平均占有的粮食一直在六百斤上下，近两年来稍有提高，也仅比一九五二年提高百分之十二的样子。有些经济作物按人口平均的产量还有所降低：一九七八年平均每人占有的棉花为四点五二斤，比一九五二年降低百分之零点二；平均每人占有的油料为九点五斤，比一九五二年降低百分之二十六点九。

① 按当年物价计算的，实际增长速度要低一些。

—1—

Picture 5.2 Appendix 1: Current population problems in China (first page)

of our attention. Leaders of all ranks had recognised the importance of this work and gradually understood that population growth must move in tandem with the increase in material goods, and that "population reproduction and production of material goods should both be valued". Furthermore, family planning regulations or methods to implement family planning had been formulated in all provinces, municipalities, and autonomous regions except the Inner Mongolia Autonomous Region. Family planning work had broad public support and Academics and the media were propagating the population issues. The *Outline* noted that it was very

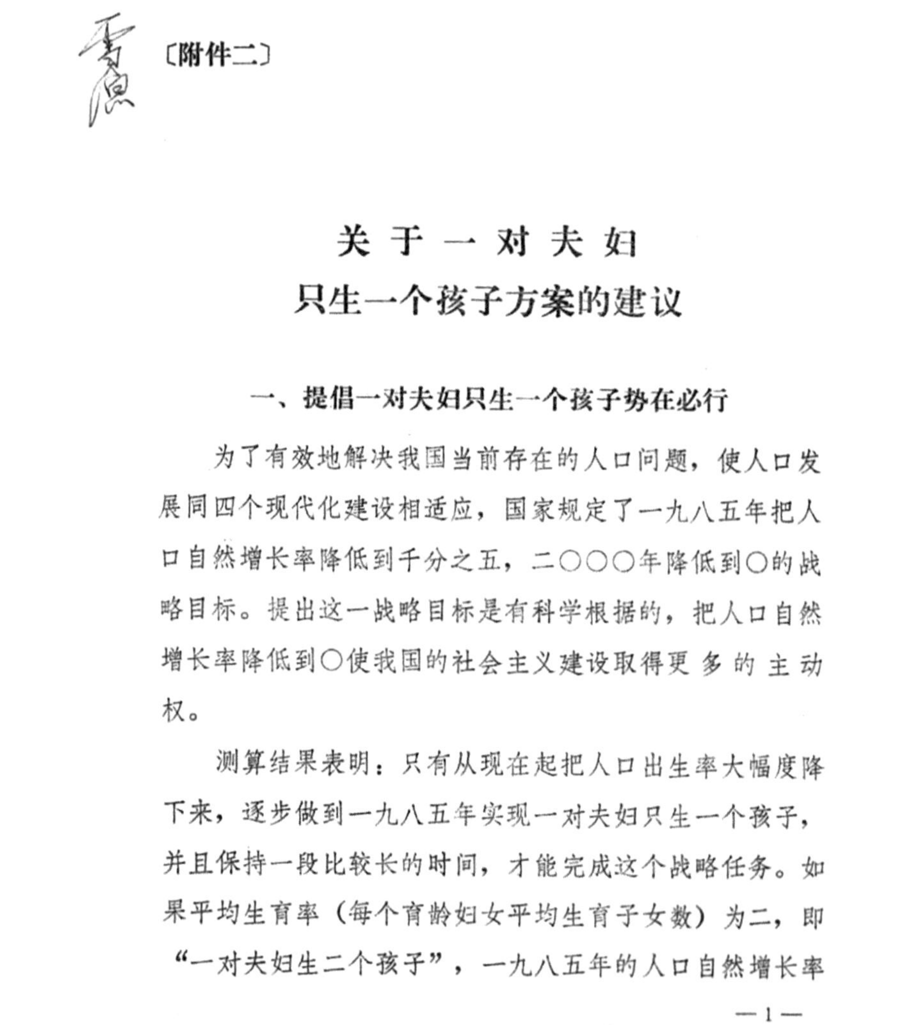

〔附件二〕

关于一对夫妇
只生一个孩子方案的建议

一、提倡一对夫妇只生一个孩子势在必行

为了有效地解决我国当前存在的人口问题，使人口发展同四个现代化建设相适应，国家规定了一九八五年把人口自然增长率降低到千分之五，二〇〇〇年降低到〇的战略目标。提出这一战略目标是有科学根据的，把人口自然增长率降低到〇使我国的社会主义建设取得更多的主动权。

测算结果表明：只有从现在起把人口出生率大幅度降下来，逐步做到一九八五年实现一对夫妇只生一个孩子，并且保持一段比较长的时间，才能完成这个战略任务。如果平均生育率（每个育龄妇女平均生育子女数）为二，即“一对夫妇生二个孩子”，一九八五年的人口自然增长率

—1—

Picture 5.3 Appendix 2: Proposal of the plan that each couple have only one child (front page)

difficult to carry out family planning work. Though the natural increase rate in 1979 (11.66 per 1,000 people) had not met the target (of 10 per 1,000 people), great efforts had been made to achieve it.

Second, some problems were currently being confronted in the implementation of the family planning work. Officials engaged in the work still harboured many doubts about the idea of having only one child and were seriously worried about the change to population policy. People also often complained about how the policy was being implemented, and were upset at the use of coercive methods and the application of a single solution for diverse problems. It was pointed out that if inappropriate

二、"一对夫妇只生一个孩子"
的方案可以搞到二〇一〇年左右

"一对夫妇只生一个孩子"势在必行，但"一对夫妇只生一个孩子"的方案搞多长时间为宜？这是人们普遍关心的一个问题。当前，人们对"一对夫妇只生一个孩子"的某些疑虑，有的纯属由于误解造成的，以为"一对夫妇只生一个孩子"要永远搞下去，会产生这样或那样的不堪设想的后果。因此，有必要清楚地告诉人民："一对夫妇只生一个孩子"不是永久之计，它是一段时间，我认为可以从一九八五年实现"一对夫妇只生一个孩子"算起，大体经过一个人口再生产周期，即到二〇一〇年左右的一个方针政策。

为什么"一对夫妇只生一个孩子"要搞到二〇一〇年左右？主要有以下几方面的原因：

（1）自然增长率在二十一世纪初降到〇以后，虽然总人口开始下降，但仍稍高于目前的水平，二〇一〇年总人口为十亿四千四百万。

（2）劳动适龄人口（男十八至六十四岁，女十八至五十九岁）在未来二十四年内是一直上升的，二〇〇五年以后有所下降，但二〇一〇年仍为七亿四千八百万，比目前多二亿四千五百万人。要降到目前的水平，需二〇三〇年以后（详见附表）。

（3）每个劳动力抚养的未成年人口和老年人口，即

Picture 5.4 Appendix 2: Proposal of the plan that each couple have only one child (page 1)

〔附件二〕

距目前所占比重最高的东德等国家百分之十六的水平，相差甚远。

看来，为了扎扎实实地解决我国人口问题，“一对夫妇只生一个孩子”搞到二〇一〇年是完全必要的。那时，总人口、劳动适龄人口、抚养指数、“老龄化”等都不会出现什么大的问题。

三、二〇一〇年以后平均生育率应作适当调整

二〇一〇年总人口和劳动适龄人口高于现在水平，抚养指数也低于现在水平，似乎“一对夫妇只生一个孩子”方案可以继续搞下去。可是，人口再生产和物质资料再生产有很大不同，人口再生产周期长，具有长期、累进和稳定增长的特点。因此，一种人口政策的实施，起码要顾及到二十五年以后的结果，再长一点时间更好。按照“一对夫妇只生一个孩子”方案，二〇一〇年以后将出现下列一些情况：

（1）虽然总人口高于现在的水平，但下降的趋势比

Picture 5.5 Appendix 2: Proposal of the plan that each couple have only one child (page 2)

work methods were used then family planning work might disturb social stability. In some places, doctors' skills in performing sterilization operations were not good, so medical accidents occurred sometimes, and the attitudes of some doctors to their work created concern.

The *Outline* proposed intensifying population size control and improving population quality simultaneously, thus enabling the population to improve national economic development and accelerate the realization of the "four modernizations". This guideline was based on the current population situation and the basic characteristics of the very large Chinese population. Every year, both the number of new births and the net population increase were enormous. The population grew at high speed,

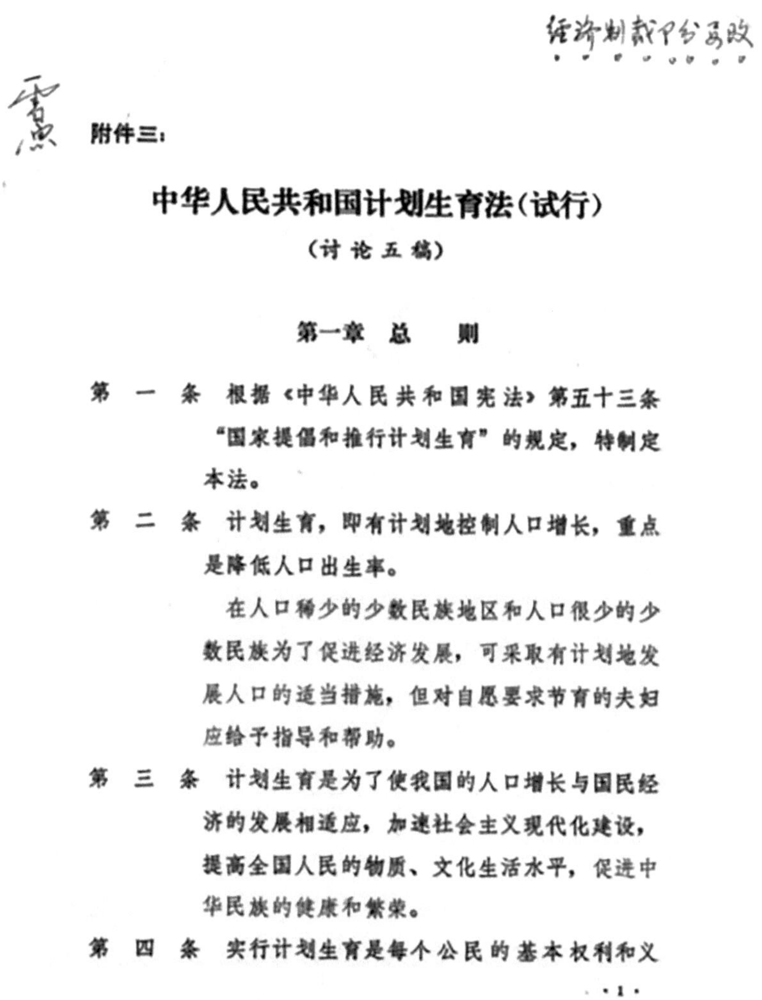
附件三：

中华人民共和国计划生育法（试行）

（讨论五稿）

第一章 总 则

第 一 条 根据《中华人民共和国宪法》第五十三条"国家提倡和推行计划生育"的规定，特制定本法。

第 二 条 计划生育，即有计划地控制人口增长，重点是降低人口出生率。

在人口稀少的少数民族地区和人口很少的少数民族为了促进经济发展，可采取有计划地发展人口的适当措施，但对自愿要求节育的夫妇应给予指导和帮助。

第 三 条 计划生育是为了使我国的人口增长与国民经济的发展相适应，加速社会主义现代化建设，提高全国人民的物质、文化生活水平，促进中华民族的健康和繁荣。

第 四 条 实行计划生育是每个公民的基本权利和义

·1·

Picture 5.6 Appendix 3: Family Planning Law in People's Republic of China (fifth draft for discussion) (first page)

resulting in a large proportion of young people. The total population did not accord with the available means of livelihood, the working age population did not fit the available means of production, and people's skills did not fit the needs of the "four modernizations". There were serious contradictions between the large population and its poor quality and national economic development. As the "four modernizations" proceeded, those contradictions grew increasingly acute.

Third, the *Outline* proposed population planning. It noted that deputy Premier Deng Xiaoping had suggested per capita GNP in China should reach $1,000 by the end of the twentieth century. To accomplish this, it was necessary to develop

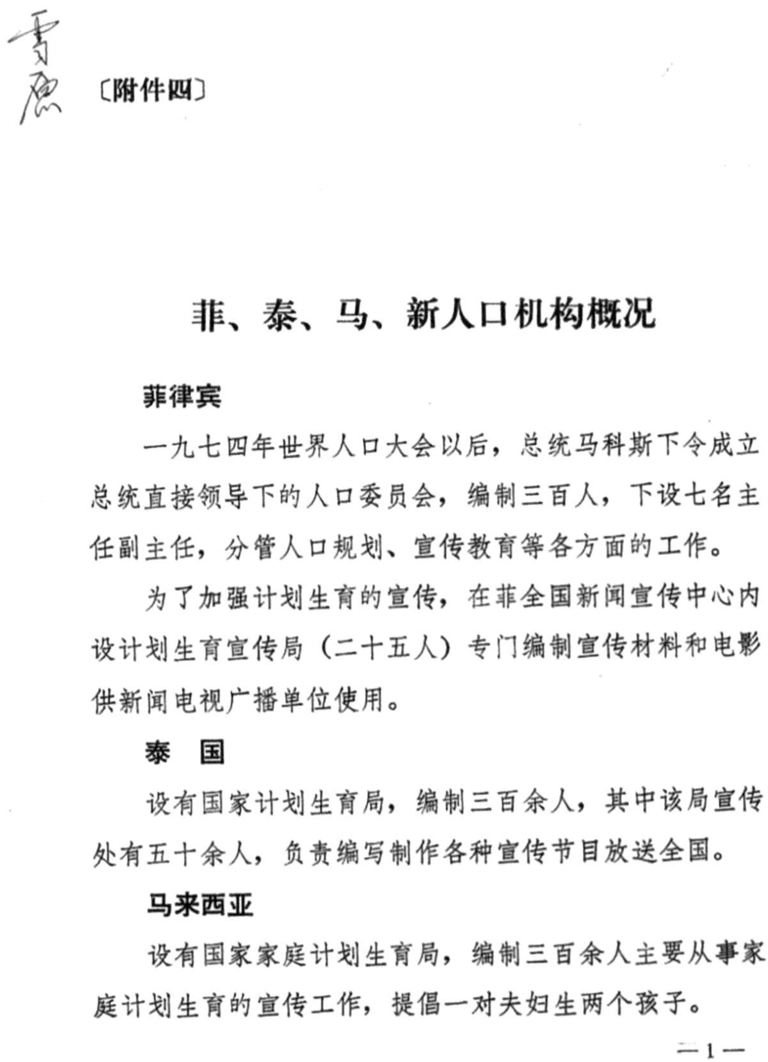

〔附件四〕

菲、泰、马、新人口机构概况

菲律宾

一九七四年世界人口大会以后，总统马科斯下令成立总统直接领导下的人口委员会，编制三百人，下设七名主任副主任，分管人口规划、宣传教育等各方面的工作。

为了加强计划生育的宣传，在菲全国新闻宣传中心内设计划生育宣传局（二十五人）专门编制宣传材料和电影供新闻电视广播单位使用。

泰　国

设有国家计划生育局，编制三百余人，其中该局宣传处有五十余人，负责编写制作各种宣传节目放送全国。

马来西亚

设有国家家庭计划生育局，编制三百余人主要从事家庭计划生育的宣传工作，提倡一对夫妇生两个孩子。

—1—

Picture 5.7 Appendix 4: A survey of population agencies in the Philippines, Thailand, Malaysia, and Singapore (first page)

production and control population growth simultaneously. The *Outline* made three predictions about population growth. According to the most conservative prediction, in 2000, the natural population increase rate would be 1.87 per 1,000 people, and in 2030, it would be down to -5.87 per 1,000 people. According to the medium prediction, in 2000, the natural population increase rate would be 6.17 per 1,000 people and down to -0.45 per 1,000 by 2030. According to the most pessimistic prediction, in 2000, the natural population increase rate would be 15.86 per 1,000

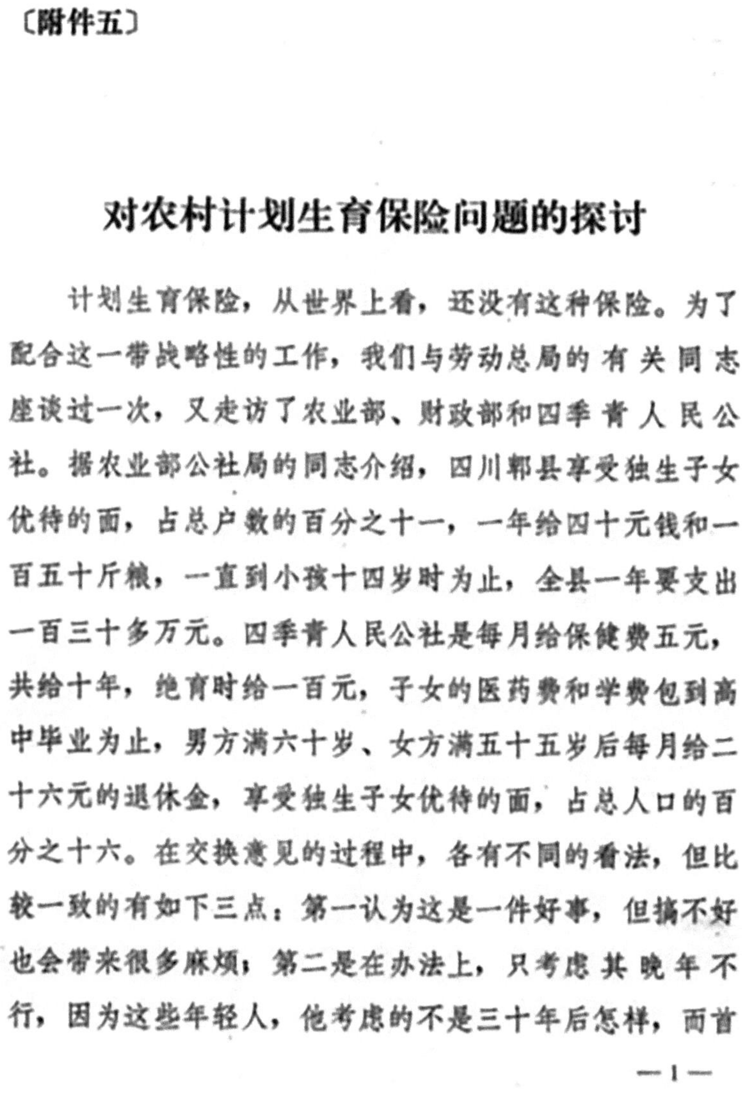

〔附件五〕

对农村计划生育保险问题的探讨

计划生育保险，从世界上看，还没有这种保险。为了配合这一带战略性的工作，我们与劳动总局的有关同志座谈过一次，又走访了农业部、财政部和四季青人民公社。据农业部公社局的同志介绍，四川郫县享受独生子女优待的面，占总户数的百分之十一，一年给四十元钱和一百五十斤粮，一直到小孩十四岁时为止，全县一年要支出一百三十多万元。四季青人民公社是每月给保健费五元，共给十年，绝育时给一百元，子女的医药费和学费包到高中毕业为止，男方满六十岁、女方满五十五岁后每月给二十六元的退休金，享受独生子女优待的面，占总人口的百分之十六。在交换意见的过程中，各有不同的看法，但比较一致的有如下三点：第一认为这是一件好事，但搞不好也会带来很多麻烦；第二是在办法上，只考虑其晚年不行，因为这些年轻人，他考虑的不是三十年后怎样，而首

— 1 —

Picture 5.8 Appendix 5: Exploration of social security problem in the rural areas (first page)

people and in 2030 remain high at 14.16 per 1,000. In order to achieve Deng's goal, 1.5 children per woman was set as the internal control index.

Fourth, several measures were proposed for carrying out the family planning policy. To achieve the population development plan, the key was to increase the portion of only children among new births and to eliminate third births as soon as possible. The goal of an average 1.5 children per woman could probably be achieved

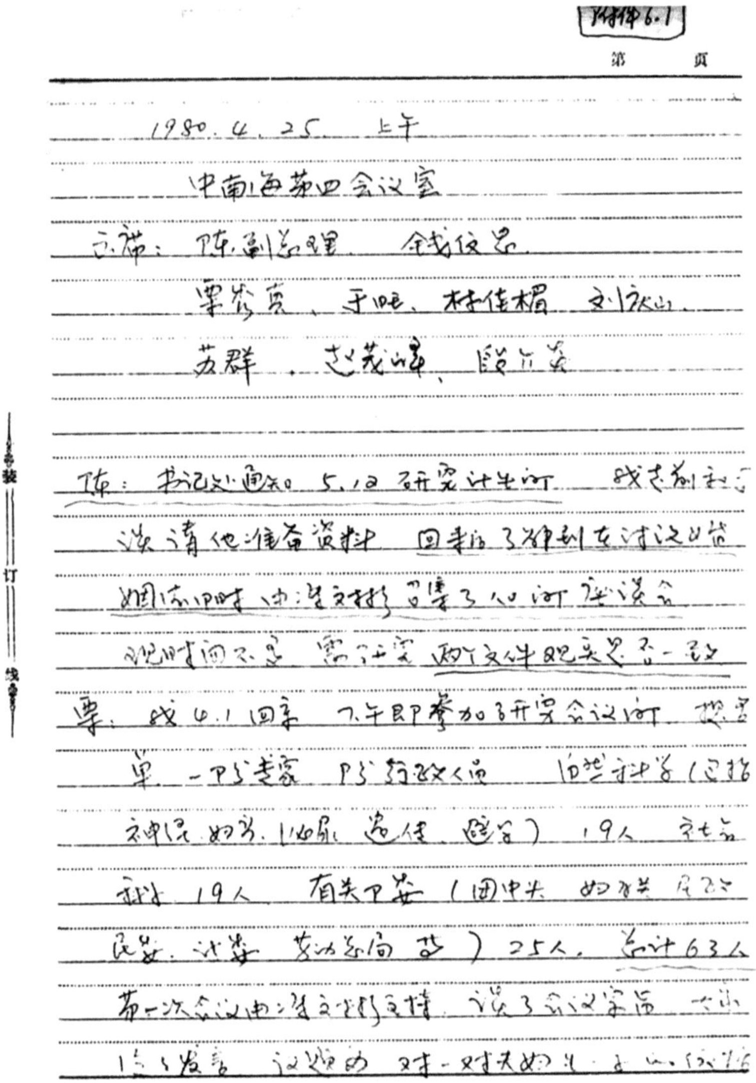

附件6.1

第 页

1980.4.25 上午

中南海第四会议室

出席：陈副总理 钱信忠

Picture 5.9 Minutes of the meeting of the Leading Group of Family Planning Work under the leadership of the State Council (25 April 1980) (Page 1)

by rewarding those who had only one child and punishing those who had a third. Seven measures were proposed.

The first was to propagate the population policy widely. The significance of having only one child, improving population quality, and reinforcing prenatal and postnatal care should be popularized and known to every household.

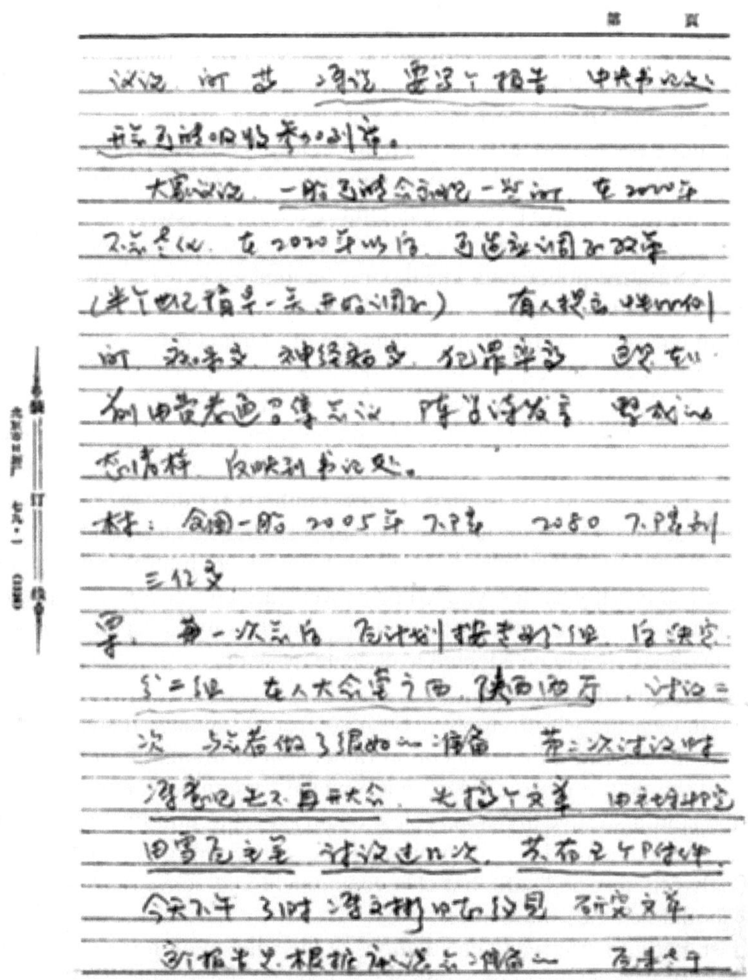

Picture 5.10 Minutes of the meeting of the Leading Group of Family Planning Work under the leadership of the State Council (25 April 1980) (Page 2)

The second measure was to reward those who had only one child and punish those who had a third. Steps should be taken to ensure that there were consistent standards for the duration of the system of rewards, for the amount of the bonuses, and for the size of the fine for third children.

The third measure was to establish the social security system in a planned way. The government should offer social security payments to the elderly in a systematic way to solve the problem of providing for senior citizens. The *Outline* pointed out specifically that in the rural areas there was no social security system, so we must address this problem and provide for the elderly who could no longer work. As the number of only children grew, relevant policies and regulations should be adapted to meet the needs of family planning work. The adaptation involved strong support and cooperation among all departments concerned.

The fourth measure concerned the minority ethnic groups, among whom the family planning policy should be applied differently. Those who belonged to certain ethnic minorities in regions with tiny populations, or in sparsely populated regions, would be permitted to have three or four children. If they lived in a city or in densely populated regions, then in principle, they should practice family planning like the Han people. Still, the policy could be flexibly applied: those who had only one child should be rewarded, and they could have three children at most.

The fifth measure referred to the promotion of prenatal care. While encouraging couples to have only one child, efforts should be made at the same time to promote prenatal care. For example, the harm caused by consanguineous marriages should be made known. Newly-wed couples should be screened for genetic diseases, and genetic disease counselling clinics should be opened to serve those in need. Prenatal and postnatal care should be promoted. New methods for delivering babies should be popularized and advice on feeding newborns should be offered to new mothers. The aim was not only to encourage people to have fewer children but also to ensure that people had healthy babies who would be well cared for. Scientific and education sectors should put more effort into studying psychology and pedagogy. Specific teaching methods should be adopted to educate these only children to improve their and the whole nation's capacities.

The sixth measure was to improve birth control techniques. It stressed that birth control should depend on contraception rather than artificial abortion. The health department should train medical personnel engaged in family planning work in facilities of the administrative divisions below county level to improve their techniques. Scientific research should also be promoted to produce new contraceptive tools that are not only efficient and safe, but also economical and convenient. Contraceptive devices should be made more accessible.

The last measure referred to calculations of the amount of money to be offered as bonuses to families who had only one child. The *Outline* pointed out that monies to be used for bonuses amounted to far less than the projected expenditures on babies that might have been born.

In part 5 of the *Outline*, it was suggested that people's doubts about the one child policy should be addressed, and the leadership should be reinforced to be able to implement the policy. There seemed to be a contradiction in people's thinking about it: on the one hand, when the total population approached 1 billion they believed that it was too big, yet for some of the couples, one child seemed not enough. To clear up the doubts, the *Outline* pointed out that the work force would keep increasing for the next 32 years, perhaps from 540 million in 1979 to 790 million by 2012. It might then decline from 2013 and not grow back to current levels until 2046. About the aging problem, the *Outline* pointed out that though the number of old people was growing quickly, the young population was declining rapidly. When the two cancelled out each other, the dependency index (the number of persons each worker must support) would keep decreasing before 1998, after which it would start to go up, though not return to current levels (slightly higher than 0.9) until 2035. Therefore, in the remaining 20 years in the twentieth century, problems such as aging, and labour force and military shortages would not appear. There was also concern about the ratio of men to women. With the strong advocacy of one child for each couple, some people would prefer sons to daughters according to the deeply embedded feudal belief that only male offspring can continue their family line. Influenced by this belief, some people would probably go to the extreme of killing baby girls or abandoning them. About concerns for the male to female ratio, the *Outline* proposed that killing or abandoning babies be prohibited by law.

At the end of the *Outline*, there was a summary: to resolve the population problems in China effectively, strong advocacy of a one child policy was imperative. To fulfill the tremendous task of controlling the size of the population and improving its capacities, Party and government organizations at all levels must reinforce their leadership in family planning work. Only thus could the policy be implemented smoothly. To deal with the existing practical problems, help should be extended to specific departments and individuals in need of it. Everyone must become familiar with the rules for implementing family planning work, and get the task done as early as possible, as carefully as possible, and as thoroughly as possible. We should not wait and act only when problems arise. We should try our best to prevent new factors of instability from appearing because of defects in family planning work. The following two images (Pictures 5.11 and 5.12) show the first and last pages of the scanned copy of *The Outline of the Presentation on Population Problems (draft)* respectively.

5.2 Essentials of the Symposia Conclusions

The symposia on population problems, commissioned by the Secretariat of the Central Committee and held by the General Office of the Central Committee in 1980, were very important meetings. The Party and government leaders, scientific experts and leaders from social organizations gathered to discuss state affairs, and participants aired their views freely on the population problems. Based on the discussion, *The Report on Symposia on Population Problems* (first draft for discussion) (henceforth the *Report*) was created. Having been assigned the task of drafting the report, I was left with deep impressions of the symposia. While informing me of the report's dominant idea, Feng Wenbin, host of the meetings, said that the chief leaders of the Central Committee already had clear perspectives on the population problem but wanted potential future problems to be identified in order to guard against them. Drafting the report was thus a serious responsibility. For me, the fundamental principle was that the report should be based on the meetings' discussions and be consistent with the spirit of the symposia. It should also stand the test of practice. With hindsight, most of the basic ideas of the report were reflected in the population policies issued later, and some ideas were neglected. It seems imperative here to discuss the report submitted to the Secretariat in its original form.

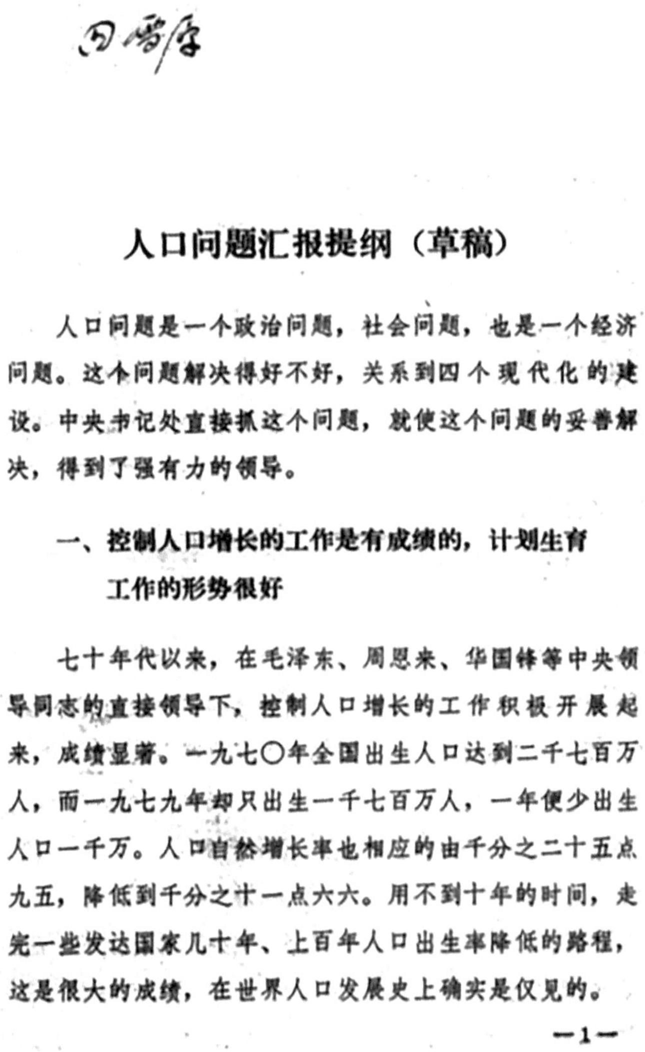

田雪原

人口问题汇报提纲（草稿）

人口问题是一个政治问题，社会问题，也是一个经济问题。这个问题解决得好不好，关系到四个现代化的建设。中央书记处直接抓这个问题，就使这个问题的妥善解决，得到了强有力的领导。

一、控制人口增长的工作是有成绩的，计划生育工作的形势很好

七十年代以来，在毛泽东、周恩来、华国锋等中央领导同志的直接领导下，控制人口增长的工作积极开展起来，成绩显著。一九七〇年全国出生人口达到二千七百万人，而一九七九年却只出生一千七百万人，一年便少出生人口一千万。人口自然增长率也相应的由千分之二十五点九五，降低到千分之十一点六六。用不到十年的时间，走完一些发达国家几十年、上百年人口出生率降低的路程，这是很大的成绩，在世界人口发展史上确实是仅见的。

—1—

Picture 5.11 The front page of the *Outline*

As the sole draftsman of the *Report*, I worked alone. When the document was rendered for discussion, some people concerned would be invited by the General Office to discuss it. The meetings mentioned in the *Report* mainly referred to the meeting held at the Hall of Diligent Administration in Zhongnanhai on 25 April and the smaller meeting of 30 April 1980. Amendments were made according to suggestions solicited from the participants. Finally, the *Report* and five further documents, attached as appendices, were submitted to the Secretariat. Why was I allowed to keep a series of documents after the symposia? As has been explained in Sect. 5.1 of this

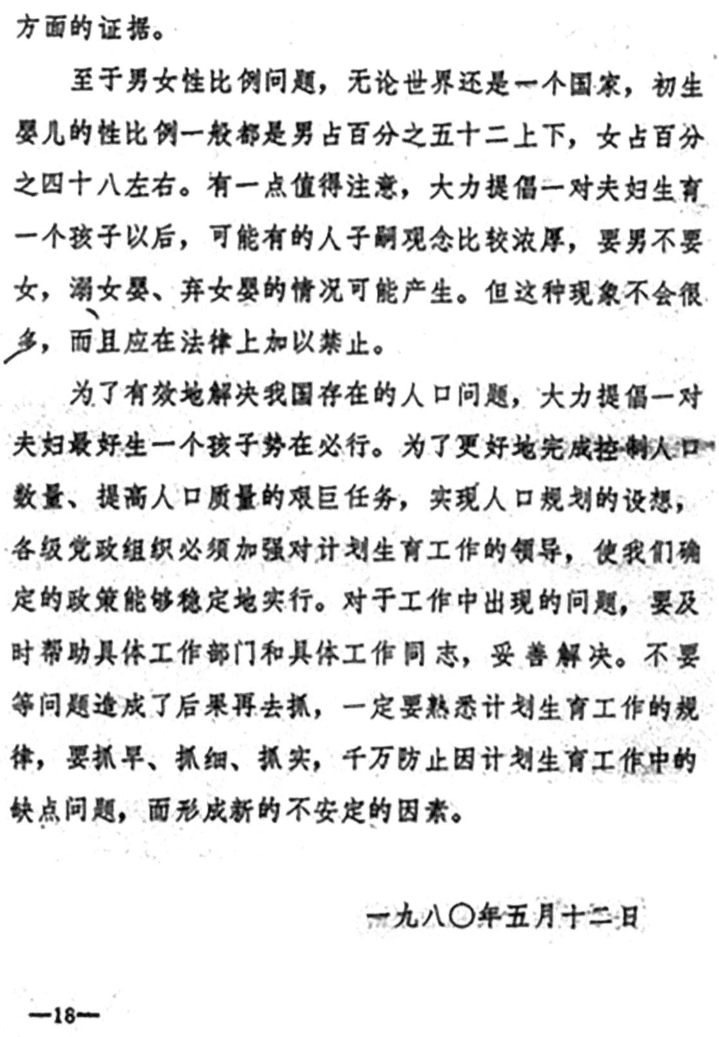

方面的证据。

至于男女性比例问题，无论世界还是一个国家，初生婴儿的性比例一般都是男占百分之五十二上下，女占百分之四十八左右。有一点值得注意，大力提倡一对夫妇生育一个孩子以后，可能有的人子嗣观念比较浓厚，要男不要女，溺女婴、弃女婴的情况可能产生。但这种现象不会很多，而且应在法律上加以禁止。

为了有效地解决我国存在的人口问题，大力提倡一对夫妇最好生一个孩子势在必行。为了更好地完成控制人口数量、提高人口质量的艰巨任务，实现人口规划的设想，各级党政组织必须加强对计划生育工作的领导，使我们确定的政策能够稳定地实行。对于工作中出现的问题，要及时帮助具体工作部门和具体工作同志，妥善解决。不要等问题造成了后果再去抓，一定要熟悉计划生育工作的规律，要抓早、抓细、抓实，千万防止因计划生育工作中的缺点问题，而形成新的不安定的因素。

一九八〇年五月十二日

—18—

Picture 5.12 The last page of the *Outline*

chapter, the host of the symposia told me to keep them for future consultations. In the interim, the Leading Group of Family Planning Work had asked me to draft *The Outline of the Presentation on Population Problems* and prepare the report (a draft version for discussion) for the third session of the fifth National People's Congress. No other leaders asked me about the symposia. Thus the most important document was the *Report*.

The Report on the Symposia on Population Problems (first draft) was submitted in early May 1980, addressed to the Secretariat. It began: "Upon the instructions of the Secretariat of the Central Committee, the General Office of the Central Committee convened a series of symposia on population problems recently." This very sentence indicates that the symposia were held by the General Office and commissioned by the Secretariat. This is significant because it was later claimed that the Central Committee had never held symposia on population problems. But this opening sentence of the *Report* clarifies the reason for the symposia, their nature, and their function. On the same page, the *Report* states who attended the meetings, the theme and the context of the symposia, and the central issues discussed there. Picture 5.1 above shows the first page of the scanned copy of the *Report* (first draft). The *Report* was composed of three parts: Part one is headed "The key to solving the current population problem in China is advocacy of the policy of only one child per couple"; Part two is headed "Discussions on some relevant questions about the one-child proposal"; Part three is headed "Several suggestions".

In addition, as mentioned, five appendices are attached: Current Population Problems in China; Proposal of the Plan that Each Couple Have Only One Child; Family Planning Law in the People's Republic of China (proposed law, fifth draft for discussion); A Survey of Population Agencies in the Philippines, Thailand, Malaysia, and Singapore, and Exploration of Social Security Problem in the Rural Areas.

The first and second attachments were written by me and the rest were provided by various government departments. Since it was required that the chief report should be no more than 6,000 words, some questions had to be elaborated in the appendices. Next, I focus on explaining the three sections of the *Report* and, where appropriate, do so by referring also to the salient appendices.

5.2.1 Part One of the Report: Advocacy of the Only One Child Per Couple Policy

This part is headed "The key to solving the current population problem in China is advocacy of the policy of only one child per couple".

Symposia participants reached a consensus that we must take China's existing population problems very seriously, for the following reasons: (a) The Chinese population was growing rapidly. Rapid population growth and slow economic growth combined sharpened the contradiction between the shortage economy and overpopulation. Since the 1970s, with the birth rate and the natural increase rate falling to some degree, the work of controlling population growth had made great progress. However, the total population still rose from 540 million in 1949 to 970 million in 1979, with a net increase of 430 million. The annual increase rate approached 20 per 1,000 people, a fast rate indeed. The current increase rate doubled that before 1949 (10 per 1,000 people) and was 2.9 times of its counterpart in the developed countries in 1979 (7 per 1,000 people). (b) China's population was huge so even though the birth rate and the increase rate dropped a bit, the absolute number of births and net population increase were still very great, resulting in a serious cumulative effect. In 1979, the total population in China (including Taiwan) was 987 million, accounting for 22.7% of the global population (4.336 billion). By 1955, with a birth rate of 32.6 per 1,000 people, the new-born population reached 19,780,000. In 1979, when the birth rate dropped to 17.9 per 1,000 (a decrease of 14.7 per 1,000), new births still amounted to 17,270,000. (c) The Chinese population was young, which meant large numbers of women of child-bearing age and a strong momentum for population growth. Inferred from a sampling survey, in 1978, there were 343 million juveniles under 15 years old, accounting for 35.8% of the total population; 264 million people were aged between 15 and 29, accounting for 27.6% of the total population. When the numbers of these two groups were put together, those under the age of 30 accounted for 63.4% of the total population; 305 million people were aged between 30 and 64, accounting for 31.8% of the total population, and just 46 million people were older than 64 years, accounting for 4.8% of the total population. The young and growing population structure would have a decisive influence on the future population changes.

These basic population characteristics resulted in "the serious population problems in China" at the time (for detailed information, refer to Appendix 1 of the *Report*). Thus it was imperative to control population growth dramatically, to lower the birth rate and natural increase rate, and to ensure gradually a reproduction rate of only one birth per couple. In the *Report*, the major population problem was summarized in three ways: first, the means of livelihood could not accommodate the size of the general population; second, the means of production could not accommodate the growing labour force; and third, the population's capacities could not accommodate the demands of public undertakings such as the development of science, education, health service, housing, and public transportation. Appendix 1, *Current Population Problems in China*, explained these three "cannot accommodates" separately based on data collected from statistics and surveys. For example, from 1953 to 1978, total household consumption in China increased 2.8 times. But with the population growing by 66%, consumption per capita only increased 1.3 times. About 58% of the increased consumption was accounted for by the newly increased population, meaning that only 42% represented an increase in consumption by the original population. As a result, improvement in people's living standards was limited (see the first six lines of Appendix 1 as shown in Picture 5.2).

Another example lies in the contradiction between a large population and the available arable land, which was getting sharper and sharper. Arable land per capita in China decreased from 2.71 *mu* around 1949 to 1.57 *mu* at the time of the *Report*. By contrast, arable land per capita in the world was 5.5 *mu* at the time. Thus the arable land per capita in China was only 28.5% of the world level.

A third example involves the education sector: there were about 5.2 million college graduates (0.5% of the total population) in China at the time, and about 210 million middle school graduates (22% of the total population). Only 5% of the senior middle school graduates could be admitted to colleges, and only 41% of the junior middle school students could attend senior middle school. Despite the situation, there were still acute shortages of teachers, buildings, and teaching aids. Since 1949, the total urban residential area had increased to 493 million square meters. As the population grew faster than available dwellings, living space per capita decreased from 4.5 square meters in 1949 to 3.6 square meters in 1977. Appendix 1 of the *Report* summarized the reasons for these situations. According to the document,

> Economically, over these years, the portion of accumulation was higher than the portion used for consumption. Of the accumulation, productive construction investment occupied a larger portion than non-productive construction investment in the likes of culture, education, health services, housing for workers, and public transportation, resulting in relatively slow progress in the development of public works. Rapid population growth since 1949 also contributed significantly to these difficulties, which cannot be denied. Here lies the heart of the population problem in China.

5.2.2 Part Two of the Report: Discussions on Questions About the One-Child Proposal

Part two was the largest part of the *Report* covering the discussions of questions raised about the one-child proposal from six angles. There was discussion about the work force and the dependent population. This was a topic that concerned many and thus was discussed frequently. There were worries that the policy might create problems such as an aging population, labour shortages and military personnel shortages, a growing dependent population, and an increase in its portion of the total population. There was also concern that it might result in households in which two labourers had to support four elderly people and one child, the so-called "four-two-one" family structure. The *Report* responded to these worries with a specific prediction: that the working-age population would keep growing in the next 24 years and then decrease, but it would not decrease to the current level until 2030, which was 50 years into the future at the time, meaning that until then, the working-age population size would not be smaller than the current one. Though there would be a big increase in the old population, there would be a simultaneous big decline in the population of juveniles. When the two cancelled each other out, the dependency index would drop over 25 years and would not rise to the (then-)current level—where one labourer had to provide for 0.91 person—until 2034. The conclusion was that in the last 20 years of the twentieth century there would be no aging problem, no shortage of work force or soldiers, and no increase in the dependent population. Nor would these problems be serious in the first 20 years of the twenty-first century. The "four-two-one" family structure—four old people, two working-age people, and one juvenile—would probably appear in some households, but would not be prevalent throughout society. Population develops over the long term, in a progressive and stable way, so if we take future possibilities into consideration when making population plan and population policy, we could guard against potential problems. The *Report* pointed out that the above-mentioned problems might appear after 2020, so it was proposed that the population policy be implemented by 2010 and by 2015 at the latest.

Appendix 2, *Proposal of the Plan that Each Couple Have Only One Child,* responded explicitly to the question of the duration of the policy:

> It is imperative to implement the "one child" policy. How long is optimum for its application? This is a question of general concern to the public. At present, people have doubts about this policy, some of which are caused by simple misunderstanding. For instance, some presume that it will be applied forever, leading to unimaginable consequences. Thus, it is necessary to state clearly that the policy will not be implemented permanently. It will last for a certain period of time only. As I[22] perceive it... this policy is a guideline that will last until around 2010, roughly a population reproduction cycle (see the first ten lines of Appendix 2 as shown in Picture 5.4 above).

Appendix 2 addressed predictions created from some data such as population increase rate, the size of the working age population, the size of aged population, and the dependency rate, around the year 2010. It stated that the policy should be adjusted after 2010. Despite slight differences between the actual indexes of population change and those predicted, population trends are consistent with those predicted several decades ago. At the time, there was a shortage of relevant data, and demography, population economics, and population sociology had only just begun. Given that situation, it was not easy to make predictions consistent with the real situation. The analysis and predictions gave clear and explicit guidance to the formulation of population policy.

There was also discussion about population quality. The *Report* pointed out that along with encouraging couples to have only one child, we should pay special attention to improving the population's various capacities. It stipulated that we should reinforce premarital sex education and perfect the premarital physical examination system. Those affected by congenital hereditary diseases would be permitted to marry but forbidden to have children, and this should be stated explicitly in the Family Planning Law or the Marriage Law. To promote population quality effectively and improve the capacities of the entire nation, research on genetics and eugenics should be strengthened. Appendix 2 took up further analysis of this issue. At the time, junior middle school and senior middle school graduates had only slim chances to continue their education, as mentioned in the first part of the *Report*, which also included more details about the Chinese population's education levels. For example, "in 1978, among every 10,000 people, only nine are college students, and 692 are middle school students. By contrast, in the US, there are over 480 college students and over 940 middle school students for every 10,000 people." So, the situation of the Chinese population's low scientific and education levels must be taken seriously.

The provision of social security to the elderly was of concern to all participants and thus was frequently discussed at the meetings, being an issue that would require urgent attention after the implementation of the one-child policy. At the time, workers in state-owned enterprises had access to retirement benefits but not the overwhelming majority who worked in enterprises of collective ownership or peasants in people's communes. The latter had to depend on their children when they became old. The one-child policy threatened to create a situation in which the child's grandparents on one side or the other would likely not be well taken care of. In the face of such a possibility, not only must the question of retirement pensions be resolved, but also

[22] Under a certain leader's instructions, I signed my own name on Appendix 2 to the *Report*, so it was considered my own proposal.

the relevant departments should investigate and plan the provision of more housing, more facilities for daily life and adequate medical services. People's concerns about these matters would be eradicated if the questions were properly resolved. Appendix 1 states that among city dwellers, there was an average of just one hospital bed per 532 people in 1977. It was common for people to have difficulties getting to see a doctor or being hospitalized when needed because of the overload of patients. Appendix 3, *Family Planning Law in the People's Republic of China* (fifth draft version) stated explicitly that the government would reward couples who had only one child. The bonus would take the form of health subsidies for their children. Those who lived in cities would receive a 5% pension bonus when they retired. Those who lived in the rural areas would receive extra economic assistance from the local government when they became old and could no longer work. These subsidies would be provided by the communes to which they originally belonged.

There was discussion about the education of the "only children". These children became the apple of their parents' eyes and the parents tended to spoil them. Some regional surveys indicated that many only children were inclined to be fastidious about food and clothing. They were found to be capricious and dependent, likely to be disrespectful to teachers, parents, and elders in their family. They did not cherish their belongings. These characteristics were common among only children and such behaviour required our attention. In discussion it was agreed that these problems could be resolved through strengthening social and family education. On the question of whether the crime rate was higher among only children, there was no clear evidence yet, but Chinese society should be concerned about that possibility and ensure the necessary education to prevent it.

There was also discussion about contraceptive methods and devices. Along with advocating that people have only one child, research into contraceptive tools must be enhanced. There was at the time a high rate of artificial abortions and mid-term induced labour. In some regions, such cases even surpassed the normal birth rate. To reduce these occurrences, the scientific branch of the family planning organization had to develop safer, more reliable, more convenient, and more efficient contraceptive tools as soon as possible. Vasectomies for men should also be encouraged.

There was also discussion about the ratio of men to women. The *Report* noted that in China and in other countries in the world, the gender ratio was about 52 male to 48 female. According to the survey conducted by the No. 1 Affiliated Hospital of Beijing Medical College, from 1975 to 1978, among 4,348 firstborns, baby boys accounted for 52.9% and baby girls accounted for 47.1%. The data were thus approaching the normal level. Against the backdrop of the one-child policy, it was feared that some couples might drown baby girls or abandon them, preferring boys to girls under the influence of traditional feudal beliefs. This phenomenon deserved our attention and should be prohibited by the law.

5.2.3 Part Three of the Report: Several Suggestions

The *Report* made seven suggestions in this regard in the third part.

(a) We should encourage family planning enthusiastically and intensify public education about it. It was believed that the one-child policy was a revolution in views of fertility, so all efforts should be made to raise people's awareness of family planning. National interests and national prosperity should be highlighted: fertility is not only a personal affair, the concern of each family but is also linked to national interests. We should grasp the rules of fertility and apply the guideline[23] proposed by Chen Muhua as early, as carefully, and as thoroughly as possible. People should be informed that controlling population growth is a strategic task and an important condition to accomplishing the goal of "four modernizations". We should stick to the proposition that the "two types of production should be strengthened simultaneously". We should spread knowledge of genetics, eugenics, and ecology and stress the importance of improving the population's capacities. We should talk about the current population policy intensively and contrive to make the slogan "one birth is encouraged and will be rewarded; a second birth is permitted according to the plan; a third birth is not permitted" strike root in the hearts of the people. The *Report* also recommended that newspapers and magazines publish columns focused on demographic theories and family planning. Demographic research institutes should be established, which would publish journals specializing in demographic research. In middle schools, courses on physiology should be created to teach sex education to students. Lectures about the importance of controlling population growth should also be delivered to students. In the economics department at colleges, a new course on demographic theories should be created. The office in charge of population and family planning should reinforce research on population policies and strengthen their investigations in order to carry out family planning work in a scientific way.

(b) A law on family planning should be passed as soon as possible. Officials both at the provincial level and communal levels urged that the law on family planning be promulgated as early as possible. The *Report* pointed out that local governments were rewarding people who complied with the population policy and punishing people who defied it according to different standards. Since there was no secure source for the bonuses, some local governments were short of the funds required for them. Because there was inadequate law to support family planning regulations, many a conflict would arise in the application of the regulations; in particular, when certain clauses contradicted regulations in other aspects, the former would tend to be considered simply "local rules", not to be implemented conscientiously. Therefore, participants of the meetings hoped that a unified national law would be issued as quickly as possible. Appendix 3, *Family Planning Law in the People's Republic of China* (fifth draft), offered a sample

[23] See details discussed in Sect. 5.2 of Chap. 3 above.

of such a law. It contained seven chapters composed of 30 articles, covering five articles on general provisions, four on basic requirements, 11 relating to the rewards and sanctions system, four on birth control measures, two about women's and children's health, another two on family planning organizations, and a final two articles covering supplementary provisions.

(c) A population commission should be created as soon as possible. The *Report* pointed out that currently the office of the national family planning work leadership was still temporary and was confined to just one aspect of the work. In a big country of around one billion people, population issues involved many dimensions. Apart from family planning matters, there were also international communications about population issues. The leading group of family planning work was not addressing the population development situation comprehensively. Many symposia participants suggested that a population commission should be created to take charge of all population issues for the purposes of coordinating the work effectively, precluding each department from acting on its own. The commission's task would be to implement a unified population policy, promote the development of demography and fulfill the strategic task of controlling population growth. Appendix 4, A Survey of Population Agencies in the Philippines, Thailand, Malaysia, and Singapore, was submitted together with the *Report* for the reference of the Central Committee.

(d) We should gradually extend the social security system to all elderly people. The *Report* said, "from now on we should take measures to solve the old people's social security [problem]". It was proposed that in the collective ownership units where there was no retirement system, the following idea should be considered. A small sum of money could be drawn from the personal income of those currently not involved in a retirement pension scheme and from the income of the collective in which they work and put into an insurance fund. Upon retirement, those workers would receive a pension from this fund as well as subsidies from the government. The proposals for a family planning security system in rural areas made by the insurance company were submitted together with the *Report* for reference. The *Report* also recommended that there be some unified planning for offering living facilities to elderly people whose children could not look after them.

(e) There was discussion about the indices concerning controlling population growth. According to the *Report,* it seemed impossible that the natural population increase rate would approach zero by the end of the twentieth century. One provision, from the draft Marriage Law made by the Standing Committee of the National People's Congress, said that men could marry only after they had reached the age of 22 and women could marry only after the age of 20. The meeting raised an objection to this because the current legal marriage ages were three or four years younger than that were recommended for a late marriage. The *Report* suggested that "the Central Committee could reconsider the question when is the optimum age for marriage. If the recommended marriage age cannot be changed, then we should take relevant measures to carry out family planning."

(f) In the context of the one-child policy, women and children must have access to good health care. Therefore, the health and education departments should improve perinatal care, popularize new methods of delivering babies, and reinforce instructions for healthy feeding of the new-borns and infants. They should also reinforce the scientific research on child psychological development and education, the construction and administration of facilities such as children's health centres, medical units, child-care centres, and kindergartens. The light industry and other relevant departments should make the proper arrangements to ensure that the infants and young children have adequate nutrition, toys, and teaching aids. All efforts should be made so that though fewer children will be born, all of them will be well cared for and educated.

(g) There were suggestions about the rewards to be given for those who have only one child. At the time, 28 provinces, municipalities and autonomous regions had formulated some tentative regulations and measures for such rewards, and to impose economic punishment on those who had three or even more children. The bonuses could be drawn from workers' welfare benefits. In the rural areas, they could be drawn from the public welfare fund of the people's commune. It had been seen that over the past year, rewarding those with only one child had promoted family planning. Yet in some enterprises, public works units, administrative units, rural communes, and production teams, bonuses could not be paid due to financial difficulties, which held back growth in the numbers of parents choosing to have just one child. It was suggested that the central and local governments offer subsidies to such units and areas in financial difficulty.

5.3 The One-Child Policy as the Pillar of the Population Policy

The symposia held by the Central Committee mainly focused on three general issues: whether it was feasible to encourage couples to have only one child; the problems that might be encountered in the implementation of the policy; how (and whether) we could solve such problems. There was no doubt expressed on the first of these matters, so there was little discussion about it during the meetings. Most discussion concentrated on the second question. The potential problems identified included those concerning population change and the development of economy, society, resources, the environment. The third matter was closely associated with the second, thus there was also plenty of discussion about it. All the questions involved consideration of how to solve the problems that might arise, if, in fact, we could.

Some symposia participants reduced the policy position "we encourage each couple to have only one child" to the "one birth" policy. In my opinion, this simplification was not an accurate reflection of the policy. Yet later the view was formed that most couples would be allowed to have only one child, thus apparently confirming the "one birth policy". The tendency was somewhat reflected in the practices in the coming days. We need some clarification of what the population policy actually said at the time.

5.3.1 Some Clarifications of the "One Birth Policy"

Policies are guidelines for action formulated by a government or political party to carry out established courses of action to fulfil certain goals. As a policy of fundamental importance for China, population policy had to be formulated by the CPC Central Committee and the State Council, for deliberation and approval by the National People's Congress (NPC) or the NPC Standing Committee. It is inappropriate, therefore, to extract one or a few elements from the policy, interpret it or them out of context, according to a particular understanding, and then to call that interpretation "the population policy". I have often heard international conference participants referring to China's twentieth century population policy as "the one-child policy". Sometimes, Chinese people call it "the one-birth policy". In reality there has never been a programme called the "one-child policy" in China. One element of the whole policy, albeit the most important element, was that "we encourage each couple to have only one child". Thus it is inappropriate and inaccurate to refer to China's population policy as a "one-child" or "one-birth" policy. Judging from the practice of "encouraging each couple to have only one child" over three to four decades, not all couples had only one child. Since people's understanding of the population and fertility policy directly concerns its nature and orientation and the evaluation of its application, in the following section, I try to clear up some misunderstandings about it.

The concept of the "only child" referred to the idea that a couple, or a woman of childbearing age, would have only one child throughout their whole lives. Accordingly, a child cannot be designated such until the couple or the woman has ended their childbearing period. Yet such an evaluation is laggard and cannot enhance the implementation of the policy. In practice in China, couples who committed to having only one child were granted an "only-child honour card", in a bid to ensure that they would have only one child in their lives, the duration of the contract. An overwhelming majority of people of childbearing age honoured the commitment to have only one child, but some did not. Thus there was a difference between the number of couples who applied for the honour card and the number who really had only one child. Moreover, there were situations in which a family had only one child but did not apply for the card. Other examples arose when couples originally had two or three children, but only one reached adulthood because the others died in infancy. Therefore, the number of couples holding the honour card did not correspond exactly to the number of families with an only child. The incidence of couples holding the honour card reflects a general situation where there is only one child in a family. A survey of the national population and family planning situation was conducted by the

National Population and Family Planning Commission in 2006. Data on the national distribution of only children from that survey is presented below[24]:

(1) The rate of families with an only child holding an only-child honour card (ROD_1) compares the portion of childbearing age women who have only one child and have claimed the honour card (NOD_1) with those who have only one child (NOC_1) (including those who have not claimed the honour card).

$ROD_1 = NOD_1 / NOC_1$
The whole-country rate of only child with an only-child honour card in 2006:
ROD_1 (whole country) = 8713/14234 = 0.6121.
The rate of only children with only-child honour cards in urban areas in 2006:
ROD_1 (urban areas) = 5126/6941 = 0.7385.
The rate of only children with only child honour cards in rural areas in 2006:
ROD_1 (rural areas) = 3587/7293 = 0.4918.

(2) The rate of only children (ROC) compares the portion of the total number of only children born of women of childbearing age (NOC) with the total number of all children born (NCT).

ROC = NOC/NCT
The whole-country rate of only children in 2006:
ROC (whole country) = 14,065/29162 = 0.4923.
The rate of only children in urban areas in 2006:
ROC (urban areas) = 6934/9599 = 0.7224.
The rate of only children in rural areas in 2006:
ROC (rural areas) = 7131/19563 = 0.3645.

This data show that both the national rate of only children and the rate of only children holding an only child honour card were between 50 and 60%. In the cities, both rates were between 72.2% and 73.9%, while in the rural areas the rates were just 36.5% to 49.2%. The data demonstrated that "only one birth" had not been practiced in China.

5.3.2 *The "One-Child Policy" as the Pillar of the Population Policy*

Though it is not quite accurate to summarize China's population policy in the last 20 years of the twentieth century as a "one-birth" policy, objectively, the "one child" element was the pillar of the population policy.

---

[24] Zhang, W. Q. (Ed.). (2008).

5.3.2.1 Family Planning as a Basic State Policy

Family planning was written into China's constitution and acknowledged as a basic state policy. Before the Central Committee symposia were held, the PRC Constitution was approved by the first session of the fifth National People's Congress, on 5 March 1978. Article 53 of the constitution explicitly stipulated that "women enjoy equal rights to men in all aspects of political, economic, cultural, social, and family life. Men and women are entitled to equal pay for equal work. Men and women are free to choose their own spouses. Marriage, family, mothers, and children will be protected by the state. The state advocates and implements family planning." This was the first time that the advocacy and implementation of family planning was written into the constitution.

On 1 September 1982, Hu Yaobang, General Secretary of the Central Committee, made a report, entitled *To Break New Ground in All Aspects of Socialist Modernization*, at the 12th Central Committee of the CPC. The report said, "the population issue remains a very crucial question in the economic and social development of China. Family planning is a basic state policy." This was the first publicly declared state policy on the question.

5.3.2.2 The Publication of the *Open Letter*

On 25 September 1980, *An Open Letter to All Party Members and Communist Youth League Members from the Central Committee of the Communist Party of China about the Control of Population Growth in China (henceforth Open Letter)* was published. The letter addressed questions such as why we should strive to control population growth, what problems might arise in doing so and how they might be resolved. It also proposed guidelines and policies to be adhered to in attempts to control population growth. These were programmatic policies formulated after careful investigation, and formally announced to the public by the Party's Central Committee They could not be summarized as a "one child" policy, because they contained many different elements.

The policy that each couple was to be encouraged to have only one child, as well as some supplementary policies, were introduced in the *Open Letter*, which reported that "the State Council has appealed to the whole nation—each couple is encouraged to have only one child. This is a significant measure that concerns the speed and future of our construction of the four modernizations, the health and welfare of our offspring, and the current and long-term interests of the entire nation. The Central Committee requires all Party members and Youth League Members, especially leaders at various levels, to be the first ones to answer the call of the State Council. They should also conscientiously propagate the policy to the public carefully and patiently." As a supplement to the major policy, the letter said, "if some people have difficulties (the situations prescribed) … they may be permitted to have two children. Yet they must not have a third child. For the minority ethnic groups, we might relax the family planning policy to some degree …. Regarding birth control measures, we suggest that contraception devices be the main means, and people are free to choose the method they like." We see from these remarks that while couples were initially encouraged to have only one child, this did not mean "all people should have only one birth"—a "one size for all" solution to diverse situations. Already at the beginning of the policy, some clear possible exceptions were noted, so that we should be aware of the needs of people with practical difficulties and of the minority groups.

The *Open Letter* also introduced some specific measures concerning controlling population growth: "to bring population growth under control, the Party and the government have decided to implement series of specific policies. For example, only-child families were to be favoured for access to nursery and school, medical doctors, and in job applications, new school enrollments, and apartment allocations in the cities or houses in rural areas. The policy that men and women should be paid equally for equal work should be carried out conscientiously. Scientific research on reproductive physiology, eugenics (to avoid the birth of babies with disabilities), and birth control techniques should be vigorously pursued. We should also train a large number of technical personnel, promote birth control techniques, and provide good health care for women and infants and good education to children. We would thus ensure the safety of birth control techniques and see the decline of babies born with congenital diseases. The relevant departments should take effective measures promptly to produce high quality contraceptive devices.

The *Open Letter* reiterated the policy of late marriage and late fertility: "while encouraging each couple to have only one child, we should underscore the importance of late marriage and late pregnancy. The marriage age prescribed by the Marriage Law is not late. For the sake of education and work, late marriage is recommended, and to some degree, late fertility should be stressed." In particular, "late marriage and late fertility are vital to slowing down the speed of population growth ... For young couples, late fertility will also bring a lot of benefits."

A time for adjusting to the major population policy—"one child policy" was addressed explicitly in the *Open Letter*. When the *Open Letter* argued for the necessity of encouraging the birth of only one child, a young population structure and the strong impetus for population growth in China was taken into consideration. It said, "at present, people aged under 30 account for nearly 65% of the total population. Henceforth, more than 20 million people on average will enter the phase of marriage and reproduction every year. If couples were not encouraged to have only one child ... In 40 years, the total population will be over 1.5 billion. This will not only create substantial obstacles to the realization of the four modernizations, but also result in serious limitations to the desired improvements in people's hard lives." The *Open Letter* suggested that the one-child policy should be recommended for the coming 30 to 40 years, especially the first 20 or 30 years. It might be said that this adjustment time was rather vague, but other remarks in the *Open Letter* defined the period very clearly. "Around 30 years from now, the currently urgent problem of population growth will not be so pressing and then we might adopt a different population policy."[25] What

[25] Peng, P.Y. (Ed.). (1997).

kind of policy would be adopted? Clearly, it would be one that included the lifting of reproductive restrictions. Thus the *Open Letter* already promised that the policy of encouraging one child per couple would be implemented for 30 years and when the time for adjustment came, adjustments would be made. It seems, however, that this point did not attract enough attention at the time, so my proposal that the population policy be adjusted after 30 years had elapsed was met with incomprehension. Some authorities asked "what background does he have?" apparently not realizing that the *Open Letter* to the whole country was my "backstage supporter".

5.3.2.3 Revisions and Amendments to the Population Policy

The population policy introduced in the *Open Letter* was both concise and general. As Chinese society developed, the policy would need to be amended and improved. In the rural areas the policy should be grounded in the specific realities there, and new revisions and amendments should be made accordingly.

On 10 September 1981, a meeting was held by the Secretariat to clarify the principle of formulating population and family planning policy. It was suggested that in the rural areas, the fertility policy should be adapted after investigation of the prevailing conditions: "we must make further investigation of the guidelines and policies of family planning according to the new circumstances and the experience of practical work in order to make policies accord with the actual situation and be easily accepted by the people. Through work it is possible to achieve the policy goal." That is to say, the meeting stressed that the fertility policy should accord with realities and thereby be feasible. About the specific policies, it was suggested that.

> From now on, in the cities we should resolutely keep encouraging each couple to have only one child. On the one hand, in the rural areas, we should tighten our grip on the family planning work based on the new conditions after adopting the household-responsibility system. On the other hand, the policy should be relaxed where appropriate. As for how relaxed it should be, there are two plans. According to Plan A, we encourage each couple to have only one child, but a second birth is permitted, and a third birth is forbidden. According to Plan B, we generally encourage each couple to have only one child but those who have practical difficulties might be permitted to have a second birth after their applications are approved.

It was also decided that "we should follow the mass line[26] and discuss the issue with experts and local officials. The Party secretaries at the provincial level will also be solicited for their opinions."

After the meeting, the General Office sent some documents to Party Committees in the provinces, autonomous regions, and the municipalities for feedback. The results were that six provinces and autonomous regions agreed with Plan A; 15 provinces, autonomous regions, and municipalities as well as the leading group of

[26] "The mass line" is the basic work route of the Communist Party of China, commonly summarized as "from the masses to the masses". Here it means listening to the opinions of the people and making decisions based on the their feedback on a certain issue.

family planning work in the army favoured Plan B; five provinces and municipalities approved what the *Open Letter* had said; and there was no feedback from three of the autonomous regions. In view of this, a symposium was held by the Central Committee, attended by all the first secretaries of Party Committees from the provinces, autonomous regions, and municipalities. After much discussion, Plan B was approved by most participants.

Based on the outcomes of all these meetings, the Central Committee and the State Council issued *the Instructions on Making More Efforts to Carry out Family Planning Work* (henceforth the *Instructions*) (also called Central Committee's Document No.11) on 9 February 1982. According to this document, the fertility policy consisted of the following:

> We will keep encouraging late marriage, late childbirth, fewer births, and better births. Specifically, among government officials, workers, and all residents of cities and towns, each couple should have only one child, except for those who have been approved to have a second child in some special cases. In the rural areas, couples are generally encouraged to have only one child. If some people have difficulties and apply to have a second child, their request could be met in a planned way after their applications are approved. A third birth is not permitted, no matter what. We also encourage couples of the minority ethnic groups to practice family planning but we can relax the policy to some degree. Specific regulations can be formulated by local governments in the designated autonomous areas, and the provincial or autonomous region governments above them, according to the actual local situations. Those regulations can be implemented after being approved by the Standing Committee of the People's Congress or the people's government at a higher level.

The *Instructions* also specified the policies concerning rewards and restrictions. For example, "those 'only children' are entitled to health subsidies. Mothers of the 'only child' are entitled to longer maternity leave and their salary will not be suspended during their leave. Neither will any pay rise or promotion be affected by maternity leave". Further, "in the rural areas where the production responsibility system has been adopted, farm output quotas for households with 'only children' might be reduced, or they may have a right to more contracted farmland." Those who do not comply with the family planning policy would suffer some degree of restriction. "If government officials, workers, or residents of cities and towns have a second child without approval, then their medical benefits and other welfare benefits that they would normally enjoy if compliant with the family planning regulations will be cancelled. They will also possibly suffer salary deductions or be denied access to financial aid and childcare subsidies. In the rural areas, there will be restrictions on members of the People's Commune who do not comply with the policy. The extra child(ren) in the rural areas will not get the fixed output quota farmland or land allocated for personal needs. Peasants who have an extra child or extra children will get less farmland or the farm output quotas for their households will be increased."

In the application of the fertility policy, different local governments applied different standards concerning the circumstances in which a couple could have a second child, some were strict and others loose. In August 1982, the National Family Planning Commission held a national meeting at which participants learned how to enforce the Central Committee's Document No.11. After the meeting, *the Minutes of the National Conference on Family Planning Work* were submitted to the Central

Committee and the State Council. On 20 October the same year, both the General Office of the Central Committee and the General Office of the State Council released the *Minutes*. They emphasized that the goal of family planning should be combined with helping people to solve their practical difficulties and the policy should be based on real social conditions. It synthesized how family planning policy was being implemented around the country.

> Before the *Instructions* were issued, three circumstances under which couples are permitted to have a second child were prescribed by the provinces, municipalities and autonomous regions. First, if the first child is disabled because of a non-genetic birth defect and will not be able to work normally; second, if one of the parties in a newly married couple is marrying for the first time but the other has a child from his or her former marriage; and third, if a couple who could not have a child despite many years of marriage becomes pregnant after adopting a child. In the implementation of the *Instructions*, many provinces, municipalities and autonomous regions introduced several other situations in which couples might have a second child. They included the following: first, if for two or three generations in a family, there was only one male child; second, if, in a household only one man among his brothers was fertile; third, if a man settles in his wife's village and his wife was an only child; fourth, a man who is an only child marries a woman who is also an only child; fifth, if a man is a disabled army veteran; sixth, if both in the couple are overseas Chinese; seventh, if a couple living in a remote mountainous area or a coastal fishing area had special difficulties.[27]

The Report on Family Planning Work submitted by the Party Committee of the National Family Planning Commission, was instructed and released by the CPC Central Committee on 13 April 1984 (it is also known as the Central Committee Document No.7 1984). After confirming the great achievements in family planning work made in the past several years, the document noted that in the future the signs of a good job being done in family planning would comprise "carrying forward the achievements, overcoming the shortcomings, and solving the problems". It underscored that "family planning policy should be reasonable, supported by the people and convenient for officials to execute". It also stated that

> We should continue to encourage each couple to have only one child. In the meantime, we should make efforts to perfect the specific family planning regulations. At present, the major regulations being practiced are listed as follows. First, in the rural areas, we should continue to relax the policy to a larger degree yet in a controlled way. According to the prescribed situations, some couples could have a second child after approval; second, we should resolutely forbid the couples to have a second child or more without approval; third, malpractice for personal gain will be forbidden. Officials involved in corrupt practices in matters concerning family planning will be resolutely punished; fourth, for minority ethnic groups, we need to formulate appropriate regulations. The following suggestions could be considered. Among the minority ethnic groups with a population of less than 10 million, each couple could be permitted to have two children. In some special cases, they could have three children, but they will not be permitted to have four. Specific regulations could be formulated by local governments in the designated autonomous areas, and the provincial or autonomous region governments above them according to the prevailing local conditions. Those regulations could be implemented after being approved by the Standing Committee of People's Congress or people's government at a higher level.

[27] Peng, P.Y. (Ed.). (1997).

Document No. 7 also included instructions and advice on advocating eugenics, reinforcing technical instructions, the use and management of funds, and the improvement of working styles of officials involved in family planning work.[28]

On 31 March 1988, after these regulations had been applied for several years, a meeting was held of the Standing Committee of the Political Bureau of the Central Committee at which *The Outline of Report on Family Planning Work* submitted by the National Family Planning Commission was discussed. The meeting reviewed the implementation of the family planning policy. In 1980, the Central Committee had sent the *Open Letter* to the whole country, encouraging each couple to have only one child. It also pointed out "if some couples do have the difficulties prescribed by the policy, they could be allowed to have two children. Yet a third child is not permitted." In the years since, many officials had made tremendous efforts to carry out the policy. Great achievements had been made but there were also some problems. Shandong province suggested that "we might relax the policy to a small degree, but we should avoid relaxing it to a large degree" to deal with the existing problems. The Central Committee approved of their suggestion. The meeting reiterated the original policy but added a new stipulation that in the rural areas, families with a single girl could have a second child. The meeting noted that,

> the current family planning policy China is that we encourage late marriage and late childbirth; we recommend that people have fewer but better births; we encourage each couple to have only one child. We will keep advocating these measures. Specifically, among government officials, workers, and residents of cities and towns, each couple should have only one child, except for those approved to have a second child in special cases. In the rural areas, if some people have practical difficulties, including that the family has only one girl and they apply to have a second child, they may be granted permission after several years if their application is approved. A third birth is not permitted, no matter what. In the autonomous regions or areas, we also encourage people to adopt family planning. The specific requirements and regulations can be formulated by the provinces and autonomous regions concerned in accordance with local conditions.

As far as I know, this was the first time the Central Committee stated explicitly that in the rural areas a family with a single daughter could have a second child. This specific policy had a measurable influence on the subsequent population change. The meeting decided specifically that the policies must be enforced over a long period in order to control population growth. Reflecting the Central Committee's decision, Peng Peiyun, Director the National Family Planning Commission, delivered a speech at the National Meeting for the Directors of Family Planning Commissions in 1989:

> We should continue to apply and enforce the current family planning policy formulated by the CPC Central Committee comprehensively. All the local governments must implement the policy strictly. We should firmly preclude any department from acting on its own, and no extra exceptional cases should appear without being approved by the Central Committee. Whenever the policy is carried out more loosely than stipulated by the current policy approved by the Central Committee, the local government must submit an official report to the Central Committee for instructions. We should continue to enthusiastically encourage each couple to have only one child. We recommend late marriage, late childbirth, fewer births, having

[28] Peng, P.Y. (Ed.). (1997).

> healthy babies, and avoiding the births of babies with disabilities. In the cities, the suburbs, and the rural areas that are economically and culturally developed, we should consolidate and develop the achievements we have made and strive to make a greater contribution to restraining population growth for the nation. In the vast rural areas, when encouraging each couple to have only one child, we should also conscientiously apply the policy of permitting a family with a single girl to have a second child after a gap of several years.[29]

Through the next nearly ten years of practice, discussion, revision, and improvement, the population and fertility policy stabilized in China. In accordance with local realities and the policy formulated by the Central Committee, the provinces, autonomous regions, and municipal cities created their own specific regulations. According to the cases summarized by the National Population and Family Planning Commission in the book *Population and Family Planning History in China*, the implementation of the family planning policy generally fell into one of five categories.

In the first category was the application of the policy that each couple had only one child and those in special situations (10 cases had been prescribed by the local government) were permitted to have a second child. The cities of Beijing, Tianjin, and Shanghai, and Jiangsu province fell into this category.

In the second category there were various provincial and municipal guidelines according to the different prevailing local conditions. In the rural areas, couples with a single girl could have a second child; Chongqing city, and Sichuan, Zhejiang and Fujian provinces were in this category. Based on their special geographical conditions, specific guidelines were applied in Chongqing and Sichuan. Specific guidelines were also applied in Zhejiang and Fujian. Peasant couples in the mountainous areas of Chongqing and Sichuan were permitted to have two children.

In the third category the policy that in the rural areas, peasants whose first child was a girl were permitted to have a second child was applied. This was the case in Hebei, Shanxi, Liaoning, Jilin, Heilongjiang, Anhui, Jiangxi, Shandong, Henan, Hubei, Hunan, Guizhou, Shanxi and Gansu provinces, and Inner Mongolia (where it applied only to people of Han nationality) and Guangxi Zhuang Autonomous Regions.

The fourth category, involving application of the policy that in rural areas couples were generally permitted to have two children, included Guangdong (before 1998), Hainan, and Yunnan provinces, as well as the mountainous area, Bashang plateau and coastal fishing area of Hebei province, the frontier area of Heilongjiang province, the mountainous area of Anhui province, some areas in Fujian province, where there was a small population yet with sizable farmland, the frontier area of Guangxi Zhuang Autonomous Region, some mountainous areas in Chongqing city, the mountainous areas in Sichuan province, and the areas where the pilot family planning policy was implemented, including Yicheng county of Shanxi province, and Jiuquan District in Gansu province.

---

[29] The National Population and Family Planning Commission. (Ed.). (2007). *Population and Birth Control History in China. China Population Publishing House.* pp. 128-129.

The fifth category involved the specific policy formulated by autonomous regions for special application among the minority ethnic groups. This category included the Inner Mongolia Autonomous Region, where couples of the minority ethnic groups could generally have two children. Those who were both engaged in husbandry could have a third child. If the minority ethnic groups such as Dawoer, Ewenke, Elunchun requested birth control assistance, family planning technical services would be provided to them. In Guangxi Zhuang Autonomous Region, couples from minority ethnic groups with a population of less than 10 million could have two children. In Xizang Autonomous Region, officials of Zang nationality, and workers who lived in the cities and towns could also have two children. Those with special cases to plead could have three children. In the central farming and stock-breeding areas, couples were free to decide how many children they had, but were encouraged to have no more than three. In the frontiers of the farming and stock-breeding areas, and among the minority ethnic groups like Menba, Luoba, Sherpa and Dengren, we only propagated the family planning policy. They were free to decide how many children they wanted. In Qinghai province, in the rural areas, couples were generally permitted to have two children. In grazing areas, minority ethnic herdsmen were permitted to have three children. In Ningxia Hui Autonomous Region, couples in rural areas, and those composed of one citizen with an urban registration and the other a peasant, generally were permitted to have two children. In Guyuan, Haiyuan, Xiji, Longde, Jingyuan, Pengyang, Yanchi and Tongxin counties, peasants of the minority ethnic groups were permitted to have three children. This was also true in Xinjiang Uygur Autonomous Region for peasants and herdsmen of the minority ethnic groups. The other peasants and herdsmen in Xinjiang were permitted two. And those in special situations were permitted to have four children.

5.3.2.4 Varying Interpretations of the Statements "Relax the Policy to a Small Degree", and "Avoid Relaxing the Policy to a Large Degree"

There were varying interpretations of the statements "relax the policy to a small degree", and "avoid relaxing the policy to a large degree" in Central Committee Document No. 7 1984. The different understandings were supported by some of the realities. According to one opinion, the rise in fertility and birth rates in the mid-1980s was created by population structure change—the peak in numbers of women of child-bearing age was reached—rather than a result of following the advice to relax the policy "to a small degree", and to avoid relaxing it too much. According to another opinion, since birth control and family planning were carried out in China without a high level of social and economic development, there was a significant gap between the fertility policy and most citizens' wishes regarding children. Therefore regardless of the fertility policy, the fertility rate was going to rebound. In the end, the policy was relaxed to a greater degree than expected with the consequence that the fertility, birth and natural increase rates rose in the mid-1980s. Which opinion

Table 5.6 Natural population change in China 1980–1990 (per 1,000 people)[30]

Year	Birth rate	Mortality rate	Natural increase rate
1980	18.21	6.34	11.87
1981	20.91	6.36	14.35
1982	22.28	6.60	15.68
1983	20.19	6.90	13.29
1984	19.90	6.82	13.08
1985	21.04	6.73	14.31
1986	22.43	6.86	15.57
1987	23.33	6.72	16.61
1988	22.37	6.64	15.73
1989	21.58	6.54	15.04
1990	21.06	6.67	14.39

was correct? We must take the real situations into consideration when addressing the question.

Table 5.6 above shows that the birth and natural increase rates did rise to some degree, bouncing back between 1985 and 1988. From 1980 to 1984, the average birth rate in China was 20.30 per 1,000 people, and the average natural increase rate was 13.69 per 1,000 people. Then from 1985 to 1988, the average birth rate was 22.29 per 1,000 people, and the average natural increase rate was 15.54 per 1,000. By contrast, the average birth and natural increase rates in the second period were higher than in the first by 1.99 per 1,000 people and 1.85 per 1,000 respectively. The average birth rate in the second period was even higher than the average birth rate in 1989 and 1990 (21.32 per 1,000 people) by 0.97 per 1,000. The average natural increase rate in the second period was higher than the average natural increase rate in 1989 and 1990 (14.71 per 1,000 people) by 0.83 per 1,000. Since the birth rate is the ratio of the number of babies born alive in a certain year to the size of the total population in that year, it is a crude rate. The natural increase rate is the ratio of the number of babies born that year minus the number of people who died that year to the size of the total population in that year, so it too is a crude rate. Since both the birth rate and the natural increase rate are crude rates, they are affected by changes in the population age structure. Thus it is inappropriate to make a simple contrast between these crude rates. A simple contrast between the crude rates is against demographic principles so we should always standardize the population age structure when making such contrasts.

Was the rise or rebound of the birth and natural increase rates during the period (1985–1988) caused by the change of population age structure and if so, to what extent? The answer to the first question is affirmative. Table 5.7 below shows contrasts

---

[30] Department of Population and Employment Statistics, National Bureau of Statistics of China. (Ed.). (1995).

Table 5.7 Population age structure at the time of the second, third and fourth census and in 1988 (%)

Year	0–14 years old	15–64 years old	65 and over
1964	40.69	55.75	3.56
1982	33.59	61.50	4.91
1988	28.05	66.63	5.32
1990	27.69	66.74	5.57

Table 5.8 Change in the population of women of child-bearing age in the second, third, fourth census and in 1988

Year	15–49 years old (million)	20–29 years old (million)	20–29 years old (%)
1964	151.62	47.94	31.62
1982	248.49	81.30	32.72
1988	297.44	100.18	33.68
1990	306.34	112.28	36.65

in the population age structure from the second census (in 1964), the third (in 1982) and the fourth (1990) and in 1988..[31]

Data from Table 5.7, based on the census data and the national survey on fertility and birth control conducted in 1988, indicate that the ratio of the first age group (0–14 years old) showed a downward trend from 1964 and 1990. During the same period, the ratio of the second age group (15–64 years old) showed an upward trend and the ratio of those aged 65 and over increased slowly. The data also demonstrate that the population age structure in China had completed its transition from young to adult, and the shift towards an elderly structure had begun. It is clear that in the 1980s, the proportion of the second age group rose most significantly, showing that the working age population and the numbers of women of child-bearing age (15–49 years old) grew rapidly. The other important factor in the changes to the birth, mortality and natural increase rates was this increase in the population of women of child-bearing age, especially among those in the most active period of child-bearing age. Table 5.8 below shows these changes, based on data collected from the same sources..[32]

According to Table 5.8 above, the number of women of child-bearing age increased substantially in the mid-1980s. Between 1982 to 1988, it rose from 248.49 million to 297.44 million, an increase of almost 20%. In particular, those in the most active child-bearing years showed dramatic growth, from 81.30 million to 100.18 million, an increase of 23.22%. Within the same period, the ratio of those 20–29 years old women rose by 0.96%. These substantial increases, particularly among those in

[31] National Statistics Bureau of the PRC. (Ed.). (2005).;
The National Population and Family Planning Commission. (Ed.). (1990).

[32] The information was based on the calculation of data provided in *A Handbook of the Principal Data of Chinese Population and Sample Survey of Fertility and Birth Control in China: Data from All of China* and *A Handbook of the Principal Data of Chinese Population*.

Table 5.9 The change in total fertility rate (TFR) from 1981 to 1990

	1981	1982	1983	1984	1985	1986	1987	1988	1989	1990
In the whole country	2.63	2.87	2.42	2.35	2.20	2.42	2.59	2.52	2.35	2.31
In the urban areas	1.40	1.58	1.34	1.22	1.21	1.24	1.36	–	–	–
In the rural areas	3.04	3.32	2.78	2.70	2.48	2.77	2.94	–	–	–

the most active child-bearing years and the obvious rise in their ratio to the total number of women of child-bearing age, were the major reason for the rise of birth and natural increase rates. Whether it was the sole reason depended on changes in the fertility rate, because the fertility rate was not affected by changes to the population age structure with the child-bearing age women included. The changes in the total fertility rate in the 1980s are shown in Table 5.9 below.[33]

Table 5.9 above demonstrates that the total fertility rate gradually decreased from 1982 to 1985 across the country, by 0.67. From 1985 to 1990, it first rose and then dropped, increasing by 0.39 from 1985 to 1987, then decreasing between 1988 and 1990 by 0.28. It is notable that before 1985 the TFR declined and then rose for three years after 1985. The changes took the shape of a small saddle. We can explain the decrease of TFR before 1985 by two factors: the socio-economic base and the implementation of policy.

The socio-economic base mainly involves the following elements. First, with the social economic development that was promoted by the reform and opening-up policy, the economic foundations for a decrease in the total fertility rate had improved. As discussed in Chap. 1 above, the theory of population transition and child cost–benefit theory reveal that the fertility rate is inversely proportional to the level of economic development and average income per capita. As a country with an under-developed economy at the time, the general rule applied to China. From 1981 to 1985, calculated at current prices, Gross Domestic Product (GDP) in China surged from 486.24 billion yuan RMB to 896.44 yuan RMB, an increase of 84.4%. Calculated at comparable prices, GDP grew by 58%, an annual increase of 12.1%. Calculated at current prices, the GDP per capita rose from 489 yuan RMB to 853 yuan RMB, an increase of 74.4%. Calculated at comparable prices, GDP per capita grew by 49.4%, at an annual rate of 10.6%.[34] Although economic development at the time was still low, it had improved noticeably as the material basis of population reproduction.

Second, increases in the rate of urbanization had facilitated the decline in the total fertility rate. As shown in Table 5.9, in the early 1980s, the average fertility rate in the urban areas was 1.35 while in the rural areas it was 2.86. The average TFR in the urban areas was just 47.2% of that in the rural areas. Thus increased urbanization meant a decline in the TFR. Urbanization in China was not rapid. The portion of urban people was only 17.9% of the total population in 1978 when the

[33] Department of Population and Employment Statistics, National Bureau of Statistics of China. (Ed.). (1995).

[34] National Statistics Bureau of the PRC. (Ed.). (2005).

reform and opening-up policies commenced. These reforms gave momentum to the urbanization process, greatly accelerating its progress. Between 1981 and 1985, the urban population rose from 191.4 million to 250.94 million, with its portion of the total population increasing from 20.2% to 23.7%, at a yearly rate of 0.9%.[35]

Third, the improvement of education levels was also favourable to the decline of the TFR. Investigations in China and many other developing countries have indicated that the fertility rate is inversely related to the population's education level. So improvement of people's academic achievements meant a decline in the TFR. From 1981 to 1985, college enrolments grew from 1,279,000 to 1,703,000, a 33.2% increase. Secondary school enrolments grew from 50,146,000 to 50,936,000, an increase of 1.6%. In the same period, the number of students who enrolled in universities rose from 279,000 to 619,000, an increase of 121.9%, and the number of graduates rose from 140,000 to 316,000, an increase of 125.7%.[36]

Before the Central Committee's Document No. 7 was transmitted to the local governments, generally speaking, the early 1980s fertility policy had featured clear planning indices and strict requirements, and family planning work was carried out strictly. As mentioned earlier, after the *Open Letter* was published in September 1980, Hu Yaobang made a report to the 12th CPC National Congress. He noted that the population question would remain a crucial issue and that implementing family planning was a basic state policy in China. In the same year, the General Office of the Central Committee and the General Office of the State Council transmitted the *Minutes of the National Conference on Family Planning Work* (the *Minutes*). According to the *Minutes*,

> Implementing family planning is a basic state policy in China. It is a major issue that concerns the construction of a high degree of material and cultural-ethical advancement in a socialist country. Whether we can fulfill the goal of building "a moderately well-off society" is strongly related to the control of population growth. Party Committees and people's government offices at all levels should clearly recognize the utmost importance of this strategic issue and intensify the leadership in carrying out family planning work. Currently, we should meticulously study Comrade Hu Yaobang's speech on family planning work in the Report delivered at the 12th CPC National Congress. Leaders at various levels should deepen their understanding of family planning work. The chief leaders of the Party Committee and people's government at all levels should take care of the family planning work themselves and take the lead in propagating it. Leaders should conscientiously oversee the work of their subordinates and strive to do a good job in carrying out family planning work.[37]

Generally, in this period, the family planning policy was scrupulously followed, and family planning work was carried out rigorously, constituting the political and social reason for the continuous decline in the fertility rate.

Document No. 7 was released on 13 April 1984. So why did the TFR not drop before 1985? As far as I know, following the release of this document, local governments in some provinces, autonomous regions and municipal cities did not quite understand the instruction to "relax the policy to a small degree but avoid relaxing it

[35] National Statistics Bureau of the PRC. (Ed.). (2005).

[36] National Statistics Bureau of the PRC. (Ed.). (1986).

[37] Peng, P.Y. (Ed.). (1997).

to a large degree" and indeed they refused to relax the policy. At the time, the leader of the National Family Planning Commission explained the adjustments to family planning policy to the provincial leaders in charge of family planning work and officials in local family planning commissions, helping them to analyse the population situation in their areas. A year elapsed before the adjusted policy was implemented and in a few provinces, it was not thoroughly applied. In the economically prosperous rural areas, the rule that each couple be encouraged to have only one child persisted. By contrast, in the places where the policy according to Document No. 7 was conscientiously implemented, there was no obvious rise in the fertility rate. Behind it was an understanding and effective implementation of the policy as laid out in Document No. 7.

The document was originally known as *the Leading Party Group of the National Family Planning Commission's Report on Family Planning Work* instructed and transmitted by the CPC Central Committee. It reported on family planning work in six aspects.

First, family planning was a basic state policy in China and an important issue that affects the prosperity of the Chinese nation. After confirming the achievements made in the previous several years, the document said that great efforts must be continued in the implementation of family planning in the future. Doing a good job meant "continuing to make achievements, overcoming shortcomings and solving problems". All local governments should carry out the work according to such requirements.

Second, to be effective, family planning policy should be reasonable and win the support of the people and be feasible for officials to apply. It reiterated that the goal for family planning was to keep the Chinese population under 1.2 billion by the end of the twentieth century, and to continue to encourage each couple to have only one child. It also introduced four specific policies: in the countryside, we should continue to relax the policy in a controlled way; we must avoid relaxing it too much; malpractices for personal gains in the execution of family planning work must be forbidden; regulations should be relaxed for the people of minority ethnic groups, to appropriate degrees.

Third, we should vigorously advocate an understanding of eugenics and the Marriage Law, in order to prevent consanguineous marriages.

Fourth, we should strengthen the techniques of family planning and provide those who committed to family planning a sense of security. We should also introduce advanced family planning techniques from other countries and improve the technological knowledge of Chinese officials engaged in family planning work. People must remain free to decide whether to undergo a sterilisation operation. Their choices must be respected and surgical accidents be guarded against.

Fifth, it underscored that the expenditures for family planning should be used appropriately. We should value the actual effects and avoid boasting about achievements.

Sixth, it commended the assiduous work of a large number of local officials. It also stated that some had failed to understand comprehensively the Central Committee's guidelines and policies. Some officials had even resorted to coercion when doing their job. Such officials should renounce the mistaken idea that coercion is unavoidable.

Instead, they should carry out civic education conscientiously to promote social stability and unity, trying to create a new atmosphere for family planning work.

In sum, Document No. 7 required the continued implementation of the state family planning policy, stating the guidelines, work methods and specific regulations to be followed. The report submitted by the leading Party group of the National Family Planning Commission made some self-criticism of the officials executing the daily work, including that they had not had a comprehensive understanding of the Central Committee guidelines, and had not applied them or improved them in accordance with prevailing real conditions. They had failed to improve work methods and relations between the Party and the public. They had not taken effective measures to rectify local malpractices such as the use of command and coercion in family planning work, the violation of laws and breach of work rules, and so on. The report also mentioned that in 1982, second births and cases of third and fourth children in a family constituted 24.2% of all new-borns respectively. That is to say, the portion of only child just accounts for around 50% of all new-borns. In view of this, the leading Party group approved the suggestion that "we might relax the policy to a small degree but avoid relaxing it to a large degree". That meant we could approve more applications by families with only one child to have a second but second and third births without permission should be avoided. Originally, one in 20 families was allowed to have a second child but that figure could now be increased to one in 10. The decrease in the number of families with more than two children meant that we could consider allowing more families with one child to have a second.

These suggestions were clear, and sensible. If the decision to "relax the policy to a small degree but not to a large degree" were conscientiously implemented, it would be conducive to controlling population growth and perfecting the fertility policy. Yet, an objective fact should not be ignored: over 300 million people born during the second baby boom from 1962 to 1973 turned 11 to 22 years old in 1984, meaning that a large cohort was about to enter the stage of high fertility, presenting a period of potentially high productivity that would last as long as 12 years. At such a time, in the face of the gigantic task of controlling population growth, all efforts had to be made to inhibit a new baby boom and to reduce the potential crest value as much as possible, rather than introduce incomprehensible or ambiguous regulations such as relax the policy a little but not too much. It can be said that such ambiguity was associated to a certain extent with the rising birth rate and natural increase rate in the mid-1980s, and the slight increase of the total fertility rate from 1985 to 1988. Mainly, the rise in birth, natural increase and total fertility rates was caused by the transformation in the population age structure, specifically the surge in women of child-bearing age. But at the same time, it was also an effect of timing of the policy, the different interpretations of "relax the policy to a small degree but not to a large degree", and ineffective corrections of deviations from the policy in its implementation. However, this situation did not last for long. By the end of the 1980s, the fertility policy had become clearer.

References

Department of Population and Employment Statistics, National Bureau of Statistics of China. (Ed.). (1995). *A Handbook of the Principal Data of Chinese Population*. p. 42

National Statistics Bureau of the PRC. (Ed.). (1986). *China Statistical Yearbook of 1986*. China Statistical Publishing House.

National Statistics Bureau of the PRC. (Ed.). (1990). *China Statistical Yearbook of 1990*, China Statistical Publishing House.

National Statistics Bureau of the PRC. (Ed.). (2005). *China Statistical Yearbook of 2005*. China Statistical Publishing House. p. 95

National Statistics Bureau of the PRC. (Ed.). (2018). *China Statistical Yearbook of 2018*. China Statistical Publishing House. pp. 31–32

Peng, P.Y. (Ed.). (1997). *A Complete Record of Family Planning in China*. China Population Publishing House.

Population Research Institute of Chinese Academy of Social Sciences. (Ed.). (1989). *Almanac of China's Population 1988*. Economics and Management Publishing House. p. 210

The Development and Planning Department of the National Population and Family Planning Commission, & China Population and Development Research Center. (Eds.). (2004). *A Databook of Population and Family Planning 2003*. China Population Publishing House. p. 80

U.N. Population Division. (Ed.). (2005). *World Population Prospects The 2004 Revision*. New York. pp. 20–24

Zhang, W. Q. (Ed.). (2008). *Collection of the Data on Population and Family Planning from Survey Conducted in 2006*. China Population Publishing House. p. 67

Chapter 6
Population Control Policy in 2000

The 1980 policy encouraging each couple to have just one child was closely associated with the population control target for 2000—to keep the total Chinese population below 1.2 billion. To fulfill it, implementation of the regulations on population growth had to be toughened and the "only child" rate had to be increased. It was a tough goal and there was a strong chance that the number could only be kept to about 1.25 billion. The population policy had therefore to go through some slight adjustments and fluctuations. The one-child policy was the pillar of the population policy but several flexible elements existed alongside it.

6.1 Origins of the Population Target for 2000

After the series of symposia, the *Open Letter* was published on 25 September 1980; its opening sentence including the words "in order to keep the total population in China below 1.2 billion by the end of the twentieth century".[1] This was the first time that this signal was sent to the Chinese public and to people outside the country. Henceforth, that was the target for all officials involved in population and family planning matters. It was also one of the century goals that the entire Party and the whole nation would strive to achieve.

How was the goal set? There are different answers to that question. Yu Wang had served as member of the National Family Planning Commission, director of the Department of Comprehensive Planning, and vice chairman of the Chinese Family Planning Association. According to his book *Thirty Years of Practice and Theoretical Explorations of Family Planning*, the goal of "keeping the total population in China below 1.2 billion by 2000" first appeared in a speech to the National Symposium

[1] Peng (1997).

X. Tian, *Population Policies in the People's Republic of China*,
https://doi.org/10.1007/978-981-96-9658-1_6

of Directors of Family Planning Commissions, delivered on 10 October 1980. Yu wrote:

Population policy should be formulated according to the population development goal set by the Central Committee. The population policy is the foundation for population planning. The goal is very clear: we should keep the total population of our country below 1.2 billion by the end of the 20th century. We should strive to fulfill that goal in our family planning work. A certain comrade said, "1.2 billion is the general goal for the whole country, yet I do not know the exact goal for my province. I need a population plan with specific data designated for each province or municipal city." This requirement is reasonable. However, what I want to tell you is we do not have such a plan at present. I want you, the people present at the meeting, to do the planning yourselves. Each province or municipal city should make its own plan. If we do not develop a plan, it will be impossible to achieve the goal. If the different administrative units at the provincial, county, and local levels do not make their own plans, then the general goal of keeping the total population below 1.2 billion will be a dead letter. For this reason, we require that every province, municipal city, and autonomous region make a plan in accordance with the guidelines and policy formulated by the Central Committee, in the light of their local conditions … The figure 1.2 billion is a goal set for the whole country, calculated through many a prediction based on the population situation across China over a long period of time. Setting this goal involved many organizations, including the Family Planning Office, the Demographic Research Centre, the Ministry of Public Security, the National Statistics Bureau, the State Development Planning Commission, and personnel from the Chinese Academy of Social Sciences and the Seventh Machinery Industry Department as well as experts and scholars from other departments. Six different predictions concerning fertility were developed on the basis of the current population increase rate and the fertility index … Briefly, the first proposal was that each couple could have 2.3 children on average. The second proposal was that each couple could have two children on average. According to the other suggestions, each couple could have an average of one child, 1.6, 1.7, or 1.5 children … After investigation and discussion the Central Committee adopted the proposal to keep the total population below 1.2 billion by 2000.

According to Health Minister Qian Xinzhong, that target assumes that each couple has an average of 1.7 children. This plan would apply to the whole country, yet we must consider the exception that couples among the minority ethnic groups will be permitted to have more children. Taking some other factors into consideration, Han couples could have about 1.5 children on average. Why is the figure not exact? Because though the calculation itself is correct, the basic data we depend on are not very precise or reliable. For example, we do not know the exact numbers within the various age groups nor do we know exactly how many couples will marry or how many women are of child-bearing age. Our prediction is calculated using data from the 1964 census and the 1978 surveys of certain regions with populations no greater than 100 million. As far as I know, some people were not included in this data … For example, when we made the predictions, women were permitted to marry at 23. According to the new marriage law, they are permitted to marry at 20. Thus there will be 30 million new couples in the child-bearing age group. Another factor is that some people have not registered. It has since been learned that in mining areas in Shanxi province and some other places, quite a few people are not registered residents. So, after the census is carried out, we will probably find that China's total population is greater than earlier believed. When some comrades hear that one couple might have 1.7 children, they think we should round up the number to two. That is, allow each couple to have an average of two children. But this certainly will not work. If we allowed all couples to have two children, there would be at least 1.25 billion people by 2000 and the Central Committee's goal for controlling population growth would not be achieved.[2]

Yu Wang, deputy director of the office of the Leading Group of Family Planning Work in 1980, knew the context and the process behind the setting of the population goal. The passage just quoted reflects the analysis of family planning at the time. The Central Committee decision and the speeches delivered by certain leaders were reflected in other documents. As early as 1979, the *Resolution on Some Issues concerning Accelerating Agricultural Development Made by the CPC Central Committee* was passed by the Third Plenary Session of the 11th CPC Central Committee, in which it was suggested that the population growth rate should decrease each year and was expected to fall to about five per 1,000 people in 1985. If the goal were achieved and the rate continued to decline, it would approach zero by 2000, by which time, GNP per capita in China was expected to reach $1,000. Some leaders including Deng Xiaoping, Li Xiannian and Chen Yun had talked about the importance of a decrease in the birth rate and population control to achieving the per capita GNP goal many times. To realize it, both economic development and control of the population growth rate had to be taken into consideration. Simple calculations indicated that if the national population were kept below 1.2 billion by 2000, then the average per annum GNP must grow by 8.1% to meet the per capita GNP goal. If the total population were larger than 1.2 billion, then GNP would have to increase more quickly. That would present greater difficulties.

To address this question, Chen Muhua, then vice premier and head of the Leading Group of Family Planning Work gave the following explanation in her report (the draft under discussion) at the third session of the Fifth National People's Congress: "We are considering whether we can make a long-term population development plan according to the following conception: in 1985, the total population is expected to reach 1,022 million, with a natural increase rate of 7.5 per 1,000 people; in 1990, the total population is expected to reach 1,068 million, with a natural increase rate of 9.4 per 1,000 people; in 2000, the total population is expected to reach 1,170 million, with a natural increase rate of 6.9 per 1,000 people. With other factors added, it is possible to keep the total population below 1.2 billion". She continued:

> We believe that this population plan is both positive and realistic. Based on this plan, apart from the women of minority ethnic groups, the portion of women of child-bearing age who have only one child should account for more than 50%. At present, those who have claimed the "only child honour card" constitute only about 6% of all women of child-bearing age.

---

[2] Yu (2007).

> If we want to raise the rate from 6% to 50% then we have no choice but to strengthen the work of encouraging couples to have only one child. It is a formidable task involving a revolution in people's attitudes towards fertility. Our main principle for family planning is that we advocate and encourage that each couple have only one child; a second birth is permitted after approval is granted; a third birth is forbidden. In the areas inhabited by the minority ethnic groups; the provinces and autonomous regions in question might make policies according to their specific circumstances. Currently, we should concentrate on the job of forbidding a third birth. More than one in four babies born last year were a third or even fourth child in their family. If we can prevent people having more than two children, then the natural increase rate will drop by a large margin. Meanwhile, we should also strive to improve the rate of only children. If we slacken our efforts, the fertility rate will rise immediately and the goal of keeping the total population below 1.2 billion by 2000 will not be accomplished.[3]

Chen Muhua's report contextualized the reasons behind the proposed goal and the paths to achieving the population control target. As the *Open Letter* was released, the population control goal became a target that the whole country strove to achieve.

6.2 Amendments to the Population Control Target

In 1970, Hagen H. Beinhauer and Ernst Schmacke's *Fahrplan in die Zukunft* inspired a new wave of research on the new millennium in various parts of the world, including in demographic studies. In 1982 *Aging 2000*, co-edited by P. Selby and M. Schechter was published. The book was an analysis of surveys conducted in 16 countries with varying degrees of economic development. The UN Centre for Social Development and Population Affairs and the Sandoz Institute for Health and Socioeconomic Research oversaw the investigations and the report was submitted to the World Conference on the Aging Problem held in Vienna in July 1982. Thus was the aging problem presented to the whole world as one of the next century's most important population issues.

China had the largest population in the world and was the developing country with the fastest growth, following the implementation of reform and opening-up policy. The question of the changes that might occur in China attracted the attention of the whole world. The goal of keeping the Chinese population below 1.2 billion by 2000 had been made known. In 1983, a research project entitled "China in 2000" was led by the Technology and Economy Research Centre of the State Council. On 30 December the same year, Ma Hong, director general of the Research Centre, presided over the first work meeting. The heads of the State Development Planning Commission, the State Economic Commission, the State Science and Technology Commission, the Chinese Academy of Social Sciences, and of some other research institutes participated. The lead unit, administrative offices, and other work organs were set up at the meeting. The project engaged hundreds of renowned experts and

[3] Chen (1980).

scholars. With its insistence on high quality research, "China in 2000" was awarded first prize for National Scientific and Technological Advancement.

The research group was composed of 21 scientific research personnel from the Technology and Economy Research Centre of the State Council, the Department of Labour and Human Resources, the Scientific and Technical Information Institute of China, the Population Research Centre, and the Youth Research Institute of the Chinese Academy of Social Sciences, the Population Research Institute of Renmin University, the Population Research Institute of Beijing College of Economics (today's Capital University of Economics and Business), and the second Academy of the Ministry of Aerospace Industry. An official of the Research Centre was put in charge of the research report on China's Population and Employment in 2000 and I was assigned to write it. Initially, it was intended to be submitted as an internal research report, prior to evaluation for possible publication for a broader public. Thus, I was not constrained by thoughts of rules and regulations when I penned the report. I wanted to produce a report that came from practices and could be applied to guide future practices. With concerted efforts over more than a year, the *Report on China's Population and Employment in 2000* was completed by the end of 1984. In the following section, I elaborate on some points concerning the 2000 population control target, using some excerpts from the report.

First, the remaining 16 years of the twentieth century was the key period within which to restrict population growth. Though an appeal had been made to the public to practice family planning before the 1970s, it had not yet been implemented on a large scale. There were significant imbalances in family planning work. In many places, the population grew unbounded. This resulted not only in a sharp surge in population growth, but also in a young population, which meant a strong momentum for population growth. From the 1970s, remarkable achievements were made. Thanks to the family planning efforts, there were tens of millions fewer births in China. More importantly, there were also changes to the population age structure. The Chinese population had begun the transition from a young (increasing) population structure to an adult (stable) population structure. Eleven years had elapsed since the fertility rate had started to decrease by a large margin in 1973. Thanks to family planning, the birth rate declined from 29.8 per 1,000 people in 1972 to 21.1 per 1,000 in 1992, and the natural increase rate decreased from 22.2 per 1,000 people to 14.5 per 1,000, which greatly shook the base of the original population structure pyramid. According to the data provided by a 10% sample from the third national census, from 1964 to 1982 the portion of children aged 0 to 14 years in China fell from 40.4% to 33.5%; those aged from 15 to 64 rose from 55.9% to 61.6%; and those aged 65 and over rose from 3.7% to 4.9% of the total population. These figures indicated that China had made a first step in the transition from a young to an adult population structure. Yet this was just the first step. To complete the fundamental shift in the population reproduction pattern, we must persist in our efforts in controlling population growth.

If the job of population control were well done, in a period of 28 years (from 1973 to the end of the twentieth century), that is, in one generation's complete reproduction cycle, the population age structure would fundamentally change, finishing its transition into an old population structure and beginning a transition from a stable to a decreasing population structure. According to the prediction that the total population could be kept below 1.2 billion by 2000, the portion of children aged 0 to 14 years would fall to 21.9%, down by 11.6% from 1982; the portion of those aged from 15 to 64 would rise to 70.9%, an increase of 9.3%; and the portion of those aged 65 and over would rise to 7.2%, an increase of 2.3% over the 1982 figure. These shifts would be the result of a continuous long-term decline in the fertility rate and it would be impossible to effect such a shift in a short period. In a predictable future, the Chinese population would depend on control of the fertility rate over the next 16 years. If it were effective in the remaining years before 2000, then efforts to control population growth in the past 10 + years would yield twice the result. It would be possible for the total population to stabilize at around one billion by the end of the next century. If we slackened our efforts and failed to do a good job in this period, our efforts would not be futile but we would achieve only half the result. It would also not be possible to stabilize the total Chinese population at about 1.5 or 1.6 million 100 years later. If that happened, then the task of controlling population growth would become even more formidable.

Second, there was an amendment to the population control target for 2000. The population must be brought under control, but how and what was the target for the year 2000? The questions aroused great controversy. Some thought it must be kept below 1.2 billion: if the Central Committee had clarified the target, then we should not compromise on that. Some thought it was such a high target that it was impossible to achieve. The report suggested, for one, that in view of the practical demographic conditions, and the economy and culture in China at the time, we should examine the factors influencing population production in the process of developing China into a modern country. For another, we must make new analyses using the newer and more reliable population data provided by the 1982 national census and make the necessary adjustments to previous predictions. In fact, population predictions made by official organizations, including the UN, are not static. Predictions are not true forever. The World Population Prospects published by the Population Division of the UN Department of Economic and Social Affairs is released every two years and it makes amendments to previously released data. Sometimes, when certain previous predictions prove close to the real situation, the previous prediction will be valid again and come back into view.

Based on the new population data collected in 1982, the research team of "China's Population and Employment in 2000" made three predictions: a low prediction, a medium prediction and a high one. They made estimates about changes in China's total population, the birth, fertility, and natural increase rates for 1985, 1990, 2000, 2020, and 2030. Tables 6.1, 6.2 and 6.3 present the relevant data.

It was found that only the low (i.e. ambitious) prediction (Table 6.1) meets the requirement of keeping the total population below 1.2 billion. To achieve this target, the total population in 1985 should not exceed 1.05 billion and in 1990 not exceed 1.1 billion. New births then must decrease from over 19 million in 1983 to around 18 million in 1990 and to over 17 million in 2000. The birth rate needed to fall from 18.6 per 1,000 people in 1983 to 16.5 per 1,000 in 1990 and to 14.1 per 1,000 people in 2000. The annual average population increase had to be maintained at 0.95%.

Table 6.1 A low-level prediction about population changes in China (1985–2030)

Year	Total population (billion)	Total fertility rate	Birth rate per 1,000	Mortality rate per 1,000	Rate of increase (%)
1985	1.048	1.80	16.3	6.7	0.96
1990	1.099	1.60	16.5	6.6	0.99
2000	1.204	1.50	14.1	6.8	0.73
2020	1.348	2.10	13.5	9.8	0.37
2030	1.374	2.10	12.8	12.4	0.04

Table 6.2 A medium-level prediction about population changes in China (1985–2030)

Year	Total population (billion)	Total fertility rate	Birth rate per 1,000	Mortality rate per 1,000	Rate of increase (%)
1985	1.050	2.10	19.0	6.8	1.22
1990	1.119	1.90	19.1	6.6	1.25
2000	1.248	1.60	14.5	6.6	0.79
2020	1.425	2.10	14.7	9.4	0.53
2030	1.467	2.10	12.8	11.7	0.11

Table 6.3 A high-level prediction about population changes in China (1985–2030)

Year	Total population (billion)	Total fertility rate	Birth rate per 1,000	Mortality rate per 1,000	Rate of increase (%)
1985	1.056	2.50	22.5	6.9	1.56
1990	1.136	2.10	20.6	6.5	1.41
2000	1.280	1.70	15.1	6.5	0.88
2020	1.483	2.10	15.3	9.1	0.62
2030	1.537	2.10	12.8	11.2	0.16

The total fertility rate was expected to drop to 1.80 in 1985, and to 1.60 in 1990. According to this low prediction, the total population in China would reach its peak (1.374 billion) in 2030 then would fall slowly to around 1.3 billion in 2070. Clearly, this most ambitious prediction favoured population growth control but it would be very hard to meet the requirements because of several potential difficulties.

First, the birth rate had already fallen to a very low level and been decreasing over the previous 10+ years and the natural increase rate had declined from 2.2% in 1972 to 1.2% in 1983. In 12 years, it had almost decreased to half of the original level, a rare occurrence in the world's population history. Among the developing countries, China's natural increase rate was already low, making it harder for it to continue to decline. In some regions, the natural increase rate had already fallen below one percent and in some places, it had reached zero. Thus there was little room for the birth rate to drop further.

Second, due to the age structure formed during the baby boom from the mid-1960s to the early 1970s, there was potential for another baby boom between the early 1980s and the mid-1990s. On average, more than 11 million women annually would enter their child bearing years, incomparable with any previous period. Hence despite a substantial decline in the birth rate, new births and the net increase in the number of people would not necessarily go down. They might even increase, resulting in a continuous growth of the total population.

Third, the urban–rural population distribution in China at the time was highly skewed. In the early 1980s, the rural population accounted for nearly 80% of the total population, with the birth and natural increase rates in rural areas much higher than in urban areas. For example, in 1981, the natural increase rate in the cities was 1.13% while in the countryside it was 1.5%. When the household responsibility system was adopted in the rural areas, there were changes to the functions of families. To a large degree, rural families resumed their production function, which laid the foundation for relatively big families, stimulating their desire for more children.

Moreover, the education level of the population was not high. The areas with high birth rates were mainly in remote mountainous areas, where there was little possibility of significantly reducing the birth rate. These factors combined multiplied the difficulties of continuing to reduce the birth rate. Thus it was not impossible that the current birth rate would be maintained across the country in the remaining 16 years of the century. In 1982, the population increase rate was 1.5%. With an upper population increase limit of 1.3% (the high prediction), then the total population would be 1.056 billion in 1985, 1.136 billion in 1990, 1.216 billion in 1995, and 1.28 billion in 2000. The total fertility, birth, and natural increase rates before 1995 were similar to the current levels, and slightly lower after 1995. So while in this scenario, population control would not be as satisfactory, the indexes of the high prediction level would not be surpassed. Yet, according to this prediction, the total population would not stop growing even when it reached 1.54 billion in 2070 and so we should try our best to avert this situation.

Family planning work had been carried out over many years and been established as a key state policy. Modernization, urbanization, and improvement in education levels were all conducive to controlling population growth. The report thus suggested that the medium prediction (Table 6.2) was closest to the real situation. According to this scenario, in 1985, the total population would be around 1.05 billion. In 1990, it would be around 1.119 billion, around 1.191 billion in 1995, and around 1.248 billion in 2000. If this prediction were realized, the TFR would decrease from 2.10 in 1985 to 1.90 in 1990, to 1.75 in 1995, and to 1.60 in 2000. Overall, it would decline to some degree, but not rapidly. New births would remain above 200 million per annum before 1995, the birth rate would remain above 18 per 1,000 people, and the natural increase rate would remain above 1.15%, were the potential baby boom to occur in the following 10 + years. After the baby boom was over, the natural increase rate would fall at a faster pace. The medium prediction not only reflected the general requirement to strengthen population controls through continuous implementation of the family planning regulations but also showed the difficulties in controlling population growth in the future and the characteristics of the changes in population age structure. In a word, the medium prediction represented the most realistic plan. It appeared less effective than the most optimistic prediction in terms of restraining population growth because it predicted that the total population would continue to grow and not stabilize until around 2070, by which time the total population would be about 1.439 billion. But according to the report, this result would be good enough if it could be achieved.

6.3 Policy Options for Achieving the 1.25 Billion Target

The three prediction scenarios indicated that it was impossible to keep the Chinese population below 1.2 billion by the end of the twentieth century. To avoid the highest prediction (1.28 billion) and to ensure the medium target (1.25 billion), current population conditions had to be taken into account and feasible measures and countermeasures formulated on the basis of past experiences. The *Report on China's Population and Employment in 2000* presented some major policy choices.

First, it said, we must continue with the policy of encouraging each couple to have only one child. Achieving the 1.2 billion target would demand a high proportion of families with one child. Only by intensifying the "one child" efforts could we achieve it. To reach the medium prediction (1.25 billion), the proportion of one-child couples also had to be high. Even to achieve the high prediction of 1.28 billion, the total fertility rate from 1995 to 2000 would need to fall below replacement level and some women would have to have only one child. Therefore, there was no alternative but to strongly advocate for one child for each couple. Yet there had to be an explicit strategic principle: the policy of encouraging one child for each couple could not be a one-size-fits-all solution to the population problem. It would be impossible and it was not permissible to apply only that policy. The basis for the recommendation was to bring one generation's fertility rate under control; the policy was neither an expedient measure, nor a permanent one. It was a measure that would be applied mainly in the twentieth century. In view of this, it should be clearly stipulated that any couple in which both husband and wife were only children themselves must be permitted to have a second child, and the fertility policy should undergo adjustments after being applied to one generation. This would give people hope and win them over. It would also alleviate and even avert the social problems associated with only children that would be incurred were the policy applied over a long period. As for the family planning requirements, different measures should be taken in different situations. In this way, family planning work would be reasonable, supported by the people, and implementable.

Second, the *Report on China's Population and Employment in 2000* said, new measures should be introduced to bring population growth under control. Many factors affect the fertility rate, therefore we must broaden our horizons and create conditions conducive to reducing it.

For example, education levels are closely associated with the number of children people have. Generally, the more highly educated a nation is, the more likely people are to pay attention to the capacities of their children rather than to the number of them they have, and vice versa. Thus different education levels produce stark differences in fertility rates: in 1982, among the cohort of 40 year-old Chinese women, illiterate women had 4.7 children on average; those with primary school education had 4.1 children on average; women with junior high-school education had 3.4 children on average; those with senior high-school education had 2.7 children on average; and women with university education had an average of two children. Among 50 year-old women, the same education levels meant averages of 5.9 children, 4.8, 3.7, 2.8, and 2.1 respectively. Since the fertility rate is generally inversely proportionate to education achievements, improved schooling levels, especially among women, would mean a fall in the fertility rate. Therefore, increasing investments in the intellectual improvement of the population, including vigorous development of the education system, are not only required to improve the population's education attainments and scientific capacities but are also a guarantee of a lower fertility rate.

At the time, there was a wide gap between the fertility rate in cities and in rural areas. Thus restrictions on peasants coming to the cities should be lifted where appropriate, particularly if they are moving to towns near their own villages. This measure would be crucial to the development of industry and commerce in cities, and would facilitate economic development in rural areas. It would also be a practical step to lowering the fertility rate. In addition, we should promote the development of medical and healthcare services, and practice eugenics and sound child rearing in the interests of developing the economy, which in turn will contribute to a decline in the fertility rate, directly or indirectly. In sum, it should be clear that family planning work is the key to achieving the population control target by the end of the twentieth century.

Third, we should gradually create a social security system for the old people. As explained earlier, an old-age pension system had been in place for workers in state-owned enterprises and for public servants, but in the vast rural areas and in most collectively owned enterprises, there was no pension system. There were still objective reasons then behind the traditional desire to have children to provide for one's old age, which constituted a solid justification for some families to insist on having more than two children. To solve this problem fundamentally and to resolve the difficulties for childless couples, the report suggested that a social security system for old people be created in all the collectively owned enterprises, including in agriculture. The system could be applied in one of two ways. One approach would be to develop retirement pension funds as the collective economy grew, as a few such enterprises had already done. Since most collectively owned enterprises would not be able to do that in a short period, the other approach was to establish elders' funds along the lines of practices in other countries. That is, the insurance companies could withhold a certain portion of an employee's monthly income, say 5%–10%, and deposit that in a fund for their retirement. The money so deposited would be invested and thus earn returns. Such money must only be used for this specific purpose and for specific people only. When the employee retires, the capital and interest would be returned to them through payments made on a fixed date each year. The state, enterprises, and public service units should also contribute to these pension funds. With such systems in place, funds would be accumulated in order to support elderly people after their retirement. It would be an effective way to reduce the fertility rate, as it is in many countries. The feasibility of such a scheme should be investigated and similar measures adopted.

Fourth, precautions should be taken against potential problems as soon as possible. Their key aspects should be understood, and appropriate adequate measures developed.

Those key problems involve the aging population. With the falls in fertility and birth rates, the percentage of young people decreases and the adult and old population increases. The aim is to control population growth effectively and at the same time to avoid an excessive aging problem. Thus a choice must be made between a surge in population growth induced by a high fertility rate and a serious aging problem caused by a low fertility rate. This is the primary reason for choosing the medium-level prediction. According to the *Report*, if we choose the low fertility rate, it should not be lower than that suggested by the low prediction. And we could thus ensure that in 2000 the percentage of people aged 65 and over would not surpass 7.2%. In 2040, when the aging population will approach a peak, the situation would not be too serious. If we choose the high fertility rate, it should not exceed that suggested by high prediction, which would ensure that the total population would not be over 1.28 billion in 2000 and would not exceed 1.56 billion when approaching the peak in 2045. The nature of the population problem in China at the time was that the population was hindering productive forces. There was a contradiction between a large population with fast growth and an underdeveloped economy. It was thus rational and practical to choose the fertility rate close to that suggested by the most ambitious prediction, which would allow us to achieve the medium, more modest prediction.

We must also try to maintain a balanced gender ratio. According to the 1982 census, the national gender ratio was 106.28 males to 100 females. Among infants less than 12 months old, the ratio was 107.15 males to 100 females. This ratio was only slightly higher than the normal range (102~107 males to 100 females). With 2000 years of feudal society behind Chinese culture, people were still attached to the traditional belief that a family should have a son to carry on its line. When applying family planning measures and encouraging couples to have just one child, special precautions would be required to prevent the killing of baby girls. New customs should be promoted to eradicate the notion that "there are three forms of unfilial conduct, of which the worst is to have no male descendants". We would have to crack down on all illegal activities involving abusing mothers and daughters and forbid the killing or abandoning of baby girls. The legal rights of women and children must be protected.

In addition, the health of newborn babies must be promoted. As already noted, the family planning laws stipulated that if a first-born child is diagnosed with a non-genetic disease that will prevent him or her from working as an adult, then the parents will be permitted to have a second child. The problem is that a limited number of children have been diagnosed with genetic diseases, because of backward technology and testing methods. As a result, a number of parents who have a child with a genetic disease have already had a second child, increasing the proportion of mentally retarded children in certain regions. And a phenomenon of "adverse selection" forms. Therefore, when a couple's first child is diagnosed with any newborn infant disease, we should be strict rather than lenient when considering whether to give them permission to have a second child. Meanwhile, great efforts should be made in research into genetics and eugenics, establishing clinics for genetic consultation, developing the means to monitor whether some unborn babies have carried genetic diseases, and enacting eugenic protection laws in order to reduce the births of unhealthy babies and promote the growth of healthy ones.

Here, it is worth mentioning that *China in 2000* was a research report for internal circulation, and it had a decisive influence on policy formulation. The Central Committee leaders in charge of family planning work were involved in the research project, yet still more channels were needed to present the latest research results in timely fashion. Before the report was finished, the Technology and Economy Research Centre of the State Council reported the latest research results in its internal publication *The References* (Issue 4, 1984) in a piece entitled "Controlling Population Size and Improving Population Capacity—China in 2000" (Series 4), which presented the views discussed above. The article said: "By 2000, population in our country might reach 1.2 billion to 1.28 billion. It will be difficult to keep the population from surpassing 1.2 billion. We should be aware that it might … reach 1.25 billion or 1.28 billion." According to my literature search, this is the earliest research report submitted to the CPC Central Committee and the State Council to suggest that the total national population might reach 1.25 billion to 1.28 billion. The following image shows the front page of the issue of *The References* containing the report in question. Note the underlined sentences (Picture 6.1).

The leading group of the project "China in 2000" decided that parts of the report would be made available to the public through summary articles. On 4 November 1985, an article entitled "China's Population and Employment in 2000" was published in the *Economic Daily*, the content of which was the same as the report published in *The References*. Then the *Report on China's Population and Employment in 2000*, which contained the recommendation that the total national population should be held at around 1.25 billion by 2000, was submitted as an internal report together with other research papers by the Technology and Economy Research Centre to the CPC Central Committee and the State Council. This population target for the year 2000 was released to the public through the publication of this report. The information aroused wide attention.

In the interim, academics and other interested sectors had also raised doubts about the original 1.2 billion population target. In July 1984, Ma Yingtong and Zhang Xiaotong of the National Population and Family Planning Commission coauthored a report to the Central Committee in which they argued that it was not practical to try to keep the population below 1.2 billion by 2000. Leaders of the Central Committee made comments on their report.[4] Some local governments believed that the fertility indices were unachievable. On 25 March 1985, when Deng Xiaoping met a group of visitors from the US press, he said, "by then (the end of the twentieth century) the

[4] The National Population and Family Planning Commission (2007).

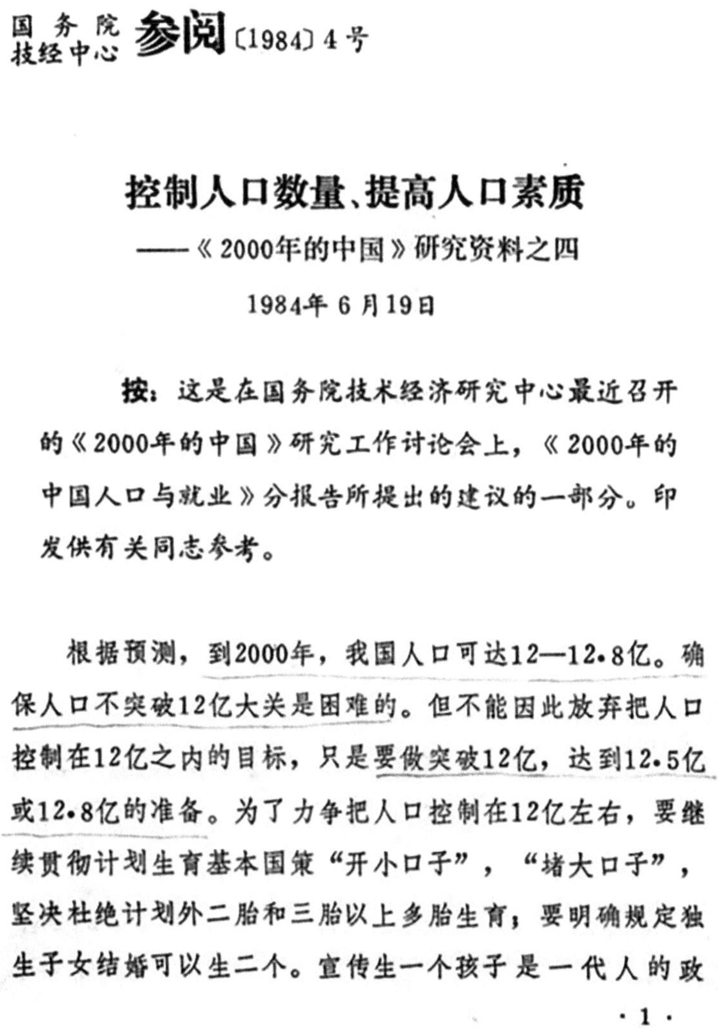

国务院
技经中心 **参阅**〔1984〕4号

控制人口数量、提高人口素质

——《2000年的中国》研究资料之四

1984年6月19日

按：这是在国务院技术经济研究中心最近召开的《2000年的中国》研究工作讨论会上，《2000年的中国人口与就业》分报告所提出的建议的一部分。印发供有关同志参考。

根据预测，到2000年，我国人口可达12—12.8亿。确保人口不突破12亿大关是困难的。但不能因此放弃把人口控制在12亿之内的目标，只是要做突破12亿，达到12.5亿或12.8亿的准备。为了力争把人口控制在12亿左右，要继续贯彻计划生育基本国策"开小口子"，"堵大口子"，坚决杜绝计划外二胎和三胎以上多胎生育；要明确规定独生子女结婚可以生二个。宣传生一个孩子是一代人的政

·1·

Picture 6.1 The front page of the *"Report"* published in *The References* (19 June 1984)

population in our country will have reached around 1.2 billion".[5] From that moment, in all newspapers, magazines, and news reports, the population control target was cited as "around 1.2 billion in 2000", a flexible target.

Not only was the change from "below" 1.2 billion to "around" 1.2 billion an amendment to the original target, in common practice, an "around" figure can be rounded up or down. Accordingly, the report's suggestion that the population be kept to "around 1.25 billion" implies that it could exceed 1.25 billion. As I remember it, in

[5] The National Population and Family Planning Commission (1999).

1984, during discussions about the *Report on China's Population and Employment in 2000*, an old scholar raised doubts about the data suggested by the three predictions. He reasoned that there was a big difference (50 million) between the low/ambitious prediction and the medium one. There were only 16 years left before the year 2000 and such a big difference between the two predictions seemed impossible. He suggested that the medium prediction could be 1.22 billion or 1.23 billion. But his suggestion was not accepted for two reasons. The first was that the low prediction was intended to interpret the control target in 2000 declared by the Central Committee. We had to explain how the birth, mortality and natural increase rates changed according to the target set, whether such changes were reasonable and whether they could be achieved. In fact, when the target of 1.2 billion appeared in the low prediction, it had been implied that it would be difficult to achieve. In demography, usually, the medium prediction is the version selected. Another reason was that the medium prediction of around 1.25 billion was the plan recommended by our research team as viable—realistic and achievable. The difference of 50 million between the low and medium predictions was a technical issue, what mattered most was what was practical. Reality proved that the medium prediction was the most practical. Subsequent data showed the total population in China in 2000 was 1.267 billion.

References

Chen, M. H. (1980). *Striving to Carry out Family Planning in Order to Control Population Growth—the Report at the Third Session of the Fifth National People's Congress* (the draft under discussion, photocopy). National People's Congress.

Peng, P. Y. (Ed.). (1997). *A Complete Record of Family Planning in China.* China Population Publishing House. p. 16.

The National Population and Family Planning Commission. (Ed.). (1999). *The Outline of Deng Xiaoping's Population Thoughts.* People's Publishing House. p. 38.

The National Population and Family Planning Commission. (Ed.). (2007). *Population and Birth Control History in China.* China Population Publishing House. p. 138.

Yu, W. (2007). *Thirty Years of Practice and Theoretical Explorations of Family Planning.* China Population Publishing House. pp. 57–58.

Chapter 7
Population Strategy Supported by the Basic National Policy

From the introduction of the reform and opening-up policy, and especially from 1980, the one-child policy was the pillar of fertility policy in China. Overall, it remained so until a new policy permitting a married couple to have two children if either one of them were an only child was introduced in 2013. Aside from a short period of fluctuation in the mid- to late 1980s, the message that population growth should be firmly restricted through encouraging couples to have only one child remained unchanged. Meanwhile, the family planning policy was promoted to a higher political rank, chiefly reflected in the following facts: from 1991 to 2005, during the annual National People's Congress and the Chinese People's Political Consultative Conference, the Central Committee convened symposia on family planning work. (Later, the meetings became symposia on population and environment, and then on population, resources, and environment). At those meetings, control of population growth and the implementation of family planning were promoted to the primary national policy and were later gradually incorporated into the Scientific Outlook on Development and the Sustainable Development Strategy.

7.1 Promotion of the Basic National Policy on Population

As detailed in Chapter 5 above, in early 1980, the General Office of the Central Committee convened five symposia on population, a rare occurrence in the Central Committee's decision-making processes. On 25 September 1980, *An Open Letter to All Party Members and Communist Youth League Members from the Central Committee of the Communist Party of China about the Control of Population Growth in China* was released, also a rare occurrence in CPC history. These actions indicate the importance that the Central Committee attached to China's population problems and their solution. The first basic national policy, "we should restrain population

X. Tian, *Population Policies in the People's Republic of China*,
https://doi.org/10.1007/978-981-96-9658-1_7

growth and implement family planning" verifies this. It is appropriate, then, to analyse how that policy was instituted and promoted.

How was this key national policy created? It was associated with prevailing conditions in China in the 1970s and 1980s, circumstances at the local level, the Central Committee leaders' understanding of the population problems, and their instructions addressing those problems. The situation in local areas and other realities in China at the time have been discussed in previous chapters. Here, we discuss some local practices in family planning work. For example, in 1974, at the second national meeting on family planning work, experiences in Nangong county, Hebei province, Heishan county, Liaoning province, Shanghai city and Hebei province were exchanged. Local governments had, for instance, introduced regulations concerning the number of children to be born, the rewards to be made to couples who had one child, and the punishments to be meted out to those who had more than two. Family planning work was also associated with economic development. The Party committee and provincial government of Sichuan province required that the top leader at each level across the province apply the principle that "two kinds of production should be managed together." This was recognized as the primary experience of family planning work in Sichuan province. The other experiences involved promoting family planning policy, rewarding those who practiced family planning and punishing those who broke the regulations, formulating population plans, building stronger teams to carry out the work, advancing scientific research in the area and improving technical instructions about it. The family planning office of the State Council approved their work and promoted it. These work methods and experience went beyond the boundaries of family planning work into the field of social and economic development, opening a new stage of common leadership and comprehensive management. They also provided some typical practices and lessons for the application of the national population policy and preparing the public for the new policy.

It is often said outside China that it is a country with a strong government. While biased, it is also an objective observation in that the formulation of policy in China mainly depends on the understanding and decisions of CPC Central Committee leaders. In January 1981, Deng Xiaoping said: "Family planning work is a strategic task, so we must do a good job. We need to whip up public opinion and praise the exemplary cases." On 19 November 1985, when Deng met the premier of Pakistan, he emphasized, "If China's population grows continuously … it will reach 1.5 billion by the end of this century. … Our population policy is a major strategic policy".[1] On 1 January 1982, "The Minutes of National Meeting on Work in Rural Areas," approved and transmitted by the Central Committee noted: "Despite a large population, there is little arable land per capita in our country. Thus to control population growth and protect the arable land is our major national policy." On 1 September of the same year, Hu Yaobang, general secretary of the Central Committee presented a report entitled "Create a New Situation of Socialist Modernization on All Fronts" to the 12th National People's Congress, according to which "in China's economic and social development, population remains a tremendously important issue. The

[1] The National Population and Family Planning Commission (2007, p. 122).

implementation of family planning is a basic national policy in China. By the end of this century, we must strive to keep our population below 1.2 billion. Our population is right in a baby boom period. A fast-growing population will not only affect the improvement of per capita income but will also create serious problems in the supply of food and housing, education, and job opportunities. Even social stability might be adversely affected."[2] He went on to declare that the "implementation of family planning is a basic national policy in China." The population policy was launched then, to be adhered to by the Party and the whole country. Family planning also became the basis for formulating population policies and making them operational.

Here two additional points need to be presented. First, two decisions closely associated with the basic national population policy were declared by the CPC Central Committee and the State Council in this period. Second, the Central Committee had held more than ten symposia on family planning work, on population and the environment, and on population, resources, and the environment. All the speeches delivered by top leaders elaborated on the national population policy.

First, the Central Committee's and the State Council's two decisions: the first document, "Decision on Intensifying Family Planning Work and Strictly Limiting Population Growth" was issued on 12 May 1991 demanded that the natural increase population rate in the next 10 years not exceed 12.5 per 1,000 people so that the second and third steps in the strategy for socialist modernization could be reached. The document stressed that we should pay attention to the following points.

1. Leaders should unify their understanding and strengthen leadership in family planning work. It is imperative for the Party committee and government at all levels to attach equal importance to family planning and economic development. Party committee leaders and the government at all levels should personally concern themselves with family planning work and take responsibility for it. Party committees and government at all levels should establish leading groups to take charge of population and family planning work, with their chief leaders at the head of these organizations. Their responsibility was to coordinate the efforts of all related departments and offices. They must take responsibility for the population and family planning work in their own regions and apply and improve the target management responsibility system in the population and family planning work. Whether a good job is done in family planning work and the population planning goal is fulfilled should be an index in evaluating the performance of Party committees and government as well as assessing individual leaders' performances. Party committee and government personnel at higher levels should strengthen supervision and inspection of the work of those at the lower levels to ensure that statistics are accurate. Providing false material or concealing information is forbidden. A reward and punishment system should be established; while good leaders will be rewarded, those responsible for losing control of population growth will be punished.
2. We should resolutely implement the current policy and manage family planning work according to law.

[2] The National Population and Family Planning Commission (2007, p. 123).

3. We should concentrate on the crucial points and carry out family planning work solidly and steadily. To restrain population growth strictly, we must focus on work at the basic levels.
4. We should make concerted efforts to ensure the implementation of family planning work goals. Party committees and government at all levels should provide the funds and material resources necessary for family planning work. They should strengthen the organizational structures of the departments in charge of family planning work and assign competent officials to the work.

A second document, "Decision on Strengthening Population and Family Planning Work and Maintaining a Low Fertility Level" was issued by the Central Committee and the State Council on 2 March 2000. According to this document, controlling the size of the population and improving its skills are strategic to achieving the goals of modernization and sustainable development in China. Family planning is a basic national policy to be applied over the long term. When the nature of population reproduction has shifted, the primary task of population and family planning work will be to maintain a low fertility level and to improve the population's competencies. It is imperative that the whole Party and society keep in mind the big picture of socialist modernization in China and the country's long-term survival and development, and strive to do a better job in family planning work. Some key points of this document are summarized here.

1. Maintaining a low fertility level over the long term is a major and formidable task. Leaders and officials at all levels must understand and solve population problems from a strategic perspective, starting from actual conditions. Being responsible for the Party, the nation, children, and the good of the people, they must carry out population and family planning work conscientiously, ensuring the future targets are met.
2. The target for population and family planning work in the next decade was to keep the total population in China (excluding Hong Kong and Macau Special Administrative Zone and Taiwan province) below 1.4 billion and the average annual birth rate at 15 per 1,000 people or below by the end of 2010. The health of newborns was to be enhanced. The gender ratio of the newborn was expected to become normal. Women of child-bearing age were to be offered basic reproductive health services. New ideas about marriage and fertility were to be formed. A family planning security system was to be created, with efficient regulation, effective management, and comprehensive policies and regulations.
3. The regulations and socio-economic policies relating to population and family planning work should be perfected. Governments at all levels and relevant departments should coordinate with each other to facilitate the work when formulating systems for land, business management, medical services, social security, household registration, labour, education, and fiscal and taxation systems, and when developing reforms. Legal structures should be strengthened as well as the benefit mechanisms. Couples who have practiced family planning would be rewarded

with preferential treatment. Social security measures should be rigorously developed. The population question should be incorporated into the plan for Development of China's Western Region. We should strive to boost economic development while the population increase rate drops and the population's competencies improve.

4. A population and family planning work management mechanism that can accommodate the socialist market economy should be built. The whole society must be mobilised to apply family planning measures and to incorporate the work into community management and service systems. The so-called floating population also requires effective management, and basic knowledge about population and family planning projects must be improved.
5. Party and government leadership in this work should be strengthened, taking maximum advantage of the part-time members of the National Population and Family Planning Commission and their work units. Leading groups of population and family planning work in each province, autonomous region and municipal city must play their role. The Central Committee would hold a symposium on population work each year and population and family planning work would be incorporated into the major matters supervised by the Central Committee. Party and government officials including those in charge of family planning work should all follow the family planning policy. None should abuse their powers and have more children than permitted by the policy. We should insist and improve the population and family planning target management responsibility system. Vetoes shall be placed on the promotion and benefits of those who fail to apply the family planning policy or who do not take their family planning management responsibilities seriously.

The two Central Committee decisions first clarified the work priority into the future, then they went beyond the limitations of "carrying out family planning work for the sake of solving the population problem" through stressing its importance for "boosting economic development." Most importantly, "family planning is a basic national policy we must adhere to over a long period", thus placing population and family planning work in the category of key national policies.

Second, the symposia on population (with the sub-themes of population, resources, and the environment) also emphasized that the family planning policy must be adhered to over a long period. From 1991 to 2005, during the annual National People's Congress and the Chinese People's Political Consultative Conference, the Central Committee successively convened more than ten symposia on family planning work. Both the reports made by the director of National Population and Family Planning Commission and the speeches delivered by both the general secretary of the Central Committee and the premier, stressed that we must stick to the basic national policy of family planning and implement it conscientiously. In 1999, the symposium on population, resources, and environment made limiting population growth and the tapping of natural resources, and protection of the environment the three cardinal state policies, intended to coordinate efforts to solve problems in these three areas. Jiang Zemin, then general secretary of the Central Committee addressed the symposium saying "we should concentrate on economic construction and make all efforts to boost the national economy. To achieve this goal, we must consider the economic development, population, resources, and the environment holistically, and make coordinated plans. We should strive to restrain population growth, use resources reasonably, and to preserve the environment effectively, making sure that economy can develop successively, rapidly, and healthily and the society advance steadily."[3]

In 2006, in a meeting held by the Political Bureau of the CPC Central Committee to discuss population and family planning work, Hu Jintao, then general secretary of the Central Committee, spoke. The meeting stressed that all regions and all departments should understand the importance of intensifying population and family planning work and resolutely coordinate efforts to solve the population problem with Chinese characteristics. Family planning—the key national policy—must be implemented, and the population and fertility policy be improved and stabilized.[4] Subsequently, documents and speeches on family planning upheld this call to implement, perfect, and stabilize the population and the fertility policy.

7.2 The Relative Stability of the Population Policy

In 1980, the Central Committee decided that each couple was to be encouraged to have only one child, and this became the keystone of population and fertility policy for the following more than three decades. On 9 February 1982, "Instructions on Making More Efforts to Carry out Family Planning Work" (Central Committee Document No. 11) was issued, according to which, restraining population growth was a major strategic question in China's modernization. It pointed out that

> there are two possibilities before us. Either we choose to restrain population growth strictly and effectively, ameliorate people's living standards gradually and expand the construction scale in our country year by year, or we lose control of population growth due to ineffective measures. Consequently, the population will grow substantially and we will be able neither to improve people's lives nor to develop the economy efficiently, advance our culture and construct our national defence. We must choose between the two options; comrades of the whole Party and the entire nation must recognize the significance of restraining population growth. They must also see that effectively controlling population growth will not only solve our present difficulties but also meet the long-term interests of the descendants of Chinese nation.[5]

This Central Committee and State Council instruction reiterated the spirit of the open letter addressed 18 months earlier to all the Party members and members of the Communist Youth League of China. It demonstrated serious resolution to restrain population growth and clarified the requirements for Party and government organizations. With the establishment of the target management responsibility system and

---

[3] Zhang (2013, p. 122).

[4] Zhang (2013, p. 12).

[5] The National Population and Family Planning Commission (2007, pp. 122–123).

the vetoes on the promotion and benefits against those who fail to observe the family planning policy or to meet their family planning responsibilities, local government organizations and work units could introduce measures coherent with local realities. Population problems, control targets, policies, and regulations were placed on the agenda of the Party and government departments at all levels.

The policy was a single national one, but included some detailed specificities designed to address the various circumstances that would be met during implementation of the fertility policy. It would be impossible to ensure that all couples had only one child, and it would be difficult to exhaust all the specific regulations of permitting a couple to have a second child proposed by the different local governments. For this reason, the Central Committee Secretariat met on 10 September 1981 to discuss the issues concerning family planning policy. It was suggested by this meeting that new situations and practical experiences be taken into consideration in the promotion of family planning. The policy of encouraging couples in the cities to have only one child should remain as is. In the rural areas, "to deal with new situations after the adoption of the household contract responsibility system with remuneration linked to output, we should intensify family planning work on the one hand. But on the other hand, we can relax the policy appropriately." How to relax the policy and to what degree? The Central Committee decided to follow the mass line: the specific regulations would be framed after consultations with governments of all the provinces, autonomous regions, and municipalities. Later, in response to the feedback, it was proposed that "in the rural areas, generally, each couple is encouraged to have only one child. If some couples have practical difficulties and apply to have a second child, they might do so according to the plan after gaining approval. Regardless of the circumstances, rural couples may not have a third child." Two years later, the *Report on China's Population and Employment in 2000* proposed that by 2000, the total national population should have been kept at around 1.25 billion and the total fertility rate at 1.75. The real situation was that in the cities, couples generally had one child and in rural areas they generally had two children. In academic and other social sectors, researchers put forward some different proposals. The leading Party group of the National Population and Family Planning Commission submitted a report to the Central Committee, which was approved and released by the latter on 13 April 1984. In the document "Report on Family Planning Work", (Central Committee Document No. 7), regulations were formulated to address the varying real circumstances (for details see Sect. 5.3).

After Document No. 7 was issued, population growth fluctuated. There were different interpretations of the policy to "allow some couples to have a second child in special circumstances" (开口子). Should the established population policy be enforced? And if so, how? These different interpretations and applications of the new regulation led to a sudden increase in population in the mid- and late 1980s. The increase did not last long, and the peak was not high (see Sect. 5.3). Nevertheless, the spike indicated that policy had a significant effect on population change and hence should not be neglected.

On 31 March 1988, the standing committee of the Political Bureau of the CPC Central Committee met to discuss the consensus achieved in the *Outline of the Report on Family Planning Work* submitted by the National Population and Family Planning Commission. The meeting reiterated:

> the current family planning policy in China is that we encourage late marriage and late childbirth; we recommend that people have fewer births and better births; we encourage each couple to have only one child. Specifically, to government officials, workers, residents of the cities and towns, each couple should have only one child, except for those who have received approval to have a second child in some special cases. In rural areas, if people have practical difficulties, including being in a family with only one girl, and they apply to have a second child, they may be given permission do so after a wait of several years following approval of their application. A third child is not permitted under any circumstance. In the autonomous regions or areas, we also encourage people to adopt family planning. The specific requirements and regulations can be formulated by the provinces and autonomous regions concerned in accordance with the local conditions.

The meeting stressed that the policy must be implemented consistently and over a long period; the policy had to remain stable in order to restrain population growth. It is worth mentioning that the regulation permitting rural couples with only one daughter to have a second child subsequently affected population changes. According to calculations made by the National Population and Family Planning Commission and other experts, the ratio of children who were second-born surged to around 50% after the 1988 regulation allowing rural couples with only one daughter to have a second child.[6]

On 21 February 1989, a meeting was held by the National Population and Family Planning Commission with the directors of provincial family planning commissions. Peng Peiyun, the Commission's director addressed the meeting, saying:

> we should continue to enforce and implement the current family planning policy formulated by the CPC Central Committee comprehensively. All local governments must implement the policy strictly. We should firmly preclude each department from acting on its own, and no further exceptions should be made without approval by the Central Committee. Whenever the policy is applied more loosely than stipulated by the current policy … the local government must submit an official report to the Central Committee for its instructions. We should continue enthusiastically to encourage each couple to have only one child. We recommend late marriage, late childbirth, fewer births, healthier babies and efforts to avoid the birth of babies with disabilities. In the cities, suburbs, and rural areas that are economically and culturally developed, we should consolidate the achievements we have made and strive to make a greater contribution to restraining population growth for the nation. In the vast rural areas, when encouraging each couple to have only one child, we should also conscientiously implement the policy of permitting families with a single girl to have a second child after a wait of several years.[7]

Peng's speech effectively precluded allowing more couples with special cases to have a second child. In the meantime, the fertility policy was to be implemented according to the established regulations without any major change. This policy remained in place until 2013, when it was changed to allow a married couple that included someone who was an only child to have a second child.

---

[6] The National Population and Family Planning Commission (2007, pp. 128–129).

[7] The National Population and Family Planning Commission. (2007, p. 129).

During this period, there was no big change to the births target stipulated in the population policy. Neither was there any significant change to the reward and punishment regulations. There were, however, changes, improvements, and innovations to the management system and policy mechanisms, work methods, and connections between population and other policy fields. A series of successful approaches and experiences had been summarized in the domain of family planning work. For example, there were the so-called "three priorities", namely, in family planning work, among the methods of publicity, education and the imposition of economic restrictions, we should prioritize the former; between contraception and artificial abortion, we should also prioritize the first; and among regular activities and occasional activities, the former should be prioritized. There were also the "three combinations": combining family planning work with economic development, with helping farmers work industriously to become well-off, and with building civil and happy families. There were also three "remain unchanged" elements: the current family planning policy would remain unchanged; the population target would remain unchanged; and the principle that the top Party and government leaders at all levels were personally responsible for family planning work and should take the overall responsibility for the work would remain unchanged. These three precepts were the guiding lights of family planning work, highlighting the guiding principles, the rules to be followed, and the essentials of the work. From a certain point of view, they indicated that since the late 1980s, population policy had remained quite stable and the family planning work had been gradually intensified and deepened.

This point is clear when we look at the document "The CPC Central Committee and State Council Decision on Comprehensively Intensifying Population and Family Planning Work and Solving Population Problems with Coordinated Efforts" (Central Committee Document No. 22, 17 December 2006). Some new content was added with this document, ensuring that the strategy for solving the population problem was up-to-date and comprehensive. Document No. 22 elaborated on issues such as improving population capacity, solving the problem of the highly unequal sex at birth ratio, improving the services provided to the floating population, tackling the aging problem, and intensifying efforts to establish mechanisms that facilitated the population and family planning work. Still, the document remained focused on population and fertility policy. It stressed that we should become fully aware of the urgency of intensifying the population and family planning work, should solve the problem with Chinese characteristics, and make every effort to maintain a low fertility level. It also emphasized that "maintaining a low fertility level is the priority of population and family planning work in the new era" and "we must never waver from the basic national policy of family planning or from maintaining the current fertility policy; we must never waver in our adherence to the principle that the leaders of the Party committees and government at all levels are personally responsible for family planning work and must take overall responsibility for family planning work; we must keep the population and family planning organizations and their personnel intact; we should innovate work systems, mechanisms, approaches and methods of the population and family planning work, time and again."[8] The series of "never wavers" indicated the orientation, guidelines, and focus of population and family planning work, maintaining the current fertility policy as the core of the work in question.

7.3 Population Policy Being Integrated into Sustainable Development Strategy

In 1994, *China's Agenda in the 21st Century—A White Paper on China's Population, Environment, and Development in the 21st Century* was published, the first of its kind in the world. In the book, sustainable development was elevated to a national strategy. In the mid and late 1990s, population policy was integrated into the Scientific Outlook on Development and Sustainable Development Strategy, which was a big turn in its development. In the White Paper, one chapter was dedicated to the question of population and sustainable development. It argued that to achieve sustainable development goals, we must "take fully into account the element of population, and carefully manage the relationships among population, resources, environment, and development to provide a relatively flexible population condition for building socialist modernization. The flexible population condition is important to achieving sustainable development of the society and the economy."[9] The White Paper set out the goals and requirements for future population change and development, with a focus on controlling population size and improving its skills. It outlined a series of action plans, for controlling population growth, family planning and health care, education and improvement of the population's capacities, the aging population and human resources development, construction and improvement of comprehensive data collection and information systems, the formulation and evaluation of policy, and regional cooperation and communication. There emerged a trend towards a population policy that took into account the requirements of sustainable development strategies.[10]

This new integration of sustainable development and population goals was reflected in various ways, including in the Central Committee symposia on family planning, population, resources, and environment from 1991 to 2005. First, the change in the symposium title reflected a leaning towards the Scientific Outlook on Development and Sustainable Development Strategy. In 1991, the symposium's name was the "Central Committee Symposium on Family Planning Work", and the speeches and meeting topics were confined to population. In 1997, it was renamed the "Central Committee Symposium on Family Planning and Environmental Protection". In 1999, it was changed again, to the "Central Committee Symposium on Population, Resources, and Environment".

---

[8] The National Population and Family Planning Commission (2007. pp. 259–260).

[9] State Council of the PRC. (1994, p. 39).

[10] Idem, pp. 42–46.

Next, the content of the symposia changed from population and the basic national policy of family planning to a combination of population, resources, and environment, forming a strategic pattern of three in one. In 1999, Jiang Zemin, then general secretary of the Central Committee delivered a speech at the first symposium on population, resources, and environment. He stated that "to facilitate the sustainable development of China's economy and society, we must restrict population growth, protect natural resources, and preserve a good ecological environment while keeping the economy growing. This is the basic national policy established according to the present conditions in our country and the strategic goal of long-term development. We must understand that it is important to handle the relationship between economic development, population, resources, and environment strategically."[11] In 2005, Hu Jintao, Jiang's successor, addressed the symposium on the same theme: "Comprehensively implementing the scientific outlook on development and further adjusting economic structures and transforming the economic growth mode are the main ways to alleviate the pressures of population, resources, and environment and to pursue the comprehensive, balanced and sustainable development of the economy and society. We should accelerate the adjustment of unreasonable economic structures and completely transform the extensive growth model. In so doing, we can grow the economy on the basis of improved population skills, effective use of resources, reduction of environmental pollution, and concentration on efficiency and quality, in an endeavour to build a resource-conserving and environment-friendly society."[12] Sustainable development refers to a type of development that satisfies present needs, but does not threaten future generations' capacities to satisfy their needs. With human beings at the centre, sustainable development aims to satisfy people's physiological, psychological, and all-round development. The driving force of development is human capital and its ultimate goal is to create a natural and social environment in which there is harmony, both between people and nature and between people and society. The tenets of sustainable development were reflected in more than ten symposia on population, resources, and the environment held by the Central Committee, speeches delivered by Central Committee leaders, and reports made by the heads of relevant departments.

The 2006 Document No. 22, "The CPC Central Committee and State Council Decision on Comprehensively Intensifying Population and Family Planning Work and Solving Population Problems with Coordinated Efforts" analysed family planning since the population policy had been adapted to the Scientific Outlook on Development and the sustainable development. Document No. 22 first summarized the previous implementation of the national family planning policy, saying that "since family planning has been practiced in China, there have been more than 400 million fewer births, realizing the transformation of population reproduction pattern ahead of time. Family planning policy has effectively alleviated pressure on resources and the environment and strongly promoted economic growth and social progress. Practice has proven that the persistent implementation of family planning in China has had a

[11] State Council of the PRC (1994, p. 124).

[12] Idem, p. 128.

tremendous impact on building socialism with Chinese characteristics and achieving national prosperity and rejuvenation, playing an important part in promoting population development in the world." Then it analysed the current situation of population and family planning:

> China has a large population, yet its per capital natural resources are quite low. Economic and social development is under massive pressure from the population and there are significant tensions among population, resources, and the environment. These present prominent contradictions and problems for building a moderately prosperous society in all respects and a harmonious socialist society in China... Our major obstacles to achieving rapid and good economic development in China are all closely connected with population size, population capacity, population structures, and population distribution. Any slip in the population question will have an irreversible long-term effect on China's economic and social development. We should comprehensively solve the population problem with the aim of realizing all-around human development through turning population pressures into a human resources advantage that will provide continuous impetus for economic and social development. This is a strategic choice if we are to achieve the great rejuvenation of Chinese nation. It is incumbent on the whole Party to keep doing a good job in population and family planning work in the new era to facilitate the implementation of the Scientific Outlook on Development, to ensure that Party and state power are exercised for the public good, for the benefit of the people, to build a moderately prosperous society in all respects, and a harmonious socialist society, while taking responsibility for the future development of the Chinese nation.[13]

In the section entitled "Resolutely Solving the Population Problem with Chinese Characteristics and Coordinated Efforts", the Document emphasized that "we should comprehensively implement the scientific outlook on development, make preferential investment in all-round human development, maintain a low fertility level, improve population capacities and structures, guide the rational distribution of the population and guarantee the population security. In so doing, we can facilitate China's transformation from a country with a large population to a strong country with rich human capital, and promote the coordination and sustainable development of population, economy, society, resources, and environment." The Document contains other sections such as "Vigorously improving the health of new births", "Continuously perfecting services to the floating population", and "Positively addressing the aging problem", all of which raise questions and elaborate on solutions, and decision-making processes from the perspective of the Scientific Outlook on Development and sustainable development, to promote the implementation of a population and sustainable development strategy. The approach to solving the population problem reflected in Document No. 22 is thus very different from the former, narrower approach of limiting population issues to population and family planning. Here, the population question is integrated into questions of the sustainable development of resources, environment, economy, and society. With all these aspects in mind, guidelines and policies such as comprehensive management and coordinated development are proposed therein.[14]

---

[13] Zhang (2013, p. 17).

[14] Idem, pp. 17–21.

References

State Council of the PRC. (Ed.). (1994). *China's Agenda in the 21st Century—A White Paper on China's Population, Environment and Development in the 21st Century* [White paper]. China Environmental Science Press. p. 39.

The National Population and Family Planning Commission. (Ed.). (2007). *Population and Birth Control History in China.* China Population Publishing House.

Zhang, W. Q. (Ed.). (2013). *Population and Family Planning in China* (1997–2007). China Population Publishing House. p. 122.

Part III
Prospects for Population Policy Adjustments

Chapter 8
General Population Trends and Developments in the 21st Century

The vigorous implementation of the one-child population policy led to a sustained decline in China's fertility and birth rates. In the early 1990s, the country transited to the low birth, low mortality, low increase model, and new developments appeared in population quantity, quality, and structure. This does not mean that the population policy has completed its historical mission and can now be dispensed with; if we believe that the population policy is still required because it plays an essential role in socioeconomic development, then we must understand the general population trends of the twenty-first century, especially its first 50 years. By "the general trends", I mean the following: Population change generally refers to natural changes in the population, while population development mainly refers to the deliberate improvement of population quality, adjustments to population structure, and the various aspects of population and development. The second "general trend" refers to macro and long-term developments. Empirical scientific research should include quantitative and predictive research conducted over a certain period but for the remainder of this century, quantitative prediction can only be a study of trends. Generally, the predictions can be divided into short-term, medium-term, and long-term. The accuracy of short-term prediction is relatively high, for medium-term prediction it is lower, and long-term prediction accuracy is the lowest. Therefore, population predictions for the rest of the century, including for population policy itself, can only be a forecast of a large trend. In the words of American demographer Philip M. Hauser: population prediction is merely prediction, it is not precision. The general trend cannot be said to be ironclad and absolutely accurate, nor can it be regarded as erratic and arbitrary. The direction and nature of the general trend cannot be changed, but the specific prediction data may differ from the future reality. The error should be set within a certain range to have practical significance.

X. Tian, *Population Policies in the People's Republic of China*,
https://doi.org/10.1007/978-981-96-9658-1_8

8.1 New Population Trends and Developments

Proceeding from current realities to think about future trends is the foundation for all good policy formulation and implementation. In the early 1990s, China's population entered the stage of "post demographic transition"; after 20 to 30 years of this transition, a new population reproduction cycle has become manifest. Overall, the fertility and birth rates continue to decline, and the population growth rate has slowed; and population capacity has been greatly improved, with improvements in physical fitness and health, in science and education, and in literacy and civility, to varying degrees. Structural changes such as age, gender, urban and rural distribution, and regional distribution have shown more uneven and growth trends. These constitute fundamental demographic change that involves both the population and socioeconomic development and which has a profound impact on the current and future adjustments of population policy.

8.1.1 New Trends in Population Quantity

In 1991, the total fertility rate dropped below 2.1, and the net reproduction rate dropped below 1.0, marking China's entry into the stage of "post demographic transition". The fertility rate dropped below the replacement level thereby creating a population quantity change quite different from that of the past. This trend has now been in place for more than 20 years and has remained relatively stable. This was the first milestone towards solving China's overpopulation problem and achieving sustainable development of population, economy, society, resources, and the environment. The second milestone, zero population growth, will be reached gradually under the effect of the negative intrinsic rate of natural increase. After that, under the influence of negative growth inertia, the population size will begin to shrink, moving towards the third milestone—achieving an all-around optimum population required for sustainable development, i.e., a stable or standard population. These three milestones are closely linked: only by passing the first milestone can we pass the second and then reach the third one. At present, we are between the first and second. We can only complete the transition by maintaining a relatively stable low fertility level.

The difficulty in doing so is manifest in the fact that China's economy is still "developing", while its population change has approached the type seen in developed countries. Practically speaking, it is appropriate to evaluate the current Chinese economy as basically at the level of a developing country. China has been the second largest economic entity in the world; since 1949, and especially since the reform and opening-up period, the economy has developed considerably, its achievements praised around the world. China's GDP jumped to the second place in the world, and it is closing in on first place, while its distance from third place is increasing. But only when the share of per capita GDP goes up can China be truly described as having completed the transition from standing up to growing prosperous. There is still a certain gap between China's per capita GDP and the world average, and a large gap between China and the richest developed countries. Still, in such a populous country whose economy is basically at the level of a developing country, the development of its population has reached a level close to that of a developed country. A comparison of some indicators from 2005 to 2010 demonstrate the point.

The total fertility rate is 2.55 in the world, 1.59 in developed countries, 2.75 in developing countries, and 1.74 in China. China's rate is therefore equivalent to 68.24% of the world rate, 109.43% of the developed countries, and 63.27% of the developing countries. It is slightly higher than that of developed countries, but much lower than that of the world and developing countries—the gap has widened.

The net reproduction rate is 1.09 in the world, 0.76 in developed countries, 1.16 in developing countries, and 0.79 in China. Thus China's rate is equivalent to 72.48% of the world, 103.95% of developed countries, and 68.10% of developing countries. Again, China's net reproduction rate is slightly higher than that of the developed countries but still much lower than that of the world and developing countries. The gap has widened.

The infant mortality rate is 52 per 1,000 people in the world, 7 per 1,000 in developed countries, 57 per 1,000 in developing countries, and 31 per 1,000 in China. China's rate is equivalent to 59.62% of the world, 4.43 times that of developed countries, and 54.39% of developing countries. It shows that China's infant mortality rate is much lower than that of the world average and the developing country rate, and is still at a distance from developed countries. Here, China sits between the developing and developed countries.

Globally, life expectancy estimates for both sexes combined is 66.5 years, 76.2 years in developed countries, 64.6 years in developing countries, and 72.6 years in China. China's life expectancy is 6.1 years longer than the world average, 8.0 years higher than in the developing countries, and 3.6 years lower than in the developed countries. It is not far from the level of the developed countries.[1]

These figures show that China's economy is at a relatively high level among developing countries, while its population is close to that of developed countries. This is a great contradiction, which indicates that although there may be some uncertainty in China's low fertility level, the uncertainty is weakening, and China's population model is approaching that of the developed countries. Therefore, on the one hand, the potential momentum of a growing population is waning and there is no possibility of another baby boom like that of the 1960s and 1970s; it is almost impossible for the fertility rate to rebound above the replacement level, and zero population growth is just around the corner. On the other hand, the potential momentum of population growth has "weakened" rather than "disappeared", which means that it still exists. More importantly, as mentioned above, the decline of China's fertility rate is an objective basis of the country's economic, scientific, technological, cultural, and social development; the contribution of population policy, which, for more than 40

[1] U. N. Population Division (2009).

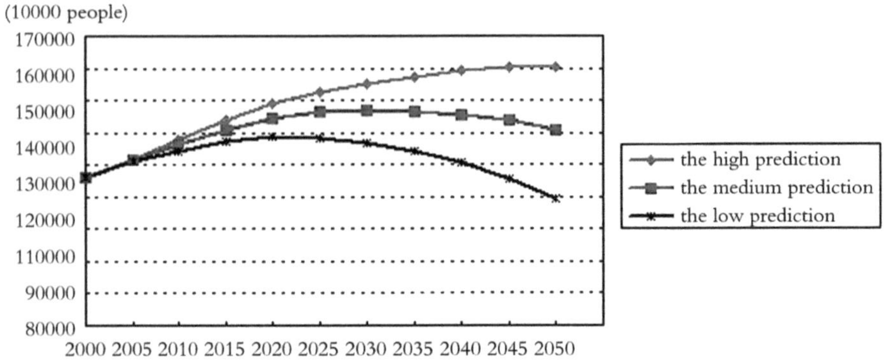

Fig. 8.1 China's population change trend in the first half of the twenty-first century

years focused precisely on reducing the fertility rate and was implemented consistently, cannot be ignored, or replaced. Therefore, in the new post demographic transition, population policy still requires careful attention. To a large extent, it will affect whether China can achieve the transition to zero growth and realize the goal of an all-round optimum population.

Figure 8.1 graphs the prediction of China's population change trend in the first half of the twenty-first century.[2]

Three possible population curves are identified in Fig. 8.1: in the low plan, the total population would have been 1.387 billion in 2021; in the medium plan, the total population will be 1.465 billion by 2030; and in the high plan, China will approach zero growth in 2050, with a total population of 1.6 billion. The population in 2050 according to the three plans is: the low plan 1.192 billion, the medium plan 1.401 billion, and the high plan 1.6 billion. In terms of the effectiveness of solving China's overpopulation and controlling population growth, the low plan should be the first choice. However, because the fertility rate of the low plan has dropped too far, and population structure problems such as aging will become prominent, the low plan must be abandoned (this is discussed later in the book). In the high plan, the national population would reach 1.6 billion in 2050 without completely stopping, and there would be slight growth until 2100. Thus, the high plan too is impractical and the medium plan best fits requirements. The change of the total fertility rate in the medium plan is: 1.75 between 2000 and 2005, 1.80 from 2005 to 2010, 1.83 from 2010 to 2020, and 1.80 between 2020 and 2050. The change of total population is as follows: 1.36 billion in 2010, 1.444 billion in 2020, 1.465 billion in 2030, and then down to 1.402 billion in 2050. This plan is based on the current fertility level and is a realistic foundation for implementation. However, given the certain gap between the current fertility level and the desire of most citizens to have more children, it is not without contradiction. In addition to external demands such as developing the economy and promoting social progress, the most important goal for the national population itself is first to stabilize a low fertility rate, even slightly dropping below

[2] Tian et al. (2007).

the replacement level within a certain period. The second goal is to stabilize the low fertility rate at a reasonable level. It will be necessary from time to time to adjust this rate to avoid problems such as extreme aging of the population.

The so-called low fertility level is less than 2.1 or a net reproduction rate of less than one child per woman. The so-called reasonable level means that the low fertility rate and the aging of the age structure that it causes should not be too serious and can be borne by the society. The above-mentioned medium plan predicted that the total fertility rate would fluctuate between 1.75 and 1.83 between 2000 and 2020, and remain at 1.80 after 2020. If the fertility rate is higher than the replacement level then the surplus population problem will not be solved. For example, the total fertility rate predicted by the high plan from 2000 to 2005 was 1.90, from 2005 to 2010 it was 2.00, which is below replacement level; but from 2010 to 2020, it would be 2.13 and from 2020 to 2050, it will be 2.16. The total fertility rate will thus be higher than the replacement level for 40 years. The total population will grow slowly and continuously in the long term and will not stop completely until 2100. It should be noted that it was not easy for China's total fertility rate to be made to drop to the current level. If the low fertility level cannot be stabilized and the fertility rate returns to above replacement level, then the opportunity to solve the problem of overpopulation will be lost.

The fertility rate should be adjusted at the appropriate time. To achieve zero population growth and long-term, all-round optimum population goals, the low fertility level must be maintained. However, this rate is not static, and it is not the simple case that the lower the fertility rate, the better. The medium plan predicted that the fertility rate would rise from 1.75 to 1.83 before 2020, and then remain at a relatively stable 1.80, with some minor ups and downs. Why the slight rebound in the early stage? First, there was a short period of fertility growth in the mid-1980s, and when the population born in that period reached parental age, between 2007 and 2012, the fertility rate would rise slightly. Second, in 2013, a new policy was introduced which allowed married couples to have two children if one parent was a single child. And in 2015, the across-the-board implementation of the two-child policy was introduced. Although its impact is not very significant, it will have some influence, and there is a possibility that the fertility rate will rebound slightly. After 2020, the impact of the policy change may be further weakened, and a relatively stable fertility rate is expected. This will establish a more reliable foundation for achieving an all-round ideal and optimum population in the future. On the question of whether policy should encourage people to have children, I believes that, at least at present, such a policy is not desirable. What is needed is a population policy linked to the strategic goal of population development, coordinated with social and economic development goals, and which promotes a long-term, all-round optimal population size.

8.1.2 New Trends in Population Capacity

There is a Chinese saying that “the poorer, the more children; the more children, the poorer”, which is not very scientific but from a cost–benefit perspective on children, is not unreasonable because the low marginal cost of children due to the poverty of parents leads to the birth of more children, which in turn hinders the improvement of population capacity, the transition to fewer and better births and the improvement of the family economy. It constitutes a vicious circle of “poverty–more births–poverty”. Transforming this into a virtuous cycle of “fewer children–better education–fewer children” is an effective way to improve the quality of the population.

Scholars have different ways of defining population capacities, which has been discussed in Sect. 5.1. I am in favour of the view that population quality contains three elements: physical fitness, education achievements and civic awareness. When China’s per capita GDP has entered the medium and high-income levels, it is necessary to advance the cultural and moral fitness of population too. It not only reflects the population’s quality in one of its aspects, but also is essential to building a modern, civilized society that is productive, affluent, and has good environmental practices. The “three elements” of China’s population quality have greatly improved, transitioning to a new stage.

8.1.2.1 Physical Health Improvements

Over the past 40 years, the physical health of the Chinese population has greatly improved. In 2008, China successfully hosted the 29th Olympic Games. The scale of the Games, the performance of the athletes, and their good management were an improvement on previous Olympic Games and won praise both at home and abroad. Chinese athletes were highly competitive and won the highest number of gold medals, in sharp contrast with its “sick man of East Asia” image of 70 years earlier. The Olympic Games are a microcosm that reflected the improvement of the physical health of all Chinese. According to estimates of the Population Division of the UN Department of Economic and Social Affairs, the world average infant mortality rate from 2005 to 2010 was 52 per 1,000 people, ranging from 7 per 1,000 in developed countries, to 57 per 1,000 in developing countries and 31 per 1,000 in China. The data of China’s sample survey has dropped below 10 per 1,000. It is more than 21 per 1,000 lower than the world average and 26 per 1,000 lower than the rate in the developing countries. The world average life expectancy estimates at birth is 66.5 years, ranging from 76.2 in developed countries, to 64.6 in developing countries, and 72.6 in China.[3] This is 2.23 years lower than the 2010 China Statistics Survey figure of 74.83, but even so is 6.1 years higher than the world average, and 8 years higher than the developing countries. In sum, the physical health of the Chinese population is not only higher than that of the developing countries but is also higher than the world average and moving closer to the level of the developed countries.

[3] U. N. Population Division (2009).

According to the 2006 UNDP's Human Development Index, China's ranking has risen significantly above the average. By the middle of the twenty-first century, the physical health of the Chinese is expected to be one step ahead of its social economy and to have reached the level of moderately developed countries.

8.1.2.2 Advances in Science and Education

To realize the transition from a big country in population terms to a country powerful in human capital, improving people's knowledge and education is key. Human capital is the sum of human knowledge, skills, experience, and health. The formation of these values cannot be separated from the improvement of the population's intelligence and scientific and education capacities. Where does the knowledge come from? Viewed from the standpoint of inter-generational transition, knowledge is the sum of the practical experiences of our predecessors. Standing on the shoulders of our predecessors, we absorb the essence of their achievements, mainly through education. Science and technology progress constantly, but if new science and technology are to be mastered by more people and the country reach the forefront of science and technology development, then this too will be realized through education. Although improvement of skills, experience and health mainly depends on practice, it is also inseparable from the development of education and science. These characteristics will be continuously improved in a "practice– theory– practice" process, and theoretical and technological innovations are the signs of the improvement of scientific knowledge and education achievements. They cannot be obtained simply through theoretical education, but must be consolidated, improved, and innovated through practical application. And after testing, identifying, and evaluating, the appropriate professional and honorary titles can be awarded. Scientific knowledge and education achievements are the core and backbone of the "three elements" of population quality. With the use of specific assessment indicators, it also receives the most attention.

Scientific and education capacities of China's population have improved significantly in the past 40 years. Comparing a 1982 census with one in 2010, for example, the number of people with some college education (all those who took up studies, whether they graduated or not) in 1982 was 6,037,279 (0.6% of the total population). In 2010, the number of people with a university education (including associate-degree students, undergraduates, postgraduate) was 118,374,897, representing 8.9% of the total population of 1,332,810,869. In 1982, the number of illiterate and semi-illiterate people over 12 was 237,720,787, 23% of the total population. In 2010, the number of people over 6 years old and not attending school was 62,136,405, 4.7% of the total population.[4] The proportion of illiterate and semi-illiterate people has dropped significantly, and the proportion of people with a higher education certificate has

---

[4] Population Census Office under the State Council and Department of Population Statistics, State Statistical Bureau, PRC (1985).

Population Census Office under the State Council, Department of Population and Employment Statistics, National Bureau of Statistics of China (2012).

risen sharply, which indicates that the population's improvement in education level in the past three or four decades is unmatched by other historical periods. The years of education per capita increased from fewer than five to more than eleven, which is close to the level of developed countries. The history of the world's backward countries catching up economically with the advanced countries shows that when the per capita GDP of the former reaches 40% of the latter, the per capita education years will reach about 70% of the advanced countries; when the per capita GDP of the less advanced countries reaches 80% of the advanced countries, the per capita length of education will reach the level of the advanced countries. Education is the foundation of this growth. With improvements in people's education levels, progress in science and technology inevitably rises.

Since the reform and opening-up policies, China's high-tech development has attracted much attention. Satellites and manned spaceships have been launched, the Jiaolong submarine operates in the deep sea; cloud computing, high-speed rail, large aircraft, hybrid rice varieties, 5G mobile phones, and so on, are among the best in the rapid development of high-tech in the world or squeeze into the advanced ranks. We can add to these some of the world's top infrastructure projects, the "big country craftsmen" and "big country heavy equipment" such as metallurgy, chemical industry, transportation, and machine manufacturing. China's foreign trade export structure gives us a glimpse of the bigger picture. Between 1995 and 2016, industrial manufactured goods exports increased nearly 15-fold from $127.3 billion to $1992.5 billion; high-tech product exports increased 58.8 times from $10.1 billion to $604.2 billion; and primary product exports increased 3.9 times from $21.5 billion to $105 billion.[5] The export growth rate of high-tech products is much higher than that of manufactured products and primary products, indicating that the transformation and upgrading of export trade is accelerating.

However, China should not be intoxicated by these achievements. There is still an obvious education, science and technology gap between China and developed countries. We have reached an advanced level and even played a leading role but we are still in transition to a so-called developed country. What can play a leading role is local; in most areas, development is still low or only at the half-way mark. Therefore, first, the fields with advanced development should play the leading role to advance towards higher goals, and second, the weak links must be strengthened, and attention paid to all-round development; scientific and education capacities must be comprehensively improved. The foremost task is to focus on basic education, then prepare for further advances.

8.1.2.3 Cultural and Moral Fitness Improvements

In breaking through the limitations of the previous definitions of political or moral awareness as discussed in Sect. 5.1 and to advance the concept of cultural and moral fitness, we make the following three considerations.

[5] National Statistics Bureau of the PRC (2018).

First, core values. Core values are those values held by a country in a certain historical period that dominate over other social values, and are recognized by the vast majority of the population. In relation to human reproduction, the core value is the coordinated development of population quantity, quality and structure—improvements not only for the good of those living now, but also development that does no harm to the development capacity of future generations. The core value is in line with the trend of historical development and is therefore a progressive one. Recognized by the majority, this value is the ideal embodiment of the people's will. This value is crucial because without it, any country or nation will have difficulty forming strong national cohesion and be able to stand on its own among the countries of the world. On the contrary, if this core value is deeply rooted in society then national cohesion will be enhanced, and the path to development will be smoother.

The second consideration is moral norms. As mentioned above, at present, China is in a new historical era, building a well-off society—per capita GDP is moving towards $10,000. It is also a new historical period in which new and old morals collide and a new morality remains to be built. People's consciousness often lags behind objective reality, and moral construction often lags behind social and economic development. In addition, corrupt people pervert the law for bribes and greedy people abandon morality for profits, and we can observe even in common everyday life numerous instances of moral decay. Examples include people talking "dirty" through mobile phones loudly in public, disregarding those around them; people driving on the roads without regard for other drivers; well-dressed people tossing litter in public places; people stepping on other people's feet then accusing them of carelessness. Such small acts reveal spirit, and subtle details reflect morality. Various social surveys show that with the change to a market economy and the reorientation of people's values, traditional concepts such as respect for the elderly, being polite and courteous, and being honest and friendly are being challenged. If we put personal gain first then we cannot correctly treat the relationship between rights and obligations, between dedication and demand, and thus we commercialize and marketize interpersonal relations. The construction of advanced moral norms that absorb the essence of traditional morality and yet also conform to modern development is key to building a harmonious society and improving the quality of the population; it is a social project that requires long-term persistence and struggle.

The third element is awareness of the rule of law. Establishing and perfecting a market economic system requires the legalization of economic operations. A harmonious socialist society requires strong democracy and the rule of law to promote social fairness and justice. To legalize and standardize political, economic, cultural, and social life, all citizens must be equal before the law, their rights and freedoms protected by the law, and they should have a sense of responsibility for maintaining and supervising judicial justice and integrity, and be constantly aware of the rule of law.

It is a long and arduous task to improve cultural fitness and civic awareness among a population. The focus is to integrate the education of national core values into the national education system and adhere to the principle of moral education first. Developing commodity production and establishing a sound market economy and an appreciation of exchange value is a challenge to traditional interpersonal relationships. This cannot but be a shock to traditional feelings, family ties, and friendship. It should be noted, too, that more than one generation has been born since 1980 when the country vigorously advocated the one-child policy. Growing up in the environment of reform and opening-up, the only child of this generation enjoys obvious advantages, including emancipation of the mind, wide horizon, and strong sense of modernization. However, with no brothers or sisters, and with much love from parents and grandparents, the only child also tends to be lonely, arrogant, and spoiled. Therefore, moral education is more important. It is necessary to sum up experience, for example, for education for the all-round development of morality, intelligence, physical fitness, aesthetics, and job skills, and education for workers in socialist consciousness and culture. It is also necessary to face the development realities of the current era, learn from the successful experience of the international community in carrying out moral education, and carefully incorporate national core values into textbooks for primary and secondary schools, so as to cultivate a new generation—with ideals a generation that loves the motherland, loves the people, respects the old and cares for the young, and is honest and friendly. To improve the cultural and moral fitness of population, priority should be given to moral education.

8.1.3 New Trends in Population Structure

The population structures most closely related to population policy are age, gender, the urban–rural distribution, regional distribution, and employment. Some of these, including the age structure, directly determine reproduction. There are clear differences in fertility and birth rates between urban and rural areas, and in population quality, especially in education levels. Therefore, the impact of the rate of urbanization on population change cannot be underestimated. The employment structure of the three industries is closely related to the urban–rural distribution of population, and is the result of change in economic structure and social structure. It also limits urbanization, economic development, and social development. Therefore, an integral part of population policy must be to examine how the above-mentioned population structures change, under what conditions, and policy choices. Unfortunately, they are generally not directly included in population policy considerations because of their relevance to other areas. Yet their role and influence cannot be ignored, because they are sometimes greater than that of the general population policies.

8.1.3.1 New Trends in Age Structure

The age structure of the total population forms a pyramid, at the top of which are children, aged 0–14 years, then adults aged 15–59 or 64 years, and at the bottom, people aged 60 or 65 years or more. The employment/unemployment structure of the three industries is formed when population is connected with labour and employment; the school population and education level structure is formed when population is connected with education; and the urban–rural and regional population structures are divided by where people live: in cities or rural areas or in which region. The combination of each population category and other categories forms a specific population structure. In this way, the population age structure has penetrated all fields of economic and social life. Various industries have set different recruitment and retirement policies based on age. There is a saying now that "age is a treasure, and a diploma is indispensable" because all industries have set the ages for starting employment and for retirement. For civil servants, it is "seven up and eight down": at the age of 57, one can still be promoted and appointed, because there are still three or four years before retirement at the age of 60, and it is not a problem to finish one term; at the age of 58, without possibility of being promoted, one can but wait for retirement. The retirement age varies greatly among different types of work, but generally, people retire at the age of 60. In industries involving heavy physical labour such as mining and construction industry, workers must retire at the age of 50. Thus changes in the population age structure affect all aspects of the social and economic structures, creating a situation of great diversity.

Trends in population age structure form the basis for the shape of other age structures. Predictions of the dynamic trends of China's population age structure in the first half of the twenty-first century are shown in Figs. 8.2–8.4.[6]

In the above-mentioned population age structures, the proportion of the juvenile, adult and elderly populations is constantly changing. There is a veritable pyramid structure in 2000, where the proportion gradually decreases with the increase in age. In 2020, the pyramid is upright, and the upper part is neither as sharp nor narrow as in 2000 but becomes blunter. By 2050, the middle and upper part of the pyramid become thicker and shorter, and the bottom half is more like a pillar, and slightly narrower. The trend in age structure can be summed up as an aging of the population. What are the predicted trends and characteristics of China's population aging in the twenty-first century?

The previous high, medium, and low predictions are shown in Fig. 8.5.[7]

Figure 8.5 shows that the aging of China's population in the first half of the twenty-first century will be relatively rapid and relatively high, characterized also by a phased and progressive nature in time, and an obvious imbalance in the spatial distribution. These are explained as follows.

First, the rate of aging is relatively fast, and the level it reaches is relatively high. Before 2050, the absolute number of those over 65 years old is fixed: i.e., it is the number of people aged from 28 to 64 years old still alive in 2013 after subtracting the annual age-specific death figure. However, the factors that affect the degree of aging include the birth rate, the number of births, and the prolongation of life expectancy. We still choose the change in the proportion of those over 65 years old to explain the speed and degree of aging. In this way, according to the above predictions, in

---

[6] Based on the data in Tian et al. (2007).

[7] Tian et al. (2007).

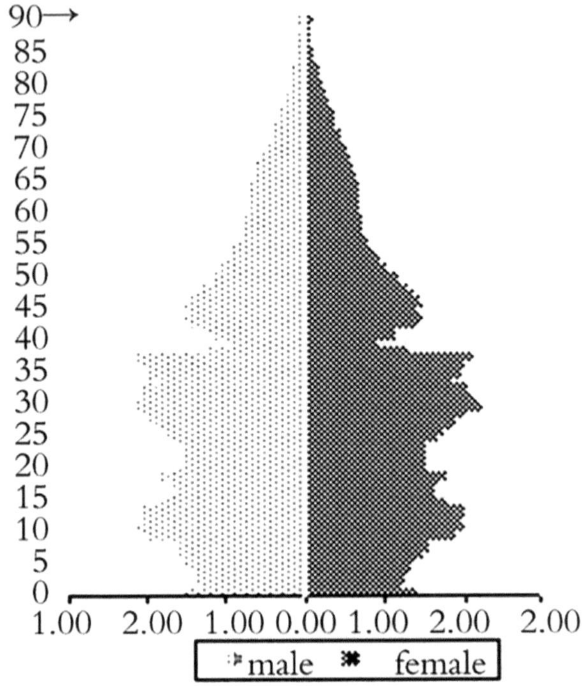

Fig. 8.2 Age structure of the Chinese population in 2000

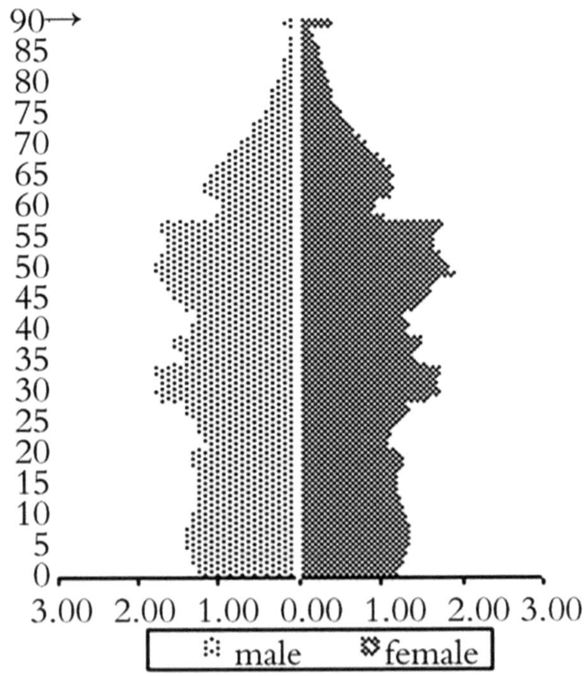

Fig. 8.3 Predicted age structure of the Chinese population in 2020

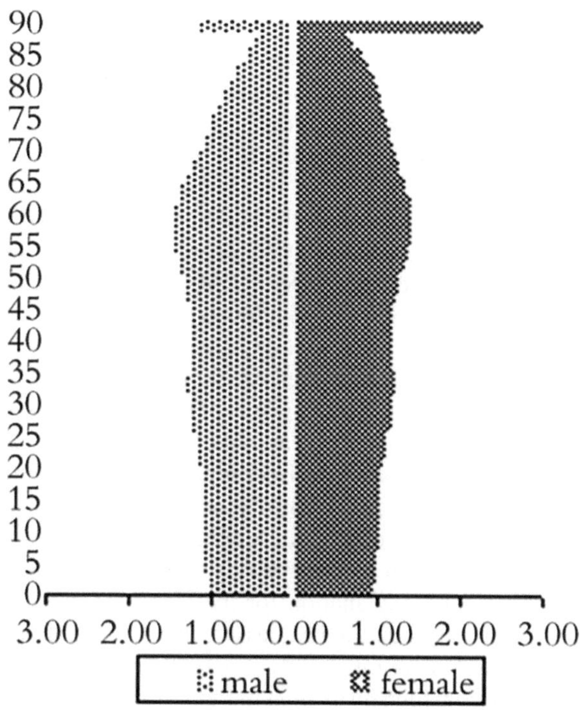

Fig. 8.4 Predicted age structure of the Chinese population in 2050

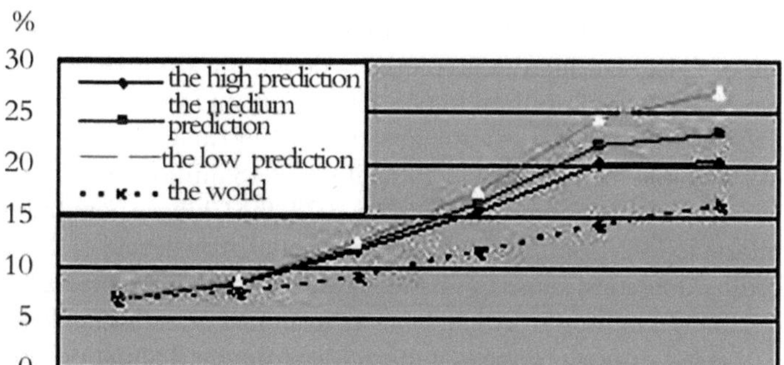

	2000	2010	2020	2030	2040	2050
the high prediction	6.92	8.5	11.66	15.36	20.1	20.23
the medium prediction	6.92	8.59	12.04	16.23	21.96	23.07
the low prediction	6.92	8.7	12.54	17.4	24.48	27.14
the world	6.8	7.6	9.3	11.7	14.2	16.2

Fig. 8.5 Trends in population aging in China and the world from 2000 to 2050

2050, the proportion of the elderly over 65 years old will reach 20.23% in the high prediction, 23.07% in the medium, and 27.1% in the low prediction. Considering that the across-the-board implementation of the two-child policy has been launched, the possibility of aging along the medium prediction is further increased. Based on this comparison, the progress of China's population aging can still be summarized as "fast" and "high". "Fast" means that from the turn of the century when population aging begins to appear, to the middle stage of aging in 2023 when the elderly reach 12.3%, will take about only 23 years while it will take more than 30 years for the world as a whole, and more than 40 years for developing countries. It will take 32 years for China's elderly population proportion to increase from 7 to 17%, and it will reach 17.5% in 2032. However, developed countries as a whole took over 50 years, until 2015, to reach this level, which is about twice as long as China. Japan is a special case. The proportion of people over 65 there increased from 7.1% in 1970 to 17.2% in 2000, which took 30 years, like China. The so-called high scenario means that the degree of aging is relatively high. In 2050, the proportion of China's population aged over 65 will reach 23.07%, nearly seven points higher than the world rate of 16.2%, more than eight points higher than in the developing countries (14.6%), and only three points lower than the overall developed country figure of level of 26.2%.[8]

The second explanation for the characteristics shown in Fig. 8.5 is that aging occurs in stages and is progressive. This is mainly caused by past natural changes in birth and death rates. Between 1949 and 1952, the population reproduction type changed from high birth, high mortality, and low increase to high birth, low mortality and high increase. During the first baby boom from 1953 to 1957, the national population increased from 574.82 million to 646.53 million, an average annual growth rate of 2.4%. During the first fertility downturn, from 1958 to 1961, the population grew from 646.53 million to 658.59 million, an average annual growth rate of only 0.5%. Population growth was negative in 1960 and 1961: in 1960, it was 10 million less than that in 1959, and in 1961, it was 3.48 million less than that in 1960. During the second baby boom from 1962 to 1973, the Chinese population went from 658.59 million to 892.11 million, an average annual growth rate of 2.6%, and in the second fertility downturn since 1974, the national population increased from 892.11 million to 1,360.72 million in 2013, an average annual growth rate down to 1.1%.[9]

It is not difficult to see that due to the different fluctuations in the birth, mortality, and natural increase rates in these five stages, an uneven age structure formed, and this will have an irregular impact on the degree of population aging in the future. Among them, the greatest part of the population age structure pyramid is the number of people born during the second baby boom from 1962 to 1973. After deducting the deaths during these years this cohort still numbers about 300 million people. This is a special phenomenon that deserves serious attention: Between 1977 to 1988, this population group passed the age of 15, and the vast majority of them joined the labour force between 1980 and 1991, creating huge pressure on the employment market. At the same time, a "golden age" of population age structure change began,

[8] For the data outside China, see U. N. Population Division (2009).

[9] Population Statistics Department of the National Statistics Bureau of the PRC (1988).

in which there was a high proportion of working-age people and a low proportion of dependents, whether old or young. It became the optimal period for "demographic profits" and "demographic dividends", which are conducive to economic development. According to the peak value of age-specific fertility rate of women in urban and rural areas of China, which is at the age of 24, there should have been a baby boom from 1986 to 1997. However, due to the continuous strengthening of population control and the implementation of the basic national policy of family planning, there has not been a baby boom that lasted more than 10 years in practice, with only a slight manifestation in the mid-to-late 1980s. Its impact is that population aging will not be too serious before the group of 300 million population has reached the age of 65; but when this large population group transits to old age, a serious "flood" of aging will arrive. This gives aging, like urbanization, the characteristic of an S-curve growth phase.

The first phase, from 2000 to 2020, is at the bottom of the S-curve, when the aging is in the slow-rise stage. The proportion of those over 65 years old increases from 6.96% to 12.04%, at an annual average rate of increase of 0.25%.

In the second phase, from 2020 to 2040, in the middle of the S-curve, the aging trend is accelerating; the proportion of the elderly increases from 12.04% to 21.96%, at an annual average of 0.5%, double the rate of the first phase.

The third phase, from 2040 to 2050, is the top of the S-curve, where aging is only slowly rising and basically stable. The proportion of the over 65 year-olds increases from 21.96% to 23.07%, at an annual average rate of 0.11%. This is the period with the slowest growth rate of the three phases. After 2050, although the proportion of the elderly may still rise, the increase is extremely limited.

The third explanation for the characteristics shown in Fig. 8.5 is that the spatial distribution of aging among urban and rural areas and regions is unbalanced. The characteristics of the aging process discussed above apply to the country as a whole. However, China is a vast territory with very different geographical conditions, and highly uneven economic, technological and social development. More directly and more importantly, the natural changes in births and deaths of the population in the past were quite different, resulting in the obvious differences in population aging between urban and rural areas and among regions.

The distribution of aging population in urban and rural areas is unbalanced. At the beginning of the 1980s, the age structure of China's population entered the period of the adult type. The 1982 national census showed that the proportion of people over 65 years old had reached 4.91%, signaling the transition to the elderly type. However, there were clear differences between urban and rural areas: the proportion of those over 65 in that year was 5% in the counties, 4.68% in the cities, and 4.21% in smaller towns. In the counties (the rural areas) their proportion was higher, and so the transition to the elderly type was happening at a faster rate. This trend continued: the 1990 census showed that the proportion of the national population aged over 65 had risen to 5.57%, to 5.64% in the counties, 5.53% in the cities, and 5.49% in towns. The gap between the locales had slightly narrowed since 1982. Data from the fifth census, in 2000, showed the proportion of the elderly at 6.96%, and an expanded gap between the counties, cities and towns: 7.74% in counties, 7.00% in cities, and

6.25% in towns. Comparing these three censuses, we see that the proportion of over 65 year-olds in counties, cities, and towns first narrowed from 1.00–0.94–0.84 to 1.00–0.98–0.97, then expanded to 1.00–0.90–0.81. The 2010 census showed that the expansion continued, in a new, uncharacteristic situation in which smaller towns outnumbered cities. The proportion of the national population aged over 65 rose to 8.91%, with 10.06% in counties, 7.97% in towns, and 7.69% in cities. The county to town to city ratio changed, expanding to 1.00–0.79–0.76.[10]

Why has the gap in the spread of the elderly population in urban and rural areas widened and the ranking of aging population in cities and towns changed? The main reason is the continuous increase of floating and migrating populations, especially since the beginning of this century. According to the 2000 census, the number of national migrants living in the local area for more than half a year reached 144,390,748. After deducting the population of other streets and towns in the county (city), the remaining 92,870,183 people are largely a floating population from other towns or cities, or rural areas of other provinces. In other words, the vast majority of these 90 million people have moved from rural areas to cities and towns. In the 2010 census, the population separated from their registered households rose to 260.94 million, a substantial increase on ten years earlier. At present, this trend continues. According to the data provided by the National Bureau of Statistics, in 2017, the number of people not living in their registered households reached 291 million, and the floating population reached 244 million.[11] When people in these floating populations enter cities and towns and settle for more than half a year, they are counted in the permanent resident population of those cities and towns, which lowers the age profile especially in cities, and widens their demographic distance from the rural areas. According to the natural population growth calculation, on 28 June 2011, the permanent resident population in urban areas exceeded that in rural areas for the first time, marking a new transition in urban–rural population structure. The direct impact on the age distribution in urban and rural areas has been to increase progressively the proportion of elderly in the rural areas. Moreover, as most of the migrant workers are young or middle-aged, and most go to cities instead of smaller towns, so population aging in smaller towns becomes higher than in the cities, in contrast to the previous situation where it was the inverse. It is predicted that when the urban population reaches 60% in the 2020s, the rate of rural migration to cities and towns will slow and the scale will be reduced. This will create conditions for narrowing the age gap among counties, cities and towns, and then the speed of aging in urban areas, especially big cities, will accelerate.

[10] Calculated according to data from: Population Census Office under the State Council and Department of Population Statistics, State Statistical Bureau, PRC (1985).

Population Census Office under the State Council and Department of Population Statistics, State Statistical Bureau, PRC (1993).

Population Census Office under the State Council and Department of Population, Social Science and Technology Statistics, National Bureau of Statistics of China. (2002).

Population Census Office under the State Council, Department of Population and Employment Statistics, National Bureau of Statistics of China (2012).

[11] National Statistics Bureau of the PRC (2018).

The distribution of aging in different regions is uneven. The above-mentioned population flows not only affect the rates of aging among cities, towns, and counties, but also have a significant impact on the age structure and the in-depth development of aging between regions. A clear sign is that the proportion of elderly has gradually shifted from the original "three plates" of the eastern, central, and western regions to a pattern of "two plates" because of the integration of the central and western regions. This change began in 2005, when the proportion of the national population aged over 65 was 9.07%, though the proportion was higher in 13 provinces and municipalities, of which seven were in the eastern coastal areas (accounting for 53.8%), three were in the central areas (accounting for 23.1%), and three were in western areas (accounting for 23.1%). There are six provinces, autonomous regions, and municipalities in the central and western regions (accounting for 46.2%). There were 18 provinces and autonomous regions below the national level, of which four were in the eastern coastal area (accounting for 22.2%), five were in the central area (accounting for 27.7%), and nine were in the west (accounting for 50%). The central and western regions account for 77.8% in total, which are areas with a low degree of aging.[12] In the 2010 census, the situation changed again: the proportion of those over 65 years old in China increased to 9.4%, and the proportion of the elderly in the central and western regions further increased. This trend continues. At present, although the gap between the eastern, central, and western regions remains, the gap between the central and western regions has narrowed. The trend of "two plates", of the eastern region on one hand and the central-western region on the other, has begun to appear, which may become more and more obvious.

8.1.3.2 New Trends in Sex Ratio at Birth

When we examine the gender structure of the population, we should pay special attention to the sex ratio at birth, which has basic and original significance for the sex ratio of the total population. The subsequent development of the sex ratio is the cumulative result of the sex ratio at birth in each year.

Sex ratio at birth refers to the ratio of male to female infants born alive within a given period. Statistics show that most countries in the world have a sex ratio at birth between 103 and 107, and only a few countries have a sex ratio lower than 103 or higher than 107. China has a long history of having a higher sex ratio at birth than this, but after the founding of the PRC, it rapidly dropped to close to the normal value, and from the 1950s to the 1970s, it fluctuated around the normal value or just slightly higher. At the beginning of the 1980s, it showed a continuous upward trend, and at the beginning of the twenty-first century, it reached a record of 120. After that, it fell back slightly and oscillated at a higher level. The rise of the sex ratio at birth and then its slow decline in recent years, coupled with the difference in the mortality rate between men and women by age, resulted in the accumulation of a certain population sex ratio in different age groups. The sex ratio of the population

[12] National Statistics Bureau of the PRC (2006).

aged 0 to 49 years provided by the third to sixth censuses showed the growth of the sex ratio of different age groups, especially the lower ones, as set out in Table 8.1.[13]

Table 8.1 shows that in 2010, the sex ratio of the population aged 20 to 29 in the period of vigorous marriage and childbirth was relatively low, with only 1.27 million more males than females, creating no need for concern. However, the sex ratio of the population aged 0 to 19 rose from 105.6 in 1982 to 107.6 in 1990, 111.3 in 2000 and 114.8 in 2010. This is an increase of 9.2 basis points in 2010 compared with 1982. In terms of the absolute number of males and females aged 0 to19, there were 12.69 million more men than women in 1982, and 22.19 million more men than women in 2010, an increase of 74.9%. These 22.19 million males will gradually reach marrying and childbearing age in the future. The first problem will be marriage squeeze. If the rise in the sex ratio at birth drops back to the average level relatively quickly then the extra 22.19 million men will reach marrying age in 20 years, which will not have much impact on the marriage gender balance. However, the fact is that the sex ratio at birth has remained high for a long time and if it continues, the problem will be become more serious and more difficult to solve.

8.1.3.3 New Trends in Urban–Rural Population Distribution

The changes in the urban–rural population structure are mainly the result of the labour migration from rural areas into urban agglomerations. Therefore the focus of research in this area should be on the new trends of urbanization.

The first new trend is the acceleration of urbanization in China. Urbanization is a common trend in China and around the world. From an economics perspective, urbanization refers to a certain spatial distribution of primary, secondary, and tertiary industries. Cities are dominated by secondary and tertiary industries, while rural areas are dominated by primary industries. Sociology treats urbanization as the reorganization of modes of production and life-styles in urban and rural areas. From a demography perspective, it refers to the distribution of the population across urban and rural areas. Regardless of the disciplinary angle, urbanization is the result of productivity development, social progress, and changes to population structures. This is a universal and common law of development and change.

Most of the highly urbanized countries in the world today are developed nation-states that industrialized early and are at the forefront of industrialization and modernization. Some of these are city-states like Singapore. Some Latin American countries have a relatively high rate of urbanization, falling into a state of blind urban development called the "Latin American Trap" of urbanization. But these are special cases. At present, the urbanization rate in developed countries such as the United

[13] Calculated according to data from: Population Census Office under the State Council and Department of Population Statistics, State Statistical Bureau, PRC (1985).

Population Census Office under the State Council and Department of Population Statistics, State Statistical Bureau, PRC (1993).

Population Census Office under the State Council, Department of Population and Employment Statistics, National Bureau of Statistics of China (2012).

Table 8.1 Sex ratio of some age groups in the 1982, 1990, 2000 and 2010 censuses (10,000 people)

Age group (years)	1982			1990			2000			2010		
	Male	Female	Sex ratio	Male	Female	Sex Ratio	Male	Female	sex ratio	male	female	sex ratio
0	1079	1002	107.6	1226	1097	111.8	746	633	117.8	746	632	118.0
0–19	23,765	22,497	105.6	22,452	20,865	107.6	20,418	18,338	111.3	17,170	14,952	114.8
20–29	8563	8330	102.8	11,775	11,228	104.9	10,817	10,401	104.0	11,485	11,358	101.1
30–49	11,740	10,562	111.2	14,747	13,556	108.8	20,768	19,554	106.2	22,730	21,821	104.2

States, the United Kingdom, and Canada exceeds 80%, and in Japan it exceeds 90%. The rate of urbanization in developing countries is relatively low, and the current average global urbanization rate is about 55%. Scholars of the history of urbanization in industrialized countries have largely reached a consensus about the paths and stages of development, and developed an S-curve, three-stage theoretical model of it. In contemporary China, the urbanization rate in the first stage is below 25%. This is called the primary stage of population urbanization, which mainly involves the transfer of rural populations to small towns. The urbanization rate in the second stage is between 25 and 65%, which can be called the intermediate stage. Here, two phenomena work in tandem: the transfer of rural populations to cities, and the transfer of small-town populations to big cities. In the early period, the main form is the transfer of rural population to cities; while in the later period, the main form is the transfer of small-town population to big cities and mega-cities. In the third stage, the urbanization rate goes above 65%, and while the overall urban population continues to increase, counter-urbanization occurs in which people living in the central areas of large cities move out to the suburbs and rural areas. This is, of course, a rough theoretical model. In reality, the urbanization process in each country is more complicated, and there may appear some repetitions and fluctuations at different stages. China's urbanization generally conforms to the three-stage theoretical model, but it also has some distinctive features due to its special national conditions. According to the three-stage model, the track of China's population urbanization process can be summarized as follows.

When the PRC was founded, the rate of urbanization was at just 10.64%. Between 1949 and 1960, it progressed smoothly, and the proportion of the urban population increased to 17.98%, an annual average increase of 0.67%. From 1960 to 1987, it first slowed down then increased, and then entered the fast lane in the reform and opening-up era. After 1960, it first retreated for 12 years, and then fell to 17.13% in 1972. This was followed by an eight-year compensatory increase, and it recovered to 19.39% in 1980, which was close to the 1960 level. Such 20 years of turbulence are rare in modern societies. After 1980, under the tide of reform and opening-up, the pace of urbanization accelerated, rising to 25.32% by 1987, thereby completing the first stage of primary urbanization. From 1960 to 1987, the urbanization rate went from 17.98% to 25.32%, an average annual increase of 0.27%; at the end of this period, from 1980 to 1987, the urbanization rate increased from 19.39% to 25.32%, an average annual increase of 0.85%. It can be said then that urbanization entered the fast lane at the beginning of the reform and opening-up period, ushering in a new era of accelerated progress.

Since then, in the 30 years from 1987 to 2017, population urbanization increased from 25.32% to 58.52%, an average annual increase of 1.11%. It became the period in which population urbanization grew fastest in China. Urbanization in China has been lower than in the developing countries for a long time, and much lower than the level across the world and in the developed countries. After 40 years of arduous efforts, it reached a new level, of 7% higher than that of developing countries and

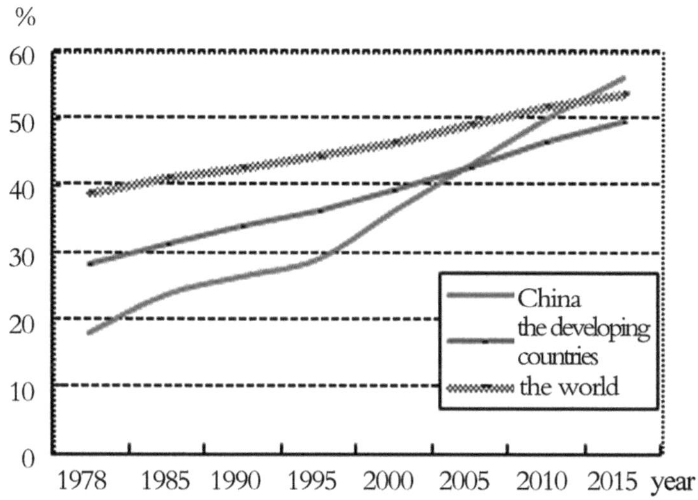

Fig. 8.6 Comparison of urbanization processes in China and the world from 1978 to 2015

3.5% higher than that of the world, placing it between the world and developed countries. See Fig. 8.6.[14]

Another new trend is featured by "big city leading". "Big city leading" here has two meanings. First, the proportion of cities that are mega-cities is on the rise, and people in small and medium-sized cities are tending to shift to big cities. Second, in the process of urbanization, big cities play an increasingly obvious central and dominant role. For a long time, China has pursued an urbanization policy that involved actively developing small towns, appropriately developing medium cities, and strictly limiting the scale of big cities. There seems little doubt that the policy was appropriate to China's realities, in line with the basic national conditions of a large rural population, a low degree of urbanization, and underdeveloped productive forces between the founding of the People's Republic to the early stage of the reform and opening-up period. In the early 1980s, of one billion people in China, the rural population was 800 million. Under such circumstances, a policy of "focusing on small cities", with less investment and quicker results, is conducive both to the transfer of rural surplus labor to small towns dominated by township enterprises, and to accelerating the process of urbanization. However, under certain conditions, all things can change direction. "Small-oriented" urbanization blossomed everywhere. While enabling rapid economic growth in urban and rural areas, it also accumulated and highlighted problems such as resource consumption, environmental pollution, and low efficiency. As a result, small towns have become big problems and this "small-oriented" urbanization is difficult to justify. By the middle of 1990s, when the urbanization rate reached

[14] Data Source:

U. N. Population Division (1989); Official website of the National Bureau of Statistics of China: *Statistical Bulletin on Social Development of the National Economy in 2018*, 2019. 03. 03.

Table 8.2 Changes in urban population structure from 1982 to 2010

Cities of different sizes (10,000 people)	Data from 1982 census	%	Data from 2010 statistics	%
Above 400	21,064,749	14.50	66,698,055	20.01
200–400	22,695,533	15.63	39,441,406	11.83
100–200	31,622,467	21.77	53,341,642	16.00
50–100	33,214,313	22.87	63,199,719	18.96
30–50	19,514,888	12.74	46,717,000	14.02
10–30	16,739,426	11.52	59,920,895	17.98
Below 10	1,401,695	0.97	4,012,822	1.20

30%, the policy was gradually changed to the "big-oriented" track. For a comparison of the urban population structure divided by non-agricultural population, see Table 8.2.[15]

Many indicators exist with which to measure the size of a city based on different levels of economic and social development. Categorization based on population size is a widely adopted measure, but there is no unified or fixed standard, due to the large differences between countries in population size, social and economic development, and the constant change in population size and scale. Combining the characteristics of China with the classification standards widely used in the international community, the following five types of city classification and corresponding indicators are proposed: small cities, with resident populations of fewer than half a million; medium-sized cities, with resident populations between 500,000 and a million; big cities, with resident populations between one and two million; large cities, with resident populations of two to four million; and super cities, with resident populations over four million. Given our original intention of distinguishing urban and rural populations based on their actual residence and the nature of the work they do, it is appropriate to focus on the resident population rather than on the division between the agricultural and non-agricultural populations.

In addition, many provinces and cities are implementing the household registration system reform, the main aim of which is to abolish the old concept of agricultural/non-agricultural population, and to introduce a unified resident identity card system for the regular population. Therefore, it is necessary to list the city scale according to the permanent resident population and to investigate the changing trend of the city scale. According to the change of city scale divided by permanent population, the population of super cities reached 99,927,015 in 2010, accounting for 15.68% of the total urban population; the population of large cities reached 77,199,472, accounting

[15] For the 1982 data (excluding the population of counties under the jurisdiction of the city), see:

Population Census Office under the State Council and Department of Population Statistics, State Statistical Bureau, PRC (1985).

For the 2010 data (non-agricultural population), see:

Department of Population and Employment Statistics, National Bureau of Statistics of China. (2012).

for 12.11%; the population of big cities reached 205,543,489, accounting for 32.25%; the medium-sized cities' populations totalled 194,173,011, accounting for 30.46%, and the number of people in all the small cities was 60,481,091, accounting for 9.49%.[16]

As we see by comparing these figures with those in Table 8.2, the population of super + large + big cities has greatly increased both absolutely and as a proportion of the urban population. Urbanization has reversed from being "small-oriented" to being "big-oriented". Why is this reversion occurring? There are many reasons. The main one is that the division between agricultural and non-agricultural populations involves methods and standards belonging to the past. The main criterion for the distinction related to whether the commercial grains were supplied to the non-agricultural population by the state government. It has long lost its practical significance, is not associated with the actual residence and employment status of residents and has often been subject to significant statistical errors. In 2017, the number of people living away from their official residence was 291 million, of which 244 million comprised a floating population, moving mostly from rural into urban areas. It is no longer reasonable to assume that the urban population is only a non-agricultural population and to exclude rural people who actually live in cities, which could lead to certain deviations.

Since urbanization in China has already entered the second stage, wherein it is dominated by big cities, some corresponding reforms to principles and policies are necessary. The direction of the reforms is to turn peasants who move into urban conglomerations into full citizens, so the new residents are entitled to the same rights as the original urban residents. That is, the division between the agricultural and non-agricultural populations will be abolished, and household registration will be carried out according to a person's current actual place of residence and work. All are registered as permanent residents, issued with resident ID cards, and the territorial management of the population is implemented according to these resident ID cards. At the same time, however, it is necessary to recognize that the leading role of big cities is an inevitable development when urbanization advances to the second stage, and give full play to the functions of big, large and super cities as much as possible. The healthy development of urbanization should be promoted.

Quite rightly, the Decision of the Central Committee of the Communist Party of China on Some Major Issues Concerning Comprehensively Deepening the Reform adopted at the Third Plenary Session of the 18th CPC Central Committee proposed to "promote the citizenization of agricultural transfer population". Indeed, this should be a top priority for urbanization and the starting point of reform.

In the first place, granting the citizenship to the migrant workers of the city where they currently live and work is a real problem to be addressed. Since the reform and opening-up period, the number of people leaving their households has been increasing, and the number of people moving out of agriculture has increased even more rapidly. The wave of internal migrant workers has become higher and higher.

[16] Department of Population and Employment Statistics, National Bureau of Statistics of China (2012).

According to the data released by the National Bureau of Statistics, in 2018, the population whose residence and place of household registration addresses did not match, and who had left the household registration address more than six months earlier, numbered 286 million, of which 241 million were a floating population. Obviously, with so many people in a state of mobility, it is not a long-term solution. In the end, people will either go back to the countryside or they will not; those who opt to stay in the cities and towns should be turned into full citizens as soon as possible.

Secondly, such reform will be required for continued development. Urbanization has an irreplaceable role in informatization, industrialization and agricultural modernization. The key to human-centered urbanization is the integration of industry and urban development. Practice has proved that if the rural migrants who are mainly agricultural laborers are permitted to stay in cities and towns and become full citizens like the original urban residents, then urbanization will drive the economy through increased consumption. In contrast, the "mobile" urbanization[17] mainly promotes economic growth by investment. After constructing buildings, roads, water, electricity and other infrastructure, many migrant workers move on to other places to do the same construction work, playing the role of builders unrelated to urban development. This is one of the main reasons why urbanization has been so important in stimulating investment since the beginning of the twenty-first century.

Thirdly, there are already relatively mature practices and experiences. In recent years, Guangdong and other places have carried out experiments to turn agricultural transfer populations into full citizens, especially in small and medium-sized cities and towns, and these efforts have been easy to implement and relatively successful. The main lesson learned is that equal treatment of citizens is essential and should be promoted and improved step by step. In this regard, *The Decision of the Central Committee of the Communist Party of China on Several Major Issues Concerning Comprehensively Deepening the Reform* provides specific policies and guidelines, and forms a reliable basis for launching and conducting reform.

8.2 Changes to the Costs and Benefits of Children

Since entering the post-demographic transition, and especially since the beginning of the 2000s, China's population has undergone tremendous changes in quantity, quality, and structure. The direct reason lies in the fundamental change to the costs and benefits of having children. The population structure change has occurred with the transition to fewer births, better care, sound child rearing, and better education. In order to understand and control the trends in population change and development

[17] The "mobile" urbanization here means the phenomenon in China that migrant workers from the rural areas only live in cities temporarily when they work there. After their work ends in a city, those workers will either return to their home in the countryside or migrate to other cities to search for new employment.

in this century, we must understand the origin and development of the child cost–benefit theory, and carefully assess the role and impact of such changes on population quantity, quality, and structure.

8.2.1 The Cost–Benefit Theory of Fertility Decisions

Changes in the fertility rate are fundamentally determined by women's biological ability to bear children. However, under normal circumstances, women's actual fertility levels do not reach their upper limit, which can be seen from women's lifetime fertility rate, total fertility rate, general fertility rate, age-specific fertility rate, and so on. On the one hand, there are human reasons for this, especially diseases that hinder fertility, physical strength, and environmental factors. On the other hand, economic development and progress in science and technology and society, accessible contraceptives and the continuous improvement in their quality, have made it easy for people to choose the number of children they want to have. This is precisely why after a small baby boom after the Second World War, the world's fertility rate has been in slow and continuous decline. In essence people have gradually shifted from pursuing children in quantity to pursuing a certain quality of child/ren.

The decline of fertility and birth rates in China since the 1970s is certainly due to the national economic development and social progress. In general, the faster socio-economic development and the higher the level of development, the faster the fertility rate and birth rate decline, and some countries and regions have achieved zero or negative growth. Low-income countries, on the other hand, still have high birth rates and high population growth. Indeed, the continuing population growth in the world today is mainly the result of the rapid population growth in low-income and developing countries. But why do economically developed countries have very low birth rates and worry about their aging populations and labour shortages while developing countries, especially the poorer ones, worry about overpopulation and surplus labour? Obviously, levels of economic, technological, and social development are linked to fertility and population growth. The connection lies in the calculated costs and benefits of having children.

The relationship between the population dynamic and socio-economic development is an interdisciplinary question that attracted attention early in economics research, seen in works from William Petty's *Theory of Taxation* of the classical school to Thomas Robert Malthus's *An Essay on the Principle of Population.* Marx and Engels made different evaluations of some relevant studies at that time, expounding a dialectical materialist relationship between population and economy as "two kinds of production".[18] However, due to the limitations of the times, especially the lack of experience of a long-term continuous decline in fertility, it was impossible to analyse further the question of fertility decline.

[18] The theory of "two kinds of production" has been discussed in Sect. 4.1.

By the end of the nineteenth century, some developed countries were taking the lead in reducing their fertility rates, ushering in a new era in which the fertility rate decreased for a long time after the fall in the mortality rate, providing new data for the development of population science. Western demography, especially population economics, has since flourished, developing population theories about fertility and mortality rates, migration, demographic transition, family, the theory of "optimum population", sociology and mathematics. In his economic theory of population growth, John Maynard Keynes argued that demand for capital depends on three factors: population, living standards and capital technology, and that a decline in the birth and population growth rates will lead to insufficient effective capital demand and hence to economic crisis. American economist Alvin Hansen and others attributed economic stagnation to stagnation of population growth and believed that changes to the population size and the labour force directly determine economic growth. Development economists such as William Arthur Lewis, Simon Kuznets, and Theodore Schultz all attached great importance to the relationship between population and economic growth.

However, before the 1950s, these and similar studies were mainly macro studies of population and economic development, while micro studies were relatively rare. Others, such as Malthusian population theory and optimum population theory, were mainly limited to the macro; some studies were mainly based on the macro, such as the demographic transition theory and the population growth theory; some included both macro and micro, such as Kuznets, Schultz and so on. After the 1950s, the research into the micro aspects of population and economic development grew and some researchers made significant breakthroughs. Among these theories, the theory of fertility rate dominated. On the one hand, fertility is the starting point of population reproduction and thus plays a more decisive role in changes in population size and structure than factors such as changes to mortality rates. On the other hand, changes in the fertility rate are closely linked with economic, technological, and social development, which demonstrates the importance of demography as a interdisciplinary discipline. The development of economics, society and mathematics greatly promoted research into fertility, and offered new theoretical frameworks and operational models. In his *Principles of Economics*, at the end of the nineteenth century, Alfred Marshall established equilibrium price theory and its equilibrium analysis methods to explain production costs and marginal costs. When Adolphe Landry and Warren Thompson developed the theory of demographic transition, they linked fertility rate changes with social development, which had a significant impact on subsequent micro-analyses of fertility. However, the most complete and influential theory in this field is that of the cost–benefit analysis of having children.

In the late 1960s, Harvey Leibenstein was professor of economics and visiting professor of demographic studies at Harvard University. His idea that a cost–benefit analysis constitutes the basis for decisions about whether to have children and how many was elucidated in such works as *A Theory of Economic-Demographic Development* and *Economic Backwardness and Economic Growth* published in the 1950s, and *Super-Economic Man* and *Super-Economic Man: Economics, Politics and Population Issues* in the 1970s. In sum, he argued that family size is determined by parental choices, and the choice whether to have a child or children depends on an assessment of the expected costs and benefits in doing so. The cost of a child are both direct and indirect. The former comprises the money spent on food and, housing, transportation, medical care, education, marriage, and so on from the time of the mother's pregnancy through to the time when the child becomes self-reliant. The indirect costs include salary loss for the mother due to pregnancy and post-natal care, as well as the income lost by both parents (though mainly by mothers) due to missed opportunities for career advancement and promotion, caused by time spent on caring for the child. Thus, the indirect costs are also called opportunity costs. Unlike direct monetary expenditure, opportunity costs occur as an indirect cost of time expenditure.

It is inevitable that families and parents will incur certain costs for having children, but this cannot constitute the motive for having them. Rather, the motive for having a child is to obtain the expected benefits. Leibenstein argues that the expected benefits that may accrue from having a child include six elements.

The first is the labour–economic benefit. Developing labour power after adolescence, the child can thereby provide labour services to the parents and larger family, or take up professional labour and thereby contribute monetary income.

The second is the elder–security benefit. In economically underdeveloped countries, because of the lack of comprehensive social security systems, economic security and personal care for the elderly are largely provided by their children. Children have the potential function of providing security for their elders.

The third is the consumption–pleasure benefit. As a piece of "durable consumer goods", a child fulfills a function of satisfying parental emotional and spiritual needs and can bring benefits such as happiness to the family.

The fourth benefit is that children can assume the risks of the family business. Not only does a child generally have a labour–economic benefit, but also the expected benefit is associated with future family economic risks. Generally, the greater the expected benefit of the child, the stronger the parental capacity to take risks; the smaller the expected benefit of the child, the weaker the capacity to take risks.

The fifth benefit is maintenance of family status. In most developing countries, the "blessings" of having more children are associated with family prosperity, including financial wealth inheritance, maintaining family status and expanding the family business, and assuming family and social obligations, so that the family maintains its social position.

The sixth benefit is to protect family safety. A large population is the basis of family prosperity. More children represent a higher safety factor for a family, while a smaller number of children means a lower safety factor.

Leibenstein showed that the cost of children is positively related to income. Direct costs rise with the increase of income, and indirect costs or opportunity costs also rise with increases in the value of time. Is the same true for the benefits of having children? Except that the change of the consumption–pleasure benefit is not clear and seems not to be strongly linked to socio-economic development, the rest generally show a downward trend. In particular, the marginal labour–economic benefit of having children is declining. With economic development, increasing family income, and the constant technological advances, the "quality" of children, especially their education levels have improved, and the marginal costs of raising a child, especially the quality cost, have greatly increased. At the same time, the labour–economic benefit of children has declined, as has the elder–security benefit. Social and family savings have increased, and societies have been able to organize more welfare programs, including old-age security, and to establish relatively comprehensive social security systems for all. Labourers are also capable of taking out part of their savings as old-age security before retirement. In practice, such social and family security schemes are often combined to form an old-age security system in which society and family complement each other, dwarfing the elder–security benefit provided by children. Other benefits, such as bearing the risks of family business, maintaining family status, and protecting family safety, also show a downward trend, to varying degrees.

Leibenstein also examined the role of social factors in the decline of fertility rates, including the rise in the proportion of nuclear families, urbanization, declining mortality rates, improvements in women's education, a declining parental preference for boys, a weakening of traditional values, and the promotion and use of contraceptives. All these factors have been conducive to the weakening of the benefits that children provide to parents and family. However, the most decisive factor, he argued, is economic development. By distinguishing different levels of socio-economic groups according to average household income, he concluded that developing countries were changing from agricultural to industrial, and the direct and indirect costs of having children were rising with this transition. Except for the consumption–pleasure benefit of children, other benefits were declining with the increasing costs, with the labour–economic and elder–security benefits declining most significantly.

On the basis of Leibenstein's research, Gary S. Becker took the cost–benefit theory of having children further, establishing the model of mutual substitution of the number of children and the quality of children. Together with Milton Friedman and Theodore Schultz, Becker constituted the influential Chicago School of economics. He applied economic theories and methods to population questions and became renowned for his work on human capital.

In works such as *Economic Analysis of Fertility*, *Human Capital*, *Family Theory*, he expounded his economic analysis of fertility. He places family labour power in the market and adopts the theory of consumption demand to explain family fertility decision-making. He applies indifference curve and consumer equilibrium theories, among others, to the study of families' needs for children. A child's net cost, he argues, is equal to the direct cost the family is expected to pay to have and raise a child as well as the indirect cost of the parents' time (via shadow prices), minus the present value of the expected income and labour to be provided by the child.[19] He argues that if the net cost of a child is positive, then the benefit provided by the child to parents and family is not enough to compensate for it. Such a child is equivalent to ordinary durable consumer goods to parents, and what parents obtain from the child is psychological benefit, which cannot create added value. If the net cost of a child

[19] Luo (1991).

is negative, then the child will provide benefits greater than the costs to parents and family. Such a child is equivalent to a durable product to parents, which can create value proliferation. Through analysis of the established models, he showed that it is not the former that parents and other family need, but rather the latter. In a dynamic sense, if the net cost of a child increases, then demand for the number of children decreases; if the net cost of a child decreases, then the demand for children increases. With economic development and an increase in per capita income, the market wage rate, especially for women's labour, increases significantly, and the shadow price of non-market activity time can be approximately equal to the market wage rate, which leads to increases in the net cost of children.

With increased agricultural modernization, the time that children in rural areas spend in school has lengthened, which also directly leads to a rise in a child's net cost. Since children's quality is more elastic than their quantity, people tend to prefer quality, and there is a substitution transition in which high quality replaces large quantity. The decisive factor in this substitution is contraceptive technology, which makes it possible for people to choose when to have a child. When family income is fixed, there is a negative correlation between the quantity and quality of children and in order to maximize utility, families choose to increase the quality cost of children instead of their quantity cost, so the fertility rate will drop significantly with economic development and higher incomes.

Thus Leibenstein and Becker use cost–benefit theories to explain people's reproductive behaviour and the number of children they have. Richard Easterlin added two variables, the supply of children and the cost of birth control, and developed a supply and demand theory of family planning decisions. The variable of the supply of children refers to the number of children born to parents without the influence of external factors, that is, births determined by the natural fertility rate of parents and the survival rate of the children. The birth control variable costs include the economic cost of adopting control measures and the psychological cost of overcoming mental disorders aroused by adopting birth control measures. Without considering the cost of birth control, if the supply of children is less than the demand for them, as in an agricultural society, then there is no birth control requirement. If the supply of children exceeds the demand for children, then birth control will occur. Easterlin argues that the impact of the birth control cost is a variable worthy of attention. He believes that the cost of birth control in developed countries has approached zero. After the infant and child mortality rate dropped significantly, that is, the supply of children increased due to better survival rates, birth control became popular, and the fertility rate has remained relatively low. On the contrary, the costs of birth control in developing countries are relatively high, and so is the infant and child mortality rate. The supply of children is reduced due to their relatively low survival rate. With poor birth control use, the fertility rate is often high, especially in countries and regions with low levels of economic and cultural development.

Going further with this work, John Caldwell developed a theory of "intergenerational wealth flow" in which he examined the direction and rate of wealth flows between parents and children in different types of society. Wealth, as Caldwell noted, includes not only money and property, but also labour and services. To illustrate the question, he proposed the concept of net wealth flow to discuss the direction and rate of net wealth flows between parents and children. His argument is that in pre-industrial, traditional societies, because the cost of raising children was low but had obvious labour–economic and elder–security benefits, net wealth mainly flowed from children to parents. Since industrialization, because children must be trained in certain cultural and professional skills before they are employable, parents' investments in their children have grown, mainly in education, which is expected to result in corresponding benefits. However, with the development of commodity and market economies, children's independence has also grown, and family size has been shrinking, so that wealth flows mainly from parents to children. As a result, people's preferences have shifted from the pursuit of numerous children to the pursuit of "quality" children and resulted in a continuous decline of the fertility rate.

It has been argued that Easterlin rejected Becker's theory, but it is not the case. For Easterlin and Caldwell, although their starting points, research perspectives, scope of vision, and conclusions differ in some ways, their basic theoretical framework is inseparable from the idea that the cost of a child increases with the rise of family income, technological progress and social development, and that family size and fertility rates are constrained and affected by income, price, and substitution effects. Their theories about fertility fall plainly within the framework of the cost–benefit approach to fertility decisions.

8.2.2 Theory and Practice of the Social Elements in the Costs and Benefits of Children

8.2.2.1 The Social Conditions of Fertility Decisions

The underlying premise for this Western cost–benefit theory of fertility decisions is that the decision making is autonomous family behaviour occurring in a free market economy. However, after the World War II, government intervention in reproductive behaviour in developing countries became increasingly prominent. Many countries introduced family planning policies aimed at controlling population growth and which have achieved some success. This poses a question for the Chicago school theory of the cost–benefit of children. What is the reason for the decline of fertility rate in certain economically underdeveloped countries, and what kind of theory can be used to explain it? China is a typical case. From 1971 to 1981, the total fertility rate dropped from 5.44 to 2.63, a rare occurrence in global demographic history. During that period, both the national economy and per capita income grew. According to contemporary commodity prices, per capita national income increased by 60% from 247 to 396 yuan. Adjusting for inflation, the real growth rate was 46.2%, which was less than the decline in the fertility rate.[20] At the time, China's economy was in transition from starvation to subsistence, and having a child then was a negative net cost.

---

[20] National Statistics Bureau of the PRC (1986).

Having more children had obvious economic benefits, which was not conducive to the decline of fertility rate. From 1981 to 1991, per capita national income more than doubled, from 396 yuan to 1,401 yuan, 1.1 times in adjusted terms.[21] This was the decade with the fastest economic and per capita income growth in the past 40 years. The total fertility rate dropped from 2.63 to 2.01, a decline of 23.6%—far from the growth in per capita national income.

Various interpretations of these shifts have been made in China and elsewhere. Some argue that it is an inevitable result of socio-economic development and had little to do with prevailing family planning policies. Some believe that it is inextricably related to both socio-economic development and family planning policies. Others argue that although China's economy developed rapidly since the application of the reform and opening-up policy, the overall level of development was not high enough to cause the fertility rate to drop so quickly to such a low level, therefore, it is must be explained by the implementation of a relatively strict policy of population control. However, a convincing interpretation must conform to the reality, needs the support of corresponding social practice, and then rises to theory. In other words, it must adhere to the materialist epistemology of practice–theory–practice. In accordance with the CPR/90/P06 agreement reached by the Chinese government and the United Nations Population Fund, under the leadership of the Population Research Institute of the Chinese Academy of Social Sciences, with the coordination of many units such as the Population Research Center of Hebei Academy of Social Sciences, the Family Planning Commission of Liaoning Province, and the Population Research Institute of Shanghai Academy of Social Sciences, I presided over a national sample survey of household economy and fertility in ten provinces and cities, and obtained a batch of complete original data that was analysed using components of the cost–benefit theory of fertility decisions. From this process was elucidated what I call the social elements in the cost–benefit theory of having children. The theory is expounded as follows.

The concept of "social elements in the cost–benefit theory of having children" is defined thus: in accordance with certain social norms (including population policies), the costs and benefits of having children increase or decrease depending on whether they exceed or fall short of the upper or lower limits specified in the norms (policies). The realities in China mean that regarding children, the main cost for individual families was incurred for having more than one child; the main benefit they received was the bonus for having only one child. In the world today, there are mainly two kinds of population policy. One aims to control population growth and thus increases the cost of having more children, and the other encourages population growth, which increases the benefit of having more children. The choice of population policy is determined by the nature and norms of population problems in different countries at different periods.

The "social elements in the cost–benefit theory of having children" theory addresses four situations: increasing or decreasing costs and increasing or decreasing benefits. In China, the first is the increased cost of having a second (or third, etc.)

[21] National Statistics Bureau of the PRC (1992).

child. For example, a certain fee is imposed on parents who have more children than is stipulated by the state population policy. The second situation is where a benefit accrues for a certain child, such as the one-child bonus, rewarded in cash or in kind. The third situation is where the marginal cost of a certain child is reduced, for example, the one-child bonus not only increases the marginal benefit of the child, but also reduces the marginal cost of the family of the child. Moreover, "rewards" are not only paid in currency, but can also come in supplementary forms: preferential kindergarten admission, school enrolment, and job places for single children in urban areas, and homestead allocation in rural areas. Rewards in both forms have reduced the marginal cost of children. The fourth situation is where the marginal benefit of a certain child is reduced. The imposition of an off-plan birth fee on parents of unplanned children reduces the marginal benefit of children to those families. In addition to the fees, some places also impose restrictions on kindergarten admissions and school enrolments in urban areas, as well as on homestead allocation, and household registration shifts from agricultural to non-agricultural in rural areas. All these policies affect the cost–benefit balance of having children, highlighting the *social* nature of the additional costs benefits involved.

Increasing or reducing the additional, social elements of the costs and benefits of having children is not limited to this material level, but also penetrates the spiritual level. This needs special attention. Both reward and punishment can affect the spiritual level, thus forming indirect psychological costs and benefits to having children. For example, in addition to being charged the off-plan birth fee, the parents of unplanned children also received criticism and education, and even got administrative punishment, which resulted in a certain loss of reputation and thus they can pay a high psychological cost for unplanned children. Parents who have received the official recognition certifying that they have had only one child may thereby earn better results in their work assessments, because many work units make this an important criterion in granting awards to outstanding workers. In the units where the "one-vote veto system" operates, unplanned children are the most important criterion. Workers who have not met the population plan targets will not be able to enter the advanced ranks, no matter how well they have fulfilled their other tasks.

Social costs to having children are also known as direct additional costs and additional benefits are also known as direct additional benefits. Such additional costs at the spiritual level can also be called indirect additional costs and additional benefits at the spiritual level can also be called indirect additional benefits. They are two aspects of the same question, forming a dialectical unity of interaction, complementarity, and mutual promotion. In a culture that attaches great importance to politics, morality, and reputation, additional costs and benefits at the spiritual level cannot be ignored, and their role in achieving the goals and requirements of fertility policy cannot be underestimated.

8.2.2.2 Social Practices in the Costs and Benefits of Having Children

The modified cost–benefit theory of fertility offered here is a theoretical generalization formed on the basis of the observation of social practices. Now it is time to elaborate the practices.

In the late 1980s and early 1990s, when the third cycle of the above-mentioned UN Population Fund assistance to China was about to begin, we conducted a sample survey of household economy and fertility in ten provinces and cities, which was completed in the first half of 1992. The results of this survey drew a fairly clear picture of the changes to China's family economy and fertility, and the costs and benefits of having children. They show that the social elements of the cost–benefit theory change quite often. It is also the practical starting point from which we can expound our theory of this social element.

A batch of relatively complete data were obtained from this sample survey. The data the most closely related to the current population policy are mainly the following.

The first is the composition of the benefits of having children. The national weighted aggregate proportions of the various benefits that children bring are: those who provide labour–economic benefits, mainly monetary income to the family, account for 17.6%; those who provide elder–security benefit make up 20.5%; those who provide psychological pleasure comprise 24.2%; those who provide family status maintenance and business expansion benefits account for 15.5%; those who provide inheritance–risk benefits are 8.5%; and those who provide elder–safety and defence benefits account for 13.9%. Among these, psychological pleasure and elder–security are the most important, both exceeding 20%. In the early 1990s then, the benefits to having children were still typically characteristics of developing countries.

The second conclusion we can draw from the survey data is the effect of family income. Among all the benefits provided by children, although the labour–economic benefit is not as significant as the psychological pleasure and elder-security benefits, it occupies third place, and is one of the crucial sources of family income, especially in rural areas. In the national weighted aggregate of household benefits, rural children who live independently from their parents contribute 47 yuan per annum when they belong to families with a monthly income of less than 25 yuan; among families with 51–100 yuan, they provide 90 yuan; among families gaining 101–200 yuan per month, they provide 215 yuan; children contribute 196 yuan in families with an average income of 201–300 yuan; and among families with 301–400 yuan, they contribute 348 yuan. The general trend is that the higher the economic benefit provided by the children, the higher the family income before that contribution. The benefits provided by the children are similar to the costs they invoke. Non-agricultural income is higher than agricultural income, by an average of 90 yuan in this survey. The benefits provided by children differ by gender too, with men contributing 46 yuan more than women.

The third conclusion relates to the effect of occupational structure. The survey shows that the benefits that children provide to their families are quite different depending on their occupation. For example, in the national weighted aggregate of the monetary income benefits provided to families by rural children who live independently, the greatest come from those who work in the food and beverage service industry, with an average annual contribution of 260 yuan; the second most important contribution comes from self-employed workers, at an average of 216 yuan per annum; the next is 140 yuan from scientific and technological personnel, 125 yuan from clerks, and just 50 yuan for public servants. These figures are related not only to the children's income levels, but also to their consumption patterns. Those who work in agriculture, forestry, animal husbandry and fishery on low incomes make greater monetary contributions to their parents and families.

Our fourth conclusion from the data relates to net costs of children and intellectual investment in children. The purpose of investigating and analysing changes to the costs and benefits of having children is to discover the motivations behind family reproductive decisions and thus to seek the best direction for reform. For this, we applied Becker's theory of the net costs of having a child to the data obtained from the survey. To restate, for Becker, the net cost of a child is equal to the direct monetary cost that the family expects to pay to have the child, plus the indirect cost (in time, via shadow price) spent by the parents, minus the present value of the income and labour expected from the child. If the net cost is positive, then the child is only equivalent to an ordinary piece of durable consumer goods, and the family will obtain only psychological benefits from the child. If the net cost is negative, then the child will be equivalent to a durable product, which can bring value appreciation to the family. In this way, families make reproductive decisions based on the estimation of child's net cost. Since the indirect time cost of children in our survey of ten provinces and cities was quite close to the labour benefit they bring, the time cost–benefit is omitted when we calculate the net cost in the cost–benefit analysis of marginal children. The results show that in cities, the net cost of children is already positive; in rural areas, the net cost of children has also begun to shift from negative to positive. For instance, in 1991, rural families spent 139 yuan on their children on average per annum, while children provided an average of 148 yuan per annum to their families. The net cost was negative nine yuan, which shows that the labour–economic benefit that children bring has fallen to a very low level.

Our survey data show that among rural children who live independently and who work in primary industries, a comparison between the direct monetary cost incurred by their family in a year and the monetary contribution the children make to the family shows a net cost of negative 50 yuan. This means that there is still some room for the net cost of rural children to be negative. The reason is that the technical component of agriculture, forestry, animal husbandry and fishery is low, thus the cost of training children for such work is also relatively low. And since the technical composition is inversely proportional to the elasticity of labour absorption, the lower the technical composition, the more elastic the labour absorption, and the earlier and more significant the benefit will be. This creates a fertility stimulation, which explains the high fertility rate in economically and technologically backward areas where the main economic activities are in the primary industries.

The main way to change this situation is to make economic and technological progress. On the one hand, progress increases the requirements for the cultural and technical capacities of workers, and increased family income depends chiefly on the quality rather than the quantity of workers, thus reducing a child's labour–economic benefit to the family. At the same time, economic development and increased labour productivity have increased social wealth. State and society are able to organize more old-age security schemes, and workers and families are better able to save for old-age care. The marginal child's elder–security benefit thereby also drops significantly. Other benefits, such as assuming the risks of family business, have also declined to varying degrees. On the other hand, economic development drives technological progress, and technological progress promotes further economic development, making families clearly aware that only the necessary intellectual investment in children can bring the desired benefits, so the main cost of a child is the cost of a quality education, which is rising rapidly. These changes to the costs and benefits of having children prompts families to shift from thinking about the quantity cost of investing in children to the quality cost, and correspondingly, the fertility pattern has shifted from more births to fewer births and better education, and the fertility rate has continued to decline. So far, the trajectory of economic development and fertility decline seen in developed countries speaks volumes about this. It is also confirmed by the successful shift in economic development and technological progress made in some regions in China since the reform and opening-up period. Economic development then, is a smooth path to a lower fertility rate—a universal law proved by both Chinese and other practice.

This does not mean that economic development will "naturally" solve the population problem. Because China has a large population, insufficient natural resources and low economic development, solving its population problem will not lie solely in economic development, but requires a two-pronged strategy. That is, while vigorously developing the economy and thereby creating the basic conditions for a decline in the fertility rate, a national policy of family planning is required, that is, effective control of population growth by administrative, economic and legislative means. We should vigorously promote the transition from negative to positive net costs for having children.

Based on the actual situation in the early 1990s, *China's 1992 Sample Survey Report on Family Economy and Fertility in 10 Provinces and Cities* made suggestions for changing attitudes to physical and mental labour and encouraging intellectual investment in children. In those years, there were still such phenomena as discrimination against mental labour reflected in sayings such as "it's better to take a razor than a scalpel" and "the pen of a professor can't match the scales of a peddler". The mental workers earned far less than the manual workers. Such attitudes have not only had an adverse impact on national economic development and modernization, but also are unfavourable to controlling population growth and to realizing the change of the net cost of a child from negative to positive. Our survey results show that on average, rural children's economic exchanges with their families change according to their education level: the net cost of illiterate and semi-illiterate children is negative nine yuan; that of children with primary school education is negative 20 yuan; among children with junior high school education it is negative 48 yuan; among children with senior high school education it is positive 78 yuan; and among children with college education and above it is positive 227 yuan. According to Caldwell's intergenerational wealth flow theory then, the net cost is negative before junior high school and positive after senior high school, which indicates an anomalous phenomenon that the more the family invests in a child's intellectual training, the more wealth flows from the parents to the child. This seriously hinders attempts to reduce the fertility rate. To reverse this inclination, we need a major policy change: following the principle that complex labour is a multiplication of simple labour, incomes paid for mental labour should be raised, to ensure that the costs incurred by families to educate their children will result in corresponding or additional economic benefits later. The policy should lead people to choose a quality of children rather than a quantity, and away from investing in quantity costs rather than quality costs.

8.3 Theory and Practice of the Post-Demographic Transition

These new trends in China's population, including the changes to the costs and benefits of having children, constitute a specific stage of population transition—the post-demographic transition stage. They show that China is mid-way in the post-demographic transition and will continue to advance. Post-demographic transition, together with population development strategy, constitute two key domains in population policy. Therefore, it is necessary to interpret the post-demographic transition using a combination of theory and practice.

In the early 1980s, five members of the Chinese Academy of Social Sciences visited Western Europe, North America, Japan, and other countries, as well as the United Nations headquarters and other relevant institutions for demographic research. They noted that these institutions mostly studied the population problems of developing countries, especially rapid population growth and the contradiction between population and socio-economic development in Asia, Africa, and Latin America. They believed that the developed countries had entered the post-demographic transition stage; apart from France and a few other countries that were studying how to increase their fertility rate and population size (though with little success), there seemed little to study in the developed world. Thus these scholars had little choice but to turn to the developing countries and study the problems brought about by rapid population growth, high fertility rates and population expansion, and their impact on the global population and resources and the environment, as well the solutions to the problems.

Against this domestic and international background, the Third Plenary Session of the Eleventh Central Committee of the CPC in 1978, armed with the restored ideological line of seeking truth from facts, began to set population theory right. Their investigation first focused on population change and growth trends and captured the principal contradiction at that time. The problem now is that in the three decades that followed, population policy, which was dominated by quantity control, has changed often, and the understanding of the population question has also advanced. According to UN predictions, India will replace China as the world's most populous country around 2030. Similar predictions by China and others show that when the Chinese population reaches 1.4–1.5 billion, around 2030, zero growth will have been achieved, and then there will be a downward trend. With the change of population size, the age structure is also advancing at a rapid and steady pace; the urban population is growing at a similar frequency; and population qualities, including health, scientific knowledge, education, literacy and civic-mindedness, are increasing at a similar rate. Therefore, to observe the changes and development of China's population at present and in the future, we should not always focus on the change in population quantity, but should gradually shift the strategic priority, and firstly to one driven by an overall improvement in population quality. The notion of improving population quality is an eternal one, but is more prominent in the post-demographic transition. The adjustment of age, gender, urban–rural and regional distribution, and other population structures is a normal problem, which receives more attention in the post-demographic transition. With the deepening of the practice of post-demographic transition, theoretical research has been reactivated.

Since the 1990s, several scholars in China have demonstrated that the country had entered a post-demographic transition stage. Others have questioned this, worrying that it may be misleading. The debate among them was a healthy academic discussion, without unsupported accusations or personal attacks and thereby breaking the perennially embarrassing situation of academic debate with no in-depth discussion. However, after a burst of activity, the debate seems to have stopped again, and the issues remain unclear and in need of further research. For example, how does the international community talk about the post-demographic transition? What does the concept actually mean and can it be extended? What standards are used to define it, and which direction will the transition take? Our interest lies in finding out the main points of divergence and consensus, especially on the substantive issues. To this end, we will summarise some theoretical research to clarify the origins and evolution of the post-demographic transition concept.

In 1909, in his "Three Theories of Population", the French scholar Landry made a historical investigation of the evolution of births and deaths in European populations, proposing the concept of demographic transition as regular population changes. Afterwards, the American scholar Thompson broke through the limitations of European research and conducted similar research on a larger scale, which also gave demographic transition research some theoretical colour. Frank W. Notestein matured demographic transition theory with his proposal of a three-stage demographic transition. C. P. Black then developed a five-stage demographic transition. Leibenstein, Ansley J. Coale, Caldwell and others made more extensive investigations, introducing some social and economic factors. Although there have been criticisms since demographic transition theories were first offered, the dissemination of the theory has not stopped. In many countries, the theory of demographic transition appears in demography textbooks and is taught as an important theory. The consensus is that the theory of demographic transition extracts a law of evolution from high birth, high mortality, and low increase to low birth, low mortality, and low increase from the population changes in Europe between the end of the eighteenth century and the first half of the nineteenth century. But there are different elaborations. One is from high birth, high mortality, low increase to high birth, low mortality, high increase, and then from high birth, low mortality, high increase to low birth, low mortality, low increase, that is, expounding the law of changes in birth and death rate from the perspective of "transition". Another combines "type" with "transition", distinguishing the process of transition in view of the type, and summarizing the different population transition laws as "three-stage", "four-stage" and "five-stage". Although these explanations are different, they all start with high-high-low, and end with low-low-low. The difference is the elaboration of the intermediate transition process.

Ansley Coale, who holds that high birth, high mortality, low increase is actually a state or type before the demographic transition; low birth, low mortality, low increase is actually a state or type after demographic transition; thus the transformation from high-high-low to low-low-low is a demographic transition in a true "transition" sense. In this way, if there is only one intermediate type in the demographic transition, that is, the type of high birth, low mortality, and high increase, then it is a "three-stage" demographic transition theory. If there are two or three intermediate types, that is, in addition to the high-low–high type, with the addition of a type of higher birth, lower mortality and higher increase, or further added a type of lower birth, lower mortality and lower increase, then it is the "four stage" or "five stage" theory of demographic transition. However, no matter how the intermediate transition process is divided, the result of demographic transition is to step into a low-low-low stage or type. There is a consensus that the demographic transition finally reaches the stage or type of "three lows".

So who first proposed the "post-demographic transition"? Due to the lack of authoritative materials, it is difficult to determine. However, according to available papers, the earliest discussions can be traced back to the 1970s and 1980s. In 1973, Coale put forward the theory of "three stages" before, during and after demographic transition, in his article "The Demographic Transition Reconsidered",[22] which defines "post-demographic transition". In the 1980s, when investigating population changes in East and South Asia, Richard Leete stated that the population changes in those regions had entered a period of "post-demographic transition". After that, the term appeared in newspapers and online media from time to time, gradually attracting attention. It is said that Luo Chun was the first to use this term in China and considered it the "most dazzling" stroke in his academic career. In 1992, *Population* magazine published Zhu Guohong's article "Southern Jiangsu Model: Post Demographic Transition Period". In 2000 and 2001, *Southern Population, Chinese Journal of Population Science*, and *Population Research* published articles on the post-demographic transition by Yu Xuejun and Li Jianmin. At the same time, Li Jianxin and Ye Mingde respectively published articles questioning the theory of the post-demographic transition in *Population Research* and *Chinese Journal of Population Science*. For a certain time, research on the post-demographic transition bloomed.

[22] Coale (1973).

However, the discussion in this area did not go further, and soon withered, leaving many unanswered questions, mainly listed as follows.

First, what is the concept of "post demographic transition"? Several articles have argued that China's population completed the process of the demographic transition in the traditional sense in the 1990s and began to enter a period of low fertility, that is, the so-called post-demographic transition, but did so without defining the concept. How is it similar to and different from the traditional demographic transition? Where is the transition headed, and along what path? Such questions have not been clearly explained, and some have not even been touched upon. Some authors argue that this is because no-one knows where the population will "transition" to in the future.

Second, what criteria and indicators do we have for measuring the post-demographic transition? Suggested indicators include total fertility rate, average life expectancy, birth rate, and natural increase rate, as well as the standard value of each indicator. If it exceeds or falls below the standard value, it will enter the post-demographic transition period. However, there is a lack of convincing elaboration on why these indicators are selected, how they are relevant, how the indicator system is constructed, and what the decisive signs are. Few scholars have treated the issue of measurement criteria and indicator systems.

Third, has China completed its demographic transition in the original sense? Post-demographic transition is related to the pre-demographic transition, or a population transition in the general, traditional, and demographic sense. Obviously, entering a post-demographic transition indicates the completion of the original demographic transition. But questioning this assumption, some researchers point out that, unlike in Western countries that have entered the post-demographic transition, the drop in the fertility rate in China below replacement level is not a natural formation but mainly the result of state policy. There are many unstable factors behind it, so it is not a demographic transition in the classical sense. Therefore, the announcement that China has entered a post-demographic transition period not only fails to withstand theoretical scrutiny, but may also be misleading in practice (Ye Mingde 2001).[23]

Fourth, Why did the "post-demographic transition" occur? What is the trajectory of the transition? What direction is the transition going in? Where does it end? What is the result of the transition? What impact will it have on social and economic development? To answer these questions, we require data about the realities of population transition, to discover some laws, to engage at the theoretical level, and to form post-demographic transition theory and theoretical systems. At present, we still have a long way to go; it might even be said that we have not started yet.

Fifth, After the general demographic transition enters the three-lows stage, will it automatically continue on to the post-demographic transition, or will it transfer only after a certain period of time? Several articles arguing that China has completed the general demographic transition suggest that a population that has achieved the three-lows transition has entered the post-demographic transition period. In order to link this with the indicators they designed, or to combine several indicators by identifying a certain year as the starting point of the post-demographic transition,

[23] Ye (2001).

some scholars identify 1998 or the late 1990s as the beginning. Yet this moment is more than six years after the fertility rate fell below the replacement level. Which "transition period" do this six + years belong to?

The concept of the post-demographic transition was not clearly explained after being put forward, which has a lot to do with the fact that China's population has not been in the three-lows stage for long, and the population change in all aspects has not been displayed sufficiently. However, we cannot propose a concept without explaining and defining it. Dialectics tells us that problems and solutions always occur at the same time. Analyse the origin and evolution of a problem, and we will find the method to solve it. It has been more than 20 years since China's total fertility rate fell below replacement level, in 1991. Although the data available for research is not rich enough, there has been a certain accumulation. Moreover, research on the demographic transition in other countries has been produced for more than a century now, and studies of the post-demographic transition have been conducted and published for decades. So, there are certain foundations to this research and it is possible to make further progress with some effort.

China has indeed completed its general demographic transition. But further research is needed into where the transition will lead and what kind of transition period or stage it will enter next. Although Coale's theory of the three stages of before, during and after demographic transition is a little too general, it is accurate and universally applicable as a definition. When China reached the low fertility level in 1991, it entered such a stage. Research into the post-demographic transition since the 1970s refers to the demographic transition of a country or region's overall population from the three-lows type to zero or near zero growth after a relatively long period of relative stability, including a certain degree of continuous growth or decline. Because of the different population conditions in different countries and regions, especially the different economic, technological, cultural, and social development, after the population has entered the three-lows level there are many possibilities, including staying at the original level, declining through inertia, and rising after repeated vibrations. Judging from the situation experienced by developed countries in Europe for example, the main trend is to decline along with the inertia force of the low fertility level and remain relatively stable. But other situations are possible, and indeed there are now signs of the other two scenarios. The most important thing here is to clarify the role of the factors of population and fertility policy after a country enters the three-lows type.

Demographic transition theory, like other social phenomena, is ever evolving. The demographic transition proposed by Landry in 1909 is a transition in the sense of natural population change, a regular description derived from the study of population births and deaths. Therefore, it is a demographic concept in a narrow or statistical sense, which is called demographic transition instead of population transition. Subsequent studies have enriched the demographic transition in authentic, traditional, usual, and general senses. However, the demographic transition shows considerable differences in different countries and regions, which are related to the population status of each country and region, as well as socio-economic development, cultural accumulation, and population policies. Together with the study and description of these variables, it has become the population transition in a broad demographic sense, with developmental implications. Many studies after Leete can be characterised in this way. If we are to discuss population change and development after China's entry into the low fertility level, it is not enough simply to adopt the former general demographic transition. Because the role and influence of social policy factors are on the increase, without the vigorous population and fertility policy of the past 30 to 40 years, especially the one-child policy, it will be impossible for China to enter the ranks of low-fertility countries quickly. More than 20 years have passed since China's fertility rate fell below the replacement level in 1991, and the low fertility level has passed through a generation. There is a relatively solid foundation for a "three-lows" population, and it is unlikely that there will be a rebound or a substantial increase, unless a special policy to encourage population growth is adopted. Therefore, the judgment that China is already in the post-population transition period seems reasonable.

In light of China's reality, the post-demographic transition is defined here as a population whose fertility rate is below replacement level, and remains relatively stable, including a certain decline, for at least one cycle of population reproduction. This definition includes following elements. One is the sign that the post-demographic transition has already occurred. The general demographic transition is described by changes in birth rate and mortality rate, reflecting the changing population increase rate from low to high and from high to low. This law of change is not only applicable to European countries but has universal significance. Natural population changes in other countries are also similar. The indicators of this natural change in population are the "three rates", of birth, mortality, and natural increase. The changes in these three rates also describe the trajectory of natural population change.

However, while the three rates are selected as the indicators for directly measuring the population transition, for example, some scholars choose the birth rate equal to or less than 16.5 per 1,000 people and the natural increase rate less than 10 per 1,000 people as the criteria for entering the post-demographic transition, so inexplicable difficulties will appear. The birth, mortality, and natural increase rates are the ratio of the number of births, deaths, and natural increase population in a certain year to the total population in that year, usually expressed in thousandths or percentages. Such a ratio is a "crude rate", which is largely influenced by the age structure. Strictly speaking, "crude rates" cannot be compared directly—it is necessary to standardize the age structure for comparison. For example, from 2000 to 2007, China's mortality rate rose from 6.45 per 1,000 people to 6.93 per 1,000 people. Why did the overall death rate increase while the age-specific death rate, including that for infants, decreased? It is because the proportion of the elderly population increased, and the age-specific mortality rate of the elderly population is much higher than that of adults and young people. Similarly, from 2005 to 2010, the population death rate in the US was 8.4 per 1,000 people, in Japan 8.8 per 1,000, and in France 9.6 per 1,000, all of which were higher than 6.9 per 1,000 people in China in 2007. It would be a mistake to infer that the health level of the populations in these countries is lower than in China; in fact, the age-specific death rates, including for infants, in these countries are much lower than that in China, and life expectancy is significantly higher than in China. The reason is that the death rate is a crude rate which is greatly affected by aging.

So what indicators are selected? The natural increase rate is the difference between the birth rate and the mortality rate, reflecting the changes in both, and has the nature of a dependent variable, so it should not be selected as an indicator for measurement. The mortality rate is mainly the age-specific one, which decreases with the development of social economy, improvement in quality of life, and improved medical and health conditions. However, an increase in the proportion of the elderly population will lead to an increase in the crude mortality rate. Under normal circumstances, without wars, epidemics and disasters, the overall trend of its change is relatively gentle, and it is not a very sensitive indicator. Which leaves the birth rate. Selecting it as a direct measurement has its own insurmountable disadvantages, so we must choose an indicator that has a decisive impact on the birth rate. That indicator is the fertility rate. Although the effect of the fertility rate on the birth rate is not immediate, the decisive role of the change in fertility rate on the change in birth rate cannot be filled by other indicators. If the fertility rate falls below replacement level and can be kept there for a long time, though the birth rate may be higher than the death rate, the gap between the birth and death rates will tend to narrow until the gap becomes zero and the population achieves zero growth. Therefore, the fertility rate is chosen as an indicator to measure the post-demographic transition. It is appropriate to say that the drop in the fertility rate below replacement level marks the beginning of an overall population entering the post-demographic transition stage or period. We must not forget that it will take at least one reproduction cycle (25–30 years) for the fertility rate to stabilize below replacement level.

Why does it take at least one cycle of population reproduction? Because after entering the three-lows type, after a population reproduction cycle, the new-born population that falls below replacement level reaches childbearing age and passes through the fertile period, the low fertility level may be considerably consolidated, and the realization of zero growth or near will be effectively guaranteed. Under normal circumstances, the population entering the three-lows type will pass smoothly through this transition period, showing that the post-demographic transition has strong stability and continuity. In special circumstances, the post-demographic transition may not be successfully completed, and it is possible to return to the state before the transition. The experience of more than 20 years in the three-lows population type shows that China has successfully passed this transition period and reached the goal of the early stage of post-demographic transition. In future adjustments to population policy, attention should be paid to maintaining the fertility rate at TFR $\leq$ 2.1 and NRR $\leq$ 1.0, so as to complete the historical mission of the post-demographic transition.

References

Coale, A. J. (1973). The Demographic transition reconsidered, International Union for the Scientific Study of Population (IUSSP): Proceedings of the International Conference 1973 (Vol. 1). pp. 53–73.

Coale, A. J. & Li, S. M. (1991). The effect of age misreporting in China on the calculation of mortality rates at very high ages. *Demography, (28)*, pp. 293–301

Department of Population and Employment Statistics, National Bureau of Statistics of China. (Ed.) (2012). *China's Demographic and Employment Statistics Yearbook 2011*(pp. 287–302). China Statistical Publishing House.

Keynes, J. M. (1997). *The General Theory of Employment, Interest, and Money*. (Lu, M. L. Trans.). Beijing: The Commercial Press.

Leete, R. (1987). The Post-demographic Transition in East and South-East Asia: Similarities and Contrasts with Europe. *Population Studies, Vol. 41* (21). pp. 187–206

Leete, R. (1994). The Continuing Flight from Marriage and Parenthood among the Overseas Chinese in East and South-east Asia, Low fertility in East and Southeast Asia: Issues and Policies. Korea Institute for Health and Social Affairs (KIHASA): Seoul.

Li, J. M. (2000). On Post-demographic Transition. *Population Research, 4.*

Luo, C. (1991). On Gary S. Becker's Theory of Family's Demand for Children. *Population Journal, 5.*

Luo, C. (2001). A Discussion on Post-demographic Transition. *Chinese Journal of Population Science,* (1)

National Statistics Bureau of the PRC. (Ed.). (1986). China Statistical Yearbook of 1986. *China Statistical Publishing House.* p. 53

National Statistics Bureau of the PRC. (Ed.). (1992). China Statistical Yearbook of 1992. China Statistical Publishing House. pp. 32–33

National Statistics Bureau of the PRC. (Ed.). (2006). China Statistical Yearbook of 2006. *China Statistical Publishing House.* p. 109

National Statistics Bureau of the PRC. (Ed.). (2018). *China Statistical Yearbook of 2018.* China Statistical Publishing House. p. 32; p. 668

Population Census Office under the State Council and Department of Population Statistics, State Statistical Bureau, PRC. (Eds.). (1985). *1982 Population Census of China (Results of Computer Tabulation).* China Statistical Publishing House. pp. 31–33; pp. 54–55; pp. 272–313

Population Census Office under the State Council and Department of Population Statistics, State Statistical Bureau, PRC. (Eds.). (1993). *Tabulation on the 1990 Population Census of the PRC (Volume 1).* China Statistical Publishing House. pp. 215–216

Population Census Office under the State Council and Department of Population Statistics, State Statistical Bureau, PRC. (Eds.). (1993). *Tabulation on the 1990 Population Census of the PRC (Volume 2).* China Statistical Publishing House. pp. 2–17

Population Census Office under the State Council and Department of Population, Social Science and Technology Statistics, National Bureau of Statistics of China. (Eds.). (2002). *Tabulation on the 2000 Population Census of the PRC (Volume 1).* China Statistical Publishing House. pp. 570–581

Population Census Office under the State Council, Department of Population and Employment Statistics, National Bureau of Statistics of China. (Ed.). (2012). *Tabulation on the 2010 Population Census of the PRC (Volume 1).* China Statistical Publishing House. p. 149; pp. 265–266; pp. 270–276

Population Statistics Department of the National Statistics Bureau of the PRC. (1988). China Demographic Yearbook 1988. China Statistical Publishing House. p. 198

Tian, X. Y. et al (Eds.). (2007). *Research on Chinese Population Development Strategy in the 21st Century.* Social Sciences Academic Press. pp. 439–448; pp. 440–454; pp. 448–454

Tian, X.Y., Hu, W. L., Liu, Q.M. (1995). Sample Survey Data on Household Economy and Fertility in 10 Provinces and Cities in China in 1992. Economic Press China.

U. N. Population Division. (Ed.). (1989). *World Population Prospects: The 1988 Version. New York.* p. 200, 204

U. N. Population Division. (Ed.). (2009). *World Population Prospects: The 2008 Revision (Vol. 1)*. New York. p. 48, 50, 52, 184

Ye, M. D. (2001). Doubts about "China's Entry into the Post-demographic Transition Period". *Chinese Journal of Population Science*, (1), 32–37

Yu, X. J. (2001). A Further Discussion on "China's Entry into the Post-demographic Transition Period". *Chinese Journal of Population Science*, (3)

Chapter 9
Population Policy and the Population Development Strategy

As discussed elsewhere in this book, China's current population policy is intended to serve two major goals: the post-demographic transition, and the population development strategy. In earlier periods, the population strategy had different aims, and thus its orientation and key characteristics were different. Here, we must also pay some attention to the population policy's effects outside the demographic field and efforts should be made to maximize its social benefits.

9.1 Population Policy and Strategic Focus Adjustments

The word "strategy" is used extensively, sometimes loosely. Some things that do not belong to the category of strategy are discussed as strategies, which creates confusion. So, let's first define the concept of "strategy" and discuss its unique attributes.

9.1.1 The Basic Attributes of "Strategy"

"Strategy" was originally a military term or concept. But then for thousands of years, war strategies were also applied in economic and other social fields, to form development plans for industry, agriculture, science and technology, education, culture, population, resources, environment, and national defence. Whatever the strategies, they share certain common attributes, including the following.

First, a strategy is intended for the long term. The priority of a development strategy is to reach its final goal. A development strategy requires an ultimate aim and longevity is one of its chief attributes.

Second, a strategy is global. Thus, a development strategy demands the ultimate achievement of the general goal, not a partial victory. While a global victory is

X. Tian, *Population Policies in the People's Republic of China*,
https://doi.org/10.1007/978-981-96-9658-1_9

achieved by an accumulation of partial victories, partial success cannot replace the global aim. Therefore, setbacks and failures must be faced and lessons learned from each failure, until the goal is achieved.

Finally, a strategy is phased. Usually, a broad strategy is designed to be accomplished in several phases rather than in one move. For example, in China's War of Resistance against Japanese Aggression, strategy was divided into three phases: strategic defence, strategic stalemate, and strategic counteroffensive. In each of the different phases, different strategies and tactics were applied, until complete victory was won. Presently, we are implementing a sustainable development strategy and the reinvigoration of China through science and education. Similarly, these can be divided into various stages, each with different targets, different paths and different methods of advancing.

Likewise, the population development strategy has these three essential attributes that allow us to understand it. The aim is for a balance among population size, capacity and structure, and also among population, resources, environment, economy, and social development, not only in the long term but also at the different development stages. Ideally, the characteristics of different phases consistent with the long-term global plan will be carried forward and those inconsistent with it will be limited. For example, no matter what the development stage, constant improvement in population capacity is required, without any contradiction between long-term and shorter-term interests. But sometimes, short-term interests in a certain phase might be sacrificed for the sake of long-term and global goals. For example, though control of population size is common to both long-term and general interests and interim targets, there is a contradiction between them. When we increase control over population growth in a certain period, various situations might arise. When population growth is under strict control, the population structure will change first, with the ratio of the working-age population to the whole population and its absolute numbers forming an inverted U. The population change first accommodates the economic development, and the demographic dividends provided by the "golden age" of population structure change greatly boosts economic growth. Next, the phase of population deficit and population debt ensues, with the social dependency ratio rising steadily, which starts to inhibit social and economic development. Such double-edged changes occur often in all the interrelationships, manifesting the coexistence of benefits and drawbacks.

Therefore, we must regard long-term and global aims as our final goal, and short-term interests as subordinate to it. There are two types of subordination. One type is any short-term appeal that is contradictory with the long-term and global aims, which will be prohibited regardless of their appeal in a certain phase. Only the short-term outcomes that are consistent with the long-term or global goals will stay. The second type is those short-term interests that are inconsistent with the long-term one but may be permitted under conditions in which they can promote development in a certain phase. The intensity and the timing of the contradictory element's function should be limited, to preclude it from undermining long-term and global aims. This requires us to introduce reforms and to adopt different policies in different phases. Realities show that through policy adjustments in different phases it is possible for short-term appeals to coexist with the long-term ones and for them to develop at the same time.

9.1.2 The "Three-Step" Population Development Strategy

Previous chapters have dealt with the population development strategy. Here, we mainly discuss the "three-step" population development strategy, and the strategic transfer.

The first step in the transition to a population reproduction type characterized by low birth rate, low mortality rate, and low natural increase rate, was control of population size. Given the contradiction between the shortage economy and surplus population after 1949, the choice to focus on population size control reflected a natural law of demography. Meanwhile, we also gave equal consideration to enhancing population capacity, adjusting population structure, and to the reality that population and socioeconomic development are mutually dependent and restraining. As mentioned, when the PRC was founded in 1949, there were 541.67 million people in China, accounting for nearly 22% of the world population. This young population was also a typically growing population. At the time, the total fertility rate (TFR) was above 5.0, the birth rate was 36 per 1000 people and the natural increase rate was 16 per 1000, indicating strong growth. In the next three decades, despite some progress in the transition to an older one, China was still a young population. Due to the first (1953–1957) and second baby booms (1962–1973, with the birth of 320 million people during this period), the transition to an adult population structure was postponed and the strong growth momentum persisted. With the application of family planning policy, the TFR dropped dramatically, to 2.24 in 1980, as did the birth and natural increase rates, with the former going down to 18.2 per 1000 people and the latter to 11.9 per 1000.[1] Nonetheless, the overall rate of population growth did not change.

For various reasons, China remained socioeconomically underdeveloped, manifesting a sharp contradiction between a shortage economy and surplus population. While population growth was not the major reason, there were serious economic shortages on the one hand and a surplus population on the other, thus it was clear that the problem had to be tackled from both sides: we should strive to increase economic supply in order to fix the shortage economy. And at the same time, we should also implement the birth control policy. Great efforts were made over the next 30 years to curb population increase and certain achievements were made. Yet there was also regression. For example, when Ma Yinchu's new population theory was criticised, the idea that the more people the better was very influential. The second baby boom (1962–1973) was closely related to this. By 1980, there had been a net increase of 445.39 million people, an 82.2% increase and an average annual increase of 2%.[2] This was the fastest population growth in China's history.

In these circumstances, it was essential for the CPC Central Committee to send an open letter to the Party members and communist youth league members, introducing the "one couple, one child" policy as the pillar of family planning. The

---

[1] National Statistics Bureau of the PRC (1986).

[2] The calculation is made according to the data on p. 91 of the book *China Statistical Yearbook of 1986*.

Central Committee's chief aim with this policy was to control the fertility rate of one generation. The Open Letter stated explicitly, "in 30 years, the currently worrying population growth problem will have been mitigated and a different population policy might then be adopted". But there was also something inappropriate in the letter. For example, its first sentence said, "we should strive to keep the total population in China under 1.2 billion by the end of the twentieth century", a rather demanding remark that indicated impatience for quick results. On this point, I discussed in Chapter 6 above how the target was revised, and the real situation of population increase at the time, so we need not go into detail again here. The important point is that after the ups and downs of Chinese population change over 30 years, it was wise for us to focus population policy on controlling the population size, by encouraging each couple to have only one child. We had grasped the main aspect of the contradiction. Practice proved that family planning measures were key to promoting the population transition, and facilitated economic development and social advancement through transforming social traditions. Nevertheless, the problems that occurred during the application of family planning policy are worthy of our attention. Some of them were quite serious and great efforts were required to remedy them. Further elaboration on these achievements and problems is made later in the book.

The second step in the transition to a "low" population reproduction type was a focus on the comprehensive enhancement of population capacity, strengthening human capital accumulation, and paying equal attention to population structure adjustment and size control in order to promote the coordinated development of population, resources, environment, economy, and society. Why concentrate on population capacity rather than size or structure? As has been explained, controlling the size of the population was the focus of the population development strategy for over 30 years. The aim, to control the fertility rate of one generation, was accomplished and we were now in the post-demographic transition period. Some people believe that the direct consequence of the one couple, one child policy is that there is now an aging problem and that hence, adjusting the population age structure should be the second step in the population strategy. I do not share this opinion, and here are some of my reasons.

First, we need a proper understanding of the aging problem. It represents a process wherein the ratio of people aged 60 or 65 years old or above continuously increases. So, the first task is to measure the percentage of old people among the total population. We also must know the median age of the total population: the higher the median age, the higher the percentage of elderly and the higher the degree of aging. Conversely, the lower the median age, the lower the percentage of old people and the lower the degree of aging. The ratio of old people to children refers to the ratio of people over 60 (or 65) to children from 0 to 14 years old. The higher this ratio, the higher the percentage of old people and the higher the degree of aging. The lower the ratio, the lower the degree of aging. According to the population transition theory, during the transition from the high birth rate, low mortality rate and high natural increase rate pattern to that of low birth rate, low mortality rate and low natural increase rate, the aging problem is a natural phenomenon, the inevitable result of socioeconomic development, improved medical and health services, a decrease in age-specific mortality rate, and increased life expectancy. As a reflection of social development and human progress, then, the aging problem is positive rather than negative.

This does not mean there are no negative aspects to the aging phenomenon. From a historical point of view, the "silver wave" is one that is certain to appear in the various waves of development. It is a phase that must be passed through on the way to the optimum population. Thus, we must put the aging problem in perspective, looking at it critically, in order to take advantage of its benefits and avoid its drawbacks. Providing for the elderly increases the social burden. The old-age social security system will consume a substantial amount of social wealth and a broad series of support services is needed to take care of seniors. Yet the aging phenomenon has not only negative elements but also positive ones. In the transition from a young population to an adult population, and in the early period of the transition to an old population, there will be a "golden era" during which the dependency ratio generally decreases, which brings demographic benefits[3] and dividends. In this so-called golden era, the ample labour force and light social burden are highly conducive to economic development, as was verified by China's rapid development since the reform and opening-up policy. It was also evident in Japan's economic take-off after World War II, and in the socioeconomic development of South Korea.

Next, we must assess the aging of the Chinese population carefully. Two tendencies should be opposed. The first is the failure of some to face squarely the rapid development of the aging problem and to try seriously to solve potential problems associated with it. For example, at the 1980 CPC Central Committee symposium on population question, a participant remarked that there would be no problem even if the "one child policy" were carried out for 50 or 80 years and that we need not worry about aging. In the enforcing of the one child policy, some did not pay much attention to the potential problems associated with aging, such as problems confronted by old couples who had lost their only son or daughter. A different tendency is over-worrying about aging; some believed that China was trapped in the aging problem and that it is insoluble. Neither of these two tendencies are practical or realistic. Neither sees the aging trend correctly nor how to solve the problems it creates.

The Chinese population is aging at a faster pace than in both developed and developing countries. The aging tendency in China is similar to the situation in Japan, the most rapidly aging society in the world. When aging reaches its peak in China in 2050, the ratio of people over 65 might be more than 23%, close to the level in the developed countries and the highest among the developing ones. Aging develops progressively and can be divided into different stages. The first stage (2000–2020) was the initial period, still partly in the old population structure. At this stage, the ratio of those over 65 years was predicted to increase by an annual average of 0.26%. The second stage (2020–2040) is when aging speeds up; during this period, the ratio of old people is predicted to grow by an annual average 0.5%. During the third stage (2040–), aging will slow until it basically stabilizes. Between

[3] "Demographic profits" here refers to favourable population conditions in a country (a high proportion of working-age population and a low dependency ratio).

2040 and 2050, the ratio of old people is predicted to rise by an average 0.11% per annum, being basically stable but with some fluctuation. The aging of the Chinese population is not equal in urban and rural areas or among the different regions, but aging differences will gradually become smaller.

Third, we must decide on the best ways to mitigate the aging problem. Population aging, unavoidable in the transition to a post-demographic period, is a positive and natural development. But that does not mean the faster and greater the aging, the better. Aging has an objectively negative influence on population change, economic development, social progress, and scientific and technological advances, which needs to be addressed. There are two solutions: to strengthen research into existing problems caused by aging and therein to seek solutions, and the other is to slow the rate of aging and reduce its degree. Through slowing down aging and alleviating its pressure, we can address the problem with fewer difficulties and develop better solutions. Judging from the aging trend in China and the practical problems it creates, both measures are needed. But first, we need a detailed analysis of the circumstances.

The decline of the birth rate and increased life expectancy are two primary reasons for the aging of a population. Can we find a solution in lowering people's life expectancy to reduce the proportion of old people? Of course not. Longevity is a common wish. Lowering people's life expectancy would be a crime against humanity. Increasing the fertility and birth rates would be both practical and viable. But if we raise the birth rate, those born will also (mostly) become old, creating an endlessly growing population. Based on our analysis, boosting the fertility and birth rates is necessary to resolve the aging of population age structure, yet this must be done appropriately. Done carelessly, the population will increase continuously. While this reduces the aging problem, we end up trapped in unchecked increase. This kind of population adjustment should be avoided. It should be clarified here that the purpose of boosting fertility and birth rates is to prevent aging from becoming too serious and to facilitate solutions to the population problem and socioeconomic development, rather than to eliminate aging. We should not simply boost the birth rate to offset aging. Fertility and birth rates should be raised in such a way that we can prevent aging from becoming too serious a problem. Thus, population development will not only comply with the population strategy but will also be conducive to a reasonable, comprehensive solution to the population problem.

Adjustments to gender structure as well as to the urban–rural and geographical distribution resembles adjustments to a population age structure, so they should not be the focus of the general population development strategy. Comprehensive improvement of the population's skills is an essential part of the whole population development strategy and should be the focus of the second population strategy, replacing population size control that was its first focus. Comprehensive improvement of human capacities is beneficial to population size control. In the 1990s, I conducted a survey on family economy and fertility that showed that women of child-bearing age with different education levels (tertiary, middle school and primary education), had different numbers of children on average. Children per woman decreased by 0.3 to 0.5 as the mother's education level improved. Therefore, the rise in the population's education level means a fall in the fertility rate. Nowadays, as the average number of children per woman has decreased and people have easier access to education, this difference has diminished, but still observable. The effect of education levels on urban–rural distribution and geographical distribution is reflected through population movements. People with higher education levels are more likely to migrate, which favourably alters urban–rural distribution, economic density of population, geographical population distribution, and sustainable development of people, resources and the environment.

There is a different view of what should be the focus of the second step in the population strategy. Some thought that since demographic profits and demographic dividends would decrease or even disappear when the proportion of the working age population (15–64 years old) reached its peak in 2010 and its absolute number peaked in 2013, the focus of population policy adjustment should be on increasing the supply of working age people. How to increase it? There is still only one prescription for it—increasing the fertility rate. If we raise the fertility rate and increase births, the proportion of working age people will be increased and more labourers will be available in 15 years. This is like the idea of reducing aging through raising the birth rate. Neither way can solve the population problem completely. Moreover, they both conflict with the requirements of the Chinese economic development strategy. When it transits to "high-quality development", increased productivity becomes the principal path. Yet the fall in the proportion of the working-age population and the decline in its absolute numbers will force us to abandon our dependence on augmenting the labour force. Instead, we must adopt a strategy of raising labour productivity. Thus, high quality development matches the strategic shift in population policy to comprehensive improvement of the population's capacities.

At the beginning of the 1980s reforms, when the goal was to quadruple the gross output value of agriculture and industry by 2000, it was made clear that "continuous improvement of economic benefits" was the prerequisite. At that time, transforming the economic growth pattern was discussed frequently and the proposed economic transformation models were constantly updated: from the extensive model to the intensive; from economic growth induced by production means to productivity-induced economic growth; from an investment-led economy to a consumption-led economy; from a quantity-oriented economy to a quality-oriented one, and from a period of high-speed growth to one of high-quality growth. The general tendency is a shift from economic growth due to increased production means, to growth due to productivity enhancements, and from quantity-oriented growth to quality-oriented growth. The final focus is on qualitative improvement. So why was the transformation of the economic development model not as effective in the 30 years following the introduction of the reform and opening-up policy? Most studies have analysed this question from the perspectives of weak economic foundation, the difficulty of breaking out of the old economic system, and according to traditional official standards and the evaluation of officials' performance. Such explanations are reasonable and necessary, but they have neglected the influence of population transition and working-age population change on economic development. This influence was clearly visible during the 30 years since 1978.

One reflection of the restrictions they created was seen in the growing pressure on employment created by the growing working-age population. Statistics showed that the working-age population (15–64 years old) grew from 615 million in 1980 to 763 million in 1990, from 889 million in 2000 to 999 million in 2010, a proportional increase from 59.8% (1980) to 66.7% (1990), 70.1% (2000) and 74.5% (2010).[4] As the working-age population expanded, the economy tended to expand production through increasing labour. Regression analysis indicates that from 1980 to 2010 the increased labour contributed to approximately 30% of GDP growth, and that demographic profits and demographic dividends played a significant role in the economic surge.

Another reflection of the impact of population transition and working-age population change on economic development was that the transformation of the economic development model was impeded by the population's underdeveloped capacity and inadequate human capital accumulation. I did some calculations using third, fourth, fifth and sixth national census data. The results indicated that the average years of schooling of those above six years old was 4.2 years in 1982, 5.2 years in 1990, 6.8 years in 2000, and 8.3 years in 2010.[5] It is correct that with a large population and the talents being pooled, the Chinese accomplished some high-end projects, exemplified by the nuclear bombs built in the 1960s and the artificial satellites launched in the 1970s. Yet, generally, it is difficult to build an industrial system underpinned by advanced technology with such low education levels, and a truly formidable task to transform the economic development model in such circumstances.

According to General Secretary Xi Jinping's report to the 19th CPC National Congress in 2017, China's economy was transitioning from a phase of rapid growth to one of high-quality development. How was this possible? There are various reasons, including that China has consolidated its position as the second largest economy in the world, its economic and technological structures having attained the medium–high level and its industrial structure approaching the modern one. Chinese population change is another important reason: the proportion of working-age people in China reached its peak in 2010, ending the era of an unlimited supply of labour and opening a new phase of economic development led by productivity improvement. As the ratio of working-age population declined and its number decreased, comprehensive improvement in population capacity reached a new high, with the average

---

[4] National Statistics Bureau of the PRC (2018).

[5] Population Census Office under the State Council and Department of Population Statistics, State Statistical Bureau, PRC. (Eds.). (1985). *1982 Population Census of China (Results of Computer Tabulation)*. China Statistical Publishing House. pp. 354–355;

Population Census Office under the State Council and Department of Population Statistics, State Statistical Bureau, PRC. (Eds.). (1993). *Tabulation on the 1990 Population Census of the PRC*. China Statistical Publishing House. pp. 32–51;

Population Census Office under the State Council and Department of Population, Social Science and Technology Statistics National Bureau of Statistics of China. (Eds.). (2002). *Tabulation on the 2000 Population Census of the PRC*. China Statistical Publishing House. pp. 593–602;

Population Census Office under the State Council, Department of Population and Employment Statistics. (Ed.). (2012). *Tabulation on the 2010 Population Census of the PRC*. China Statistical Publishing House. pp. 295–300.

schooling reaching more than 10 years. This is the moment at which quality-oriented economic development transition could begin and become effective. Thus we see that improvement in the population's capacities and the construction of a nation with strong human capital are requirements not only for resolving the population problem in China, but also for boosting high-quality economic growth. Therefore, it is necessary and reasonable that the focus of the second step in the population development strategy be on comprehensive improvement of population capacity.

The third step of population strategy is to set an all-round optimum population as its ultimate goal, to promote population development and sustainable development strategy. An "all-round optimum population" refers to optimal population size, capacity and structure and means that the population fits in with the development of resources, the environment, the economy and society. This is the end goal of the "three-step population development strategy", which can be achieved through the population and sustainable development strategy.

The relationship between population and natural environment has attracted much attention and there are different views on the question; one is that nature is to be conquered and dominated by humans, and another is that humans should return to nature. I believe rather that humans should seek a harmonious development with nature, wherein the former are integrated with the latter. In other words, population size should be optimal, within the capacity of the natural environment and resources. It should also meet the objective requirement of social and economic development: the population should have a high level of physical fitness, intelligence and technological knowledge, and moral and cultural fitness, which will ensure that they can preserve the environment and meet the socioeconomic requirements for human capital. The population's age, gender, regional distribution, and urban–rural structures should be reasonable, able to provide the conditions for continuous and sustainable development. The continuity and sustainability of development are relative, for different historical periods have had different requirements for the quantity, quality and structure of population and human capital. Yet continuity and sustainability of development are also absolute, for the population is generally stable, with stable change and development. We cannot imagine a population whose quantity, quality and structure remain unchanged forever. There is only a general population of relative stability and without drastic change. Such a general population is based on certain social and economic developments, so it is relatively stable. As the social-economic base changes, population size, capacity and structure will change accordingly before it regains its all-around optimal state.

9.2 Zero Population Growth

As the population development strategy proceeds to the second step, the focus of policy shifts to comprehensive improvement of population capacity. After that there should be clear strategic goals for population development. Despite some uncertainty, their orientation and tendency are clear and a general strategic goal is needed. The book *Research on Chinese Population Development Strategy in the 21st Century*, that I coauthored with several other scholars suggests that a series of goals could be achieved by 2030. They are based on our reaching zero population growth and a total national population of approximately 1.465 billion. Thereafter, the population will decline slowly, to around 1.402 billion by 2050. Population skills have improved significantly. According to UN predictions using the increment method, life expectancy estimates for both sexes combined could be 76.1 years in 2030, and 77.1 years in 2050. The infant mortality rate might drop below 10 per 1000, approaching the level in the developed countries. High school education will be accessible to all, and the average schooling of people over 15 might surpass 12 years. People's cultural and moral fitness will improve significantly, with a sharp decline in the crime rate (including youth crime) and there will be a manifest improvement in awareness of legal and moral codes. The proportion of working-age people will drop to around 67.36% in 2030 and 61.29% in 2050. In contrast, the ratio of those over 65 years will rise to 16.2% in 2030 and to around 23.1% in 2050. Urbanization will continue to proceed rapidly, reaching 72.2% in 2030 and around 75.4% in 2050.[6]

According to predictions about population change, the birth and mortality rates will not change rapidly in the first half of the twenty-first century and will remain the same around 2030. Then the population will reach zero growth and from then on, will decrease slowly, commencing a period of an all-around optimum population. As mentioned, the focus of the population development strategy at this second stage should be shifted to comprehensive improvement of skills. But the aging changes are closely linked to the fertility rate, and are the result of the interaction of fertility, birth and mortality rates. It is clear that comprehensive improvement of the population's capacities is closely associated with natural population change, each influencing and limiting the other. We should also note that as the focus shifts from controlling population size to enhancing capacity, people's understanding of the shift and their subsequent action also must change. The transition does not happen immediately. Therefore, how population size changes is a matter of general concern. According to stage two of the population development strategy, the population is expected to reach zero growth, but there are only around 10 more years to reach the goal and people are very concerned about it. Questions about why we must achieve zero population growth, the timeframe within which it should be achieved and what population policy is required to do so arise. Responses to these questions should be formulated according to the actual conditions prevailing in China.

Why should we aim for zero growth? This is an old question that has been debated in China and throughout the world since World War II, when there was a baby boom in the developed countries. Thereafter, worries about a rapidly increasing global population were amplified, and warnings were issued about the threats imposed by an ever-growing population or a population explosion. The 1972 report *The Limits to Growth*, by Donella H. Meadows and Dennis L. Meadows and two other scholars of

6 Tian et al. (2007).

the Rome Club was a key text. Following a comprehensive analysis of world population, food production, industrialization, non-renewable resources, and the environment, they warned that if the exponential growth trends in these five areas continued, the limits to growth would be reached sometime in the next 100 years. Food shortages, depletion of natural resources and pollution would become unbearable, and industrialization and socioeconomic development unsustainable. To avoid such a tragedy, the report suggested a series of measures to be taken, one of which was certain interventions to make the birth and mortality rates equal, to bring the population to zero growth.

That report was published nearly half a century ago. Judging from current practices and conditions, some of the arguments in the report have held water while others appear to have been somewhat exaggerated. This is not the place to evaluate it comprehensively, but regarding zero growth, at the end of its first chapter the authors asked: "How many people can be provided for on this earth, at what level of wealth, and for how long? To answer these questions, we must look in detail at those systems in the world which provide the physical support for population and economic growth."[7] They suggested the limits to both economic growth and population growth and that the population would achieve zero growth when the birth rate was brought down to match the mortality rate. Though they did not argue the case for zero population growth adequately, their views on the nature and limits of exponential growth were greatly influential.

In China, a few scholars conducted research on the question raised in *The Limits to Growth*. They have explained how many people China could support given its 9.6 million square kilometres of land and over three million square kilometres of sea, considering natural resources, especially fresh water, economic growth, especially the weighting effect of population growth, and environmental pollution, especially the effect of population growth. "How many people China can provide physical support for?" is a crude proposition; we would do better to ask "what is the optimal population China could sustain?" Although the early empirical studies achieved much, they were far from enough. Further research is needed and it should be conducted on the basis of existing research results from both within China and abroad, and taking into account actual changes in our population, economy, society, resources, and environmental development. Such a substantial and comprehensive research project will require collaboration among many disciplines, and an integration of theory and practice. This current book can go no further on this topic, but here I want to state the preconditions for such a project: zero population growth is an inevitable population change tendency in China as elsewhere and it is a stage we must pass through to achieve all-around optimum population.

First, let us talk about the purpose of population growth. This seems to be an easy question, yet it is usually not elucidated. Note that here "population" is a general concept whose demographic, economic, and sociological characteristics are abstract. That is, the concept refers to an ahistorical "population". Yet people engage in population reproduction with a certain purpose. It is not helpful to deny the purposefulness

[7] Meadows et al. (1972).

behind population reproduction by saying that in a class society, people are divided into classes. Regardless of the society, and people's class positions, the purpose of reproduction remains unchanged: to produce the next generation and to strive to enable them to prosper and to do better than their parents. For example, in feudal society, the popular view was that more children would bring more happiness. People generally wanted to have as many children as possible. Childless men were considered unfilial, as reflected in the saying "there are three forms of unfilial conduct, of which the worst is to have no descendants". Such a belief does not come from nowhere. It is the result of the cost–benefit advantage of children in pre-industrial times. More children, especially more sons, bring more benefits to the parents. For thousands of years this view was passed down from generation to generation, and during the process, the fertility ideology of having as many children as possible was formed. This ideology prompted the feudal ruling class to develop appropriate population and fertility policy, including related land and taxation policies. It also impacted people's fertility behaviour in general. People prided themselves on having a prosperous family through having more children, keeping several generations under one roof, and having male heirs to carry on the family line. In contrast, those with fewer children or even without a male heir were despised and ridiculed. A son-less family was seen as punishment for some monstrous crime.

In modern society, the cost of raising children has risen significantly, and at the same time, the economic benefits and old-age insurance benefits that children can provide to their parents have decreased dramatically. Likewise, the spiritual pleasure that children bring to their family and the benefits of continuing the family line have declined to varying degrees. As a result, the cost–benefit of raising children has changed dramatically, from a pattern of negative cost and positive benefits to one of positive cost and negative benefits. Though such transformations have not occurred in all families, the cost and benefits of raising children tend to be close to each other. The belief that having more children brings more happiness is fundamentally undermined. The idea has lost its impetus and been superseded by the notion that it is better to have fewer babies and to rear better children. This is a great step in human development, whose significance is on par with any other major invention. Why? Because it helps people to break away from blind procreation and promotes the view that reproduction should accord with socioeconomic development goals and a harmonious relationship between reproduction and natural resources. It projects bright prospects for a society on the path towards an all-around optimum population.

Zero population growth is raised to the status of a strategic goal, to be accomplished through social and economic development and a population policy designed to accommodate it. This is the approach that will resolve the contradictions between limited natural resources, the worsening threat of environmental pollution and infinite population growth. We do not want to state that the earth is overcrowded, but it is true that zero population growth is the solution to the increasingly crowded planet. Therefore, to create a favourable environment for Chinese population development should be the primary goal of our national population development. Its basic requirements include the following conditions: the shortage of natural resources should be relieved; environmental pollution should be reduced; there must be more space for economic growth, better housing conditions and decreased traffic congestion; the economy's dependence on a growing labour force must be shifted to improved productivity; with improved productivity, people will have more time for learning, relaxing and leisure activities; physical fitness, technological knowledge, and cultural and moral fitness will be effectively improved and the population will work, live and develop better than the previous generations. All of this is, surely, the ultimate purpose of human production and reproduction.

Deng Xiaoping once said, "poverty is not socialism." In 2012, Xi Jinping became General Secretary of the CPC Central Committee and declared, before Chinese and international journalists, at a press conference of the new Standing Committee of the Political Bureau of CPC Central Committee, that "fulfilling the people's wish for a good life is our goal". Does the Chinese population need to continue to grow endlessly, or should it try to achieve zero population growth at an appropriate time? Which would be more favourable to the realization of the nation's wish for a good life and reflect its fundamental interests? If we choose the latter option, we could achieve an all-round optimum population at a certain moment, providing favourable population conditions for realizing "people's wish for a good life". If we choose the former option, the population will keep growing. On the way to realizing "the people's wish for a good life", we will always be confronted with conflicts between population growth and social and economic development, natural resources, and the environment. Consequently, the nation will struggle to develop with heavy load. In my opinion, there are both merits and demerits in the traditional fertility view of "the more children, the better". We must make a specific analysis, because this view is often manipulated to criticize zero population growth or negative population growth indiscriminately. As mentioned, there is an economic and social basis to the dominance of the traditional view that "the more children, the better". Though in the modern society such basis no longer exists, such a belief will not disappear immediately. But as history advances, the decline of such a view is inevitable. The advent of zero population growth is unstoppable and it is approaching.

Next, we will talk about the problems with population growth. As I have argued in earlier chapters, the major population problem at the beginning of the reform and opening-up lay in the contradiction between the shortage economy and the surplus population: 1. The economy could only provide low-level products to satisfy people's physiological needs while those that could meet psychological and development needs were scarce. 2. There was a young population age structure and a high social dependency ratio. Since the accumulation was limited, the portion of wealth available for investment and increasing national economic growth was limited. 3. The increased means of production can not fit the big working age population. As a result, the economy was forced to offer more jobs with low wages. The national economy had to move into a cycle of "more jobs with low wages—low productivity—more jobs with low wages". 4. The population's capacities were insufficient for the advances in economy and technology. Low education and technological skills seriously hindered the development of high-tech industry. 5. The low rate of urbanization restrained the upgrading of industrial structures. 6. The imbalance in the regional population distribution made it difficult to change the unbalanced distribution of productive forces.

With rapid population growth and relatively slow social and economic development, the focal point of the contradiction between the shortage economy and surplus population was "whether living standards were sufficient." More than 40 years after the introduction of the reform and opening-up policy, these situations have changed greatly and some fundamental changes have occurred. With the long-term economic growth and development, supply capacity has been enhanced remarkably. With the continuous decline of the birth rate, population growth has weakened drastically. We are now in the middle of a post-demographic transition, and the relationship between population and economy has changed significantly.

So, has the contradiction between the shortage economy and the surplus population disappeared? I am afraid not. While the "sufficient living standards" contradiction is basically resolved, the question of whether living standards are good remains. The lack of high quality economic development begins to emerge. On the consumption side, the problem of adequate food and clothing has been basically solved. But after people's basic physiological needs are met, we need to offer them good food and clothing. More importantly, we need to address the high-order question of how to satisfy their psychological and development needs. The extensive production model has continued for over 30 years. Although there has been overall economic growth, stubborn problems remain, including poor product quality, inadequate technological equipment, lack of core technology and weak international competitiveness. Since the high quality development mode is put forward, there has been a noticeable improvement. At present, the contradiction between the shortage economy and a surplus population has not been resolved, but it no longer exists in the same way. The nature and manifestation of the contradiction has certainly changed. Economically, the focal point of contradiction shifted from product shortage to poor product quality. The focus of the population question has shifted from surplus population to poor population capacity that cannot enable high quality economic development. To solve the contradiction between the shortage economy and surplus population, we should first vigorously boost economic growth, concentrating on expansion of economic yield and output value. Then we should make efforts to control population growth and to complete the population reproduction transition. To offer better living standards, we must first upgrade and transform economic structures in the pursuit of quality economic development. Then we should promote population capacity and reinforce human capital accumulation. In solving the first contradiction, we should focus on controlling population growth; in dealing with the second one, we should shift the focus to improving population capacity. Comprehensive improvement in human capacities and the continuous reinforcement of human capital accumulation form the foundation for high quality economic growth. They also serve to curb population growth effectively, to achieve zero population growth and to facilitate population structure adjustment.

The final question here is how to achieve zero population growth when taking the second step of population development strategy. Population size, population capacity and population structure, being interdependent and interactive, are a unity of opposites that coordinate with each other and promote each other's development.

To advance the second step of population strategy, we must emphasise the comprehensive improvement of the population's skills. Though achieving zero population growth is not the focus of the second step of population strategy, it is an important move in the transition from the first step to the second and it concerns the overall population development strategy. So, how to accomplish this strategic goal? It must be integrated into the comprehensive improvement of population capacity, the enhancement of which could promote zero population growth. Such a view has already been implied in previous discussions in this book. Why is it essential to transform the economic development mode and adjust economic structures? It is because along with the problems of the economic and technological structures, as well as those of the management system, poor general population skills constitutes one of the important reasons. How do we strengthen our human capacities? By vigorously developing sports, education, science, technology, health and medical care, and cultural activities. Thus, people's health and physical fitness, education and technological skills, cultural and moral fitness can be effectively promoted, building a solid foundation for human capital accumulation. To accomplish this, curbing population growth and improvement of human skills must be integrated. Empirical evidence shows that improvement in capacities, especially education and technological skills, can reduce fertility and birth rates. Population size and population capacity are in opposition and yet also complement each other. Today, science and technology advance at a tremendous pace. Machines have replaced human labour in some work, sometimes doing work that humans cannot. They even can do some work that cannot be matched by humans. With wage rates increasing continuously, some enterprises, foreign investment enterprises and joint ventures, in particular, have chosen to reduce their workforces and started to replace people with robots. In so doing, they can raise productivity, reduce friction between workers and management, save costs and increase profits. In this sense, the rapid spread of intelligent tools promotes the achievement of zero population growth.

9.3 Reform as the Key to Improving Population Capacities

The key to accomplishing the second step of the population development strategy and the goal of improving population capacity across the board lies in reform of the mechanisms and systems associated with health and physical fitness, education, scientific knowledge, and moral and cultural fitness. Until now, these reforms have either started late and made slow progress or started relatively early yet barely made any progress or had any measurable effect. We need a clear picture of the situation in order to identify our shortcomings and grasp the best direction for development, to strengthen and deepen the reforms.

9.3.1 Deepening Reform of the Health and Physical Fitness Sectors

Among the many departments, public institutions and enterprises related to the Chinese population's health and physical fitness, the medical, health and sports fields are the most closely linked and thus are the key targets for reform. Since medical and health departments are bound up with each other, we analyze and elaborate on them together.

9.3.1.1 Deepening Reform of the Medical System

Reform of the medical system began early, with the beginning of the reform and opening-up process. However, after nearly 40 years, reforms to the medical system have been a failure. They mainly focused on who was to foot the bill for medical treatments, which resulted in an increase in private spending on medical services and a continuous fall in the part paid by the state and related social agencies such as insurance companies and private medical institutions and so on. The phenomenon of difficult and costly medical treatment has not disappeared and instead, it has worsened. According to official statistics, between 1978 and 2000, the proportion of government expenditure on health in the PRC fell from 32.16% to 15.47%, a decrease of 16.69%; the proportion of social spending on health fell from 47.41% to 25.55%, a decrease of 21.86%; private individual spending rose from 20.43% to 58.98%, an increase of 38.55%. As we approached the twenty-first century, the situation improved to some degree. In 2017, government health spending grew to 28.91%, a rise of 13.44% over 2000 expenditure, yet still 3.25% less than in 1978. Social spending rose to 42.32%, an increase of 16.77% over 2000, but 5.09% less than in 1978. And the proportion of individual spending on medical services fell to 28.77%, 30.21% less than in 2000, and 8.34% more than in 1978.[8]

The current medical reform plans must break with the "who pays for medical services?" question and begin some real change. Most importantly, the monopoly and semi-monopoly over medical resources, especially premium ones, must be broken. Specifically, we should rectify the urban–rural imbalance, wherein there is a high concentration of quality medical resources in cities, especially mega-cities like Beijing and Shanghai. Let the market play the decisive role in resource allocation and optimize the allocation of all sorts of medical resources and utilize them rationally. At present, an average of more than 100,000 people travel to Beijing every day from all around the country to seek special medical care in the capital's hospitals. The explanation lies in the high concentration of quality medical resources, their irrational allocation and the virtual monopoly or semi-monopoly over those resources. To promote medical system reform, we should first insist on national medical standards. Next, we should vigorously develop many kinds of medical services and

[8] National Statistics Bureau of the PRC (2018).

medical institutions, such as the services offered by the collective, private, joint-stock system, and joint-stock cooperative medical facilities. In so doing, we can relax and resolve the contradiction between medical workers and patients, the conflict of the mismatched medical supply and medical demands, and change the imbalances created by irrational resource allocation.

9.3.1.2 Deepening Reform of the Sports System

Recently, there have been growing concern and heated discussion about deepening reform of the sports system. We should establish coherent sports guidelines, from the higher levels down to the basic levels. Since 1949, we have adhered to the guideline "promote sports and strengthen people's constitutions". Both mass sports activities and competitive sports have made great progress. With the application of the sports system of "preparing for labour and defending the country" as adopted by the USSR and the adoption of the athlete hierarchy system, and the flourishing of gymnastic exercises broadcast on national radio, a popular mass sports activity, people have been the focus of sports activities, and their constitutions have improved unprecedentedly. However, some erroneous practices have emerged in the sports area. For example, while competitive sports are prioritized, mass activities are neglected. Let's look at the main focus of sports departments at different levels. It seems that each level prioritizes winning competitions and is not truly enthusiastic about mass sports activities. This phenomenon contradicts the original aspiration of developing sports and goes against the purpose of enhancing people's health and physical fitness.

This does not mean that we should do away with competitive sports, for they encourage people to achieve excellence through high spirits, courage, and strength. Winning competitions should be encouraged, with the relevant incentives, because it is an important way to promote national prestige and enterprising spirit; it is also indispensable to enhancing population capacities in an all-round way. Yet if boosting competitive sports and winning championships constitute the primary or sole aim, around which the whole cause of sports revolves, then we would be disconnected from the principle of "improving people's health and physical fitness". It is as if we have turned the whole matter upside down. So, we need to recognize clearly that competitive sports are underpinned by mass sports activities. Only when the latter flourishes can the former be promoted, with champions emerging constantly and continuously. When deepening reform of the sports system, we should set the relationship between competitive sports and mass sports activities right, shifting the focus to mass sports activities and facilitating people's health. This shift of focus highlights the essence of sports, consolidates the foundation of developing competitive sports, and sustains the vitality of sports.

9.3.2 Deepening Reform of the Education and Scientific Systems

To achieve modernization, science and technology are key, and education is the foundation. To make comprehensive improvements to the population's capacities, we need to emphasize effective improvement of general education levels and the scientific knowledge base. Since the founding of the PRC, and especially since the implementation of the reform and opening-up policy, the fields of both education and science and technology have developed rapidly and made great achievements. Yet some problems linger, chiefly in exam-oriented education system and the sector of science and technology.

9.3.2.1 Deepening Reform of the Education System

What is the current education system and what is its main problem? The answer to both those questions is the "exam-oriented" education system, hence this should be focus of education reforms. Currently, exams constitute the start and the end of education, the core of teachers' instruction and students' learning. This might sound exaggerated. But after a visit to the high schools known as "factories for college entrance examinations" or "college entrance examination military camps", the description sounds somewhat understated. The reality is more shocking.

In 2018, I went back to my old high school, where I studied more than half a century ago. My alma mater's reputation has been growing since four students (including myself) were admitted by Peking University, one of the top universities in the country, and over ten other students were admitted by other leading universities directly under the education ministry in 1959. At the time, the acceptance rate among the top universities was very low. Based on online information, my old school ranked 34th among the top 100 high schools in China in 2019. When I arrived at the school that day in 2018, the gate was closed and a sign on it read, "Closed teaching is practiced in this school, and it is not open to visitors." The doorkeeper told me that anyone who wanted to visit the school had to ask for permission from the Education Bureau under the municipal government, more than ten kilometres away. I was astounded and immediately understood that the school had become one of the "college entrance examination factories". A tour of some other nearby high schools revealed that they too had become such "factories" to a certain degree. Then I thought of some peculiar phenomena that arise at the summer camps that host high school students from both China and abroad. When these participants are evaluated according to their exam results, the Chinese students come out on top, but when their independent thinking and practical abilities are assessed, the foreign students do better. Why is this so and what led to it? The answer seems to lie in the nature of these "examination factories" schools. A good education is of great importance to students and their parents, connected as it is to their personal futures and the development of the nation. In today's China, only children account for a large proportion of offspring. As the popular saying goes, "man struggles upwards, water flows downwards". It is understandable that parents pin high hopes on their children's education. They hope that their child will go to university, a renowned university, and then pursue their studies through a postgraduate program, perhaps overseas. An intriguing question is what a bright future really means. Does it mean that a person has lofty moral character, high intelligence, and profound knowledge? Or that the person has good grades and has been granted a certificate by a famous university?

We cannot go further in our discussion without discussing the well-known "question of Qian Xuesen".[9] For a certain time, "the question of Qian Xuesen" was interpreted as "why do our schools not produce outstanding talents?" As a matter of fact, there is a misunderstanding in this interpretation. The real story was that in July 2005, then premier Wen Jiabao went to the PLA General Hospital to pay a visit to the renowned scientist. In discussion of topics such as independent innovation in the next 15 years, Qian agreed with what Wen had said. Then Qian added, "I want to talk about one more question on education. Right now, China has not fully developed, and one important reason for that is that not a single university in China has oriented itself towards training talents who excel in scientific and technological invention and innovation. Without their own unique, innovative stuff, outstanding talents will not emerge. This is a huge problem." This was Qian's last important contribution to China's scientific advancement and education development. His advice that "education should be oriented towards training talents who excel in scientific and technological invention and innovation" is equal to transforming the education system. Over the years, some great achievements have been made but the "exam-oriented" education system remains unbroken. Qian's statement that "not a single university in China has oriented towards training talents who excel in scientific and technological invention and innovation" strikes at the root of the problem of the Chinese education system. With such a different orientation, the teaching effects will be different from those in other countries and brilliant talents will not emerge. While the education's orientation is the cause, the (lack of) cultivation of talents is the effect. "Qian's question" became popularly known, but attention has been paid only to the effects and not to the cause, thus thwarting Qian's real intention in raising the matter. The orientation of the education system is essential and the key to whether we can produce outstanding and creative talents. Therefore, the orientation and organization of the system should be the focus of education reform.

In practical terms, education reform should focus on the following three points. First, the education aim "to cultivate students with an emphasis on moral fitness" should be maintained. The current policy states that "education must serve socialist modernization and the people and must be combined with productive labour and social practices. Its aim is to cultivate socialist builders and successors with an emphasis on the concurrent development of morality, intelligence, physical fitness,

[9] Qian Xuesen (1911–2009) was a prestigious Chinese scientist, an academician of the Chinese Engineering Academy. He was honored as "the Father of China's Missiles" and awarded the Medal of "Two Bombs and One Satellite".

aesthetics, and labour". This comprehensive policy is the result of several adaptations. However, its weakness is that while it is inclusive, it is less than concise and no focus is highlighted, so it is not easily grasped. Education policy must guide education; it should serve the all-round development of citizens, meeting their physiological, psychological, communication and cultural development needs; it should reinforce human capital accumulation in order to boost economic growth and social advancement; it should also promote the construction of a resource-saving and environment-friendly society, giving full play to education's basic function. Based on the above analysis, the essential element of education policy is to "cultivate students with an emphasis on moral fitness", and to cultivate workers with both integrity and professional competence. "Emphasis on moral fitness" means to advocate noble virtues and character, including our predecessors' noble virtues, and internalizing morality with characteristics of our time. It should be advocated not only for the time being but also in hundreds of years, because it constitutes the foundation required for a great rejuvenation of the Chinese nation. "Cultivating students" means training talent young people who are knowledgeable and competent and who can transform their knowledge into capabilities, instead of producing rigid students who cannot combine theory with practice. "Emphasis on moral fitness" and "cultivating students" complement each other and form a unity. We should not attend to only one element and ignore the other. In the process of modifying China's education policy, we must address the wording about cultivating "socialist builders and successors". Juxtaposing "socialist builders" and "successors" implies that the latter is also very important, yet the two terms mean the same. Maybe they need not be mentioned in the same sentence.

Second, the examination and enrollment systems should be changed. Since an exam-oriented education system deviates from the principle of "cultivating students with an emphasis on moral fitness" and the goal of training workers in a well-rounded way, it has drawbacks. The goal of our education is to train workers in the necessary skills to do their jobs, but also in morality, intelligence, physical fitness, and aesthetics, rather than to turn out young people who excel at exams and obtaining diplomas without paying attention to their performance in economy, politics, science, technology, culture, and society. As discussed in earlier chapters, the CPC Central Committee's decision to deepen reforms across the board recommended certain directions and crucial aspects for education reform. What is important is how those instructions are carried out. We must not only identify the direction and specific policies, but also be brave enough to launch pilot projects in order to learn from practical reforms. Some schools have run pilot programs. The high school affiliated to Fudan University, for example, has launched a program called the "Elite Students Project", in which some students were sent to rural areas, where they would live and learn together with local rural students for a week in difficult living conditions. They are thereby supposed to become tough and intelligent and develop an enthusiasm for physical work and cultivate their collective consciousness through savoring the daily hardships of life and studying. Other schools send students into poor places in remote areas. With their existing knowledge, they are encouraged to wire the huts they live in, install lights there, and assemble solar water heaters for showers. They must also cut wood and make fire to boil water and cook their own food. During their stays in the villages, some students have volunteered to instruct children in the local schools. These voluntary teachers sit in on each other's classes and evaluate each other, to hone their teaching skills. They also distribute the books and textbooks that they bring to the local pupils. During these processes, they experience emotional shocks and have their minds "purified" through living in the drastically different conditions.[10]

Finally, "de-administration" reforms should be carried out in universities. Although debates on de-administration still rage, there is a consensus that such reforms are needed to reduce excessive administrative intervention by the state. Presently, universities are run like government organizations in that their organizational systems, institutions, and management are based on the model of a government agency. When a development plan is being formulated for a university, there is excessive government interference in the allocation of personnel, finances, and material resources, and in the utilization and management of those elements. Universities cannot host major events unless they have been approved by the state education department. Inside the university, matters like organizational structure, the appointment or removal of leaders of certain organizations, teaching staff arrangements, and the ranking of technical staff are chiefly decided by the university administration. Until now, universities have been like mini societies, where the same administrative rules and systems in society prevail. In a market economy system, a university becomes a relatively independent entity that builds close ties with a wide range of market entities, and thus constitutes an indispensable segment of socio-economic development. However, the existing organizational structures, resource allocation and administrative rules in a university like those practiced in government, plus constant state intervention, distract the institute's attention away from its teaching and research, hinder innovations in running education independently, and impair its capacity to produce talent and to progress in its research. The "de-administration" reforms do not intend to rid the university of an administrative system or management. Instead, the latter will be incorporated into the education system and mechanisms. The key to education reform lies in system reform, giving higher education institutes more autonomy over enrolment, teaching, and research, the granting of professional titles, and operations and management. The aim is to maximize enthusiasm and creativity in running education autonomously.

9.3.2.2 Deepening Reform of the Science and Technology System

It is clear that great achievements have been made in science and technology since 1949, including in "high-end" projects like "two bombs and one satellite". Important breakthroughs have been made in research and application in domains such as computing, aerospace technology, ocean research, biology, new materials, high-speed rail, intelligent tools, and so on. Many researchers are leaders in their fields internationally, even if only one scientist, in the medical field, in China has been

[10] (2014, November 4). Tsinghua University is taking the lead in launching comprehensive reform and Peking University is following in its footsteps. *Wenhui Daily*.

awarded the Nobel prize. In contrast, overseas Chinese scientists, including Chen-Ning Yang, Tsung Dao Lee, and Samuel Chao Chung Ting have won Nobel prizes in science one after the other despite their being limited in number. Why is this so? I believe that their achievements are related to the scientific systems of the countries in which they live.

Since World War II, and especially since the beginning of the twenty-first century, the marginal, interdisciplinary and synthetic disciplines have advanced rapidly and a growing number of research projects are carried out by teams. This is undoubtedly an efficient research mode, exemplified in the "two bombs and one satellite" project. Most research done by such research teams can be categorized as applied. It is research that is either assigned by a government department or won through a bidding process (sometimes funded), and project evaluation, both of which are top-down or "passive" undertakings. Research conducted by teams has its advantages, especially in large-scale engineering projects. Nevertheless, it also has clear demerits: it might not match a scholar's personal research interests; it might not be easy to be creative or innovative in a team; and fewer major inventions and patented results have emerged from such projects. A thorough investigation of major inventions in science and technology both in China and elsewhere shows that few inventions have arisen from research groups. Instead, they tend to come from independent scholarly research. This is worth thinking about.

Why have our scientific and technological investments not brought about significant benefits? Statistics show that, between 2011 and 2015, R&D personnel in China rose from 2,883,000 to 3,759,000, a 30.4% increase; expenditure on R&D grew from 868.7 billion RMB to 1416.99 billion RMB, a surge of 63.1%; and the proportion of R&D expenditure to GDP increased from 1.78% to 2.07%. Regarding research achievements and results, published research papers increased from 1.5 million to 1.64 million, a rise of 9.3%; registered scientific and technological achievements rose substantially from 44,208 to 55,284, an increase of 25.1%; and the value of the import and export of high-tech goods rose from 1012 billion USD to 1204.6 billion USD, a 19% growth.[11] However, as these indices show, the increase in R&D personnel and expenditure outstripped growth in technological products and research results, clearly revealing that the scientific benefits are not great. In some research institutes, managers and administrative staff outnumber scientific research personnel. Such higher proportions of administrative staff impedes the quantity and quality of research, precluding the enhancement of efficiency. There is a popular Chinese saying that "tolerance is conducive to the emergence of great scholars". In the scientific community (including both natural and social sciences), an expert's personal interests, hobbies and expertise can develop better in a relaxing environment, which is essential to innovative research.

To guide the reform of the system of science and technology, which resembles the exam-oriented education system, the CPC Central Committee issued *The Decision on Some Major Issues concerning Comprehensively Deepening the Reform*, containing basic principles to adhere to. Currently, the most important task is to clarify the

[11] National Statistics Bureau of the PRC (2016).

relationship between government and market. The preference for scientific research carried out by groups reflects the decisive role that government plays in providing, allocating, and using resources. The relationship between government and market and their respective responsibilities must be clarified, and the government should be aware of what it should do and how it should do it. Practically, government should decide the direction and focus of basic, strategic, and cutting-edge scientific research and applied technological research, integrate scientific plans and resources, and create support mechanisms for them. On the other hand, the market should concentrate on cutting-edge research in certain fields, and (re)allocate research resources in line with market mechanisms to inspire enthusiasm and creativity among professional and technical personnel. Let the market guide the direction of technological research and development, route selection, factor prices, and allocation of various innovation factors. We should strive to promote a unification of production, teaching and research and vigorously advance innovation in the scientific sector; we should systematically encourage technological innovation and strengthen the application and protection of intellectual property; we should develop the technology market, and establish a standardized platform for technology transfer and transaction, and reinforce the legal management of it; we should create market mechanisms in management; and we should break the administration-led and department-segmented old system and contrive to create a new mechanism whereby resource allocation and management is decided and adjusted by market.

9.3.3 Deepening Reform in the Areas of Cultural and Moral Fitness

An all-round improvement in a population's qualities requires enhancing its cultural and moral fitness, but divergent views still prevail about the notion of cultural and moral fitness. Some see it as a population's political and moral awareness, highlighting the political connotations, while others see it as a cultural and moral quality, emphasizing conceptual and cultural connotations. So how shall we define "population quality"? First, population refers to all the people living in a certain area. Quality refers to the internal nature and basic characteristics of things. Combined, they refer to the nature and basic characteristics of a population living in a certain area. Restricting the definition to either political or conceptual notions fails to reveal its nature and characteristics. Further, quality is measurable and constantly developing, being comparable and continuous. It is difficult for the above two definitions of population quality to meet these requirements. I propose, then, the concept of the general cultural and moral fitness of a population in a certain area. Moral fitness includes beliefs, honesty, interpersonal and inter-generational relationships and so on; legal awareness, that is, observance of the law, discipline and good behaviour; and environmental awareness, that is, understanding of the importance of protecting nature and maintaining ecological balance, harmony between human beings and the natural environment, and so on. The elements of moral and cultural fitness can be measured using relevant indices. For example, legal awareness can be measured by the adult crime rate and the rate of juvenile delinquency. Moral fitness can be revealed by phenomena such as respect for the elderly and care of the young, attending to the injured, helping those in need; and commitment to public duty can be measured against certain indices through data obtained from specific surveys.

Cultural and moral fitness involve many aspects, so reform of the systems involved will be wide-ranging. For one thing, we should conduct awareness campaigns and education activities, especially about political and moral awareness. Vivid examples can tell people what we advocate and what we disapprove of. There should be regular publicity projects. We must also strengthen the legal aspects, making rules and regulations as detailed as possible, to incorporate public affairs into the legal system so that there will be real laws to abide by; and violators must be brought to justice. To conclude, reforms of mechanisms for improving the population's cultural and moral fitness should be carried out through awareness campaigns and legal construction and be integrated into the strategic plan for all-round development.

Next, we need to discuss core socialist values, because the creation and evolution of cultural and moral fitness are to a large degree determined by value judgments. The current core socialist values affect the value judgement of individuals, groups, and organizations. If the value system were accurately defined and clearly summarized, it might exert an even greater influence. If not, its effect will be limited. The current core socialist values include prosperity, democracy, civility, harmony, freedom, equality, justice, the rule of law, patriotism, dedication, integrity, and friendship. They are generally understood in three aspects. The first four items are national aspirations; the next four express goals of social development; and the last four express the values that individuals are expected to hold fast to. These twelve qualities form a unity in which each reinforces each other. They are inclusive because they cover every aspect of the value system, but these core values include so many terms and have so many connotations that they seem to lack focus, making it difficult for people to memorize and practice them.

In November 2012, General Secretary Xi Jinping delivered a speech to the first session of the 18th CPC Central Committee in which he said, "the fundamental aim of comprehensively building a moderately well-off society, carrying out reform and socialist modernization is to constantly improve people's material and cultural lives and to promote people's all-round development through enhancing social productivity." Subsequently, when members of the 18th Standing Committee of the CPC Political Bureau met with Chinese and foreign journalists, Xi declared that "meeting people's needs for a good life is the goal we should strive for". His instructions imply that we must adhere to the basic methodology of materialist conception of history; the crucial thing about the core socialist values is they are formulated from the standpoint of the people, and they insist that development is for the benefit of the people, development relies on the people, and the benefits of it are to be enjoyed by the people. Beginning with our core socialist values, further studies should be conducted to deepen and refine them to make them more cohesive, more appealing, and more influential. The important understandings are:

First, the foundation of the core socialist values is the people's collective economic interests, so when refining those values, priority should be given to people's common economic interests and their desire for a good life.

Second, politics is the concentrated expression of economy, so the ideology that guides the people, and the political system and institutions that maintain people's common economic interests should determine the content of the core values.

References

Meadows, D.H., Meadows, D. L., Randers, J. & Behrens III, W. W. (1972). *The Limits to Growth—A Report for the Club of Rome's Project on the Predicament of Mankind*. Universe Books. p. 44.

National Statistics Bureau of the PRC. (Ed.). (1986). *China Statistical Yearbook of 1986*. China Statistical Publishing House. p. 92.

National Statistics Bureau of the PRC. (Ed.). (2016). *China Statistical Yearbook of 2016*. China Statistical Publishing House. p. 638.

National Statistics Bureau of the PRC. (Ed.). (2018). *China Statistical Yearbook of 2018*. China Statistical Publishing House. pp. 33; 732.

Population Census Office under the State Council and Department of Population Statistics, State Statistical Bureau, PRC. (Eds.). (1985). *1982 Population Census of China (Results of Computer Tabulation)*. China Statistical Publishing House. pp. 354–355.

Population Census Office under the State Council and Department of Population Statistics, State Statistical Bureau, PRC. (Eds.). (1993). *Tabulation on the 1990 Population Census of the PRC*. China Statistical Publishing House. pp. 32–51.

Population Census Office under the State Council and Department of Population, Social Science and Technology Statistics National Bureau of Statistics of China. (Eds.). (2002).*Tabulation on the 2000 Population Census of the PRC*. China Statistical Publishing House. pp. 593–602.

Population Census Office under the State Council, Department of Population and Employment Statistics. (Ed.). (2012). *Tabulation on the 2010 Population Census of the PRC*. China Statistical Publishing House. pp. 295–300.

Tian, X. Y. et al (Eds.). (2007). *Research on Chinese Population Development Strategy in the 21st Century*. Social Sciences Academic Press. pp. 23–24.

(2014, November 4). Tsinghua University is taking the lead in launching comprehensive reform and Peking University is following in its footsteps. *Wenhui Daily*.

Chapter 10
Population Policy in the Future

There have been many different understandings of and opinions about population policy so far, but as always, time will be the fairest judge. The more time passes, the clearer things become, as will policy goals and orientation for the future.

10.1 Various Evaluations of Population Policy

While there are different views about everything, including population policy, in general, basic assessments usually conclude that something is right or wrong, worthy of support or to be opposed. Other views differ only by degrees, thus forming a third kind of evaluation, but even this cannot be completely neutral. Finally, each evaluation will exhibit a certain tendency. As a result, assessments can be divided into generally favourable ones and unfavourable ones, but there are no absolutely right or wrong evaluations. In addition, often certain matters are too complex to allow a positive or a negative evaluation.

10.1.1 Positive and Supportive Evaluations of Population Policy

Population policies since the founding of the People's Republic of China (PRC) fall into three stages, each of which was accompanied by different arguments.

The first stage was the first 30 years after the founding of the PRC (1949–1978). Soon after the founding of the PRC, thoughts of restraining population growth re-emerged, expounding the idea of population control. However, in the Anti-Rightist Campaign of 1957, this was severely criticized and its proponents were labelled as rightists, resulting in its limited influence. Among so many thoughts concerning

X. Tian, *Population Policies in the People's Republic of China*,
https://doi.org/10.1007/978-981-96-9658-1_10

population control, Ma Yinchu's "new population theory", the most influential population theory in China at that time, had played a very important role. As mentioned above, he expounded China's population problem, and the necessity and urgency of controlling population growth, arguing that policy suggestions should be developed with a combination of theory and practice. Using data generated by his investigations, he rejected the idea espoused in the *Textbook of Political Economy* published in the USSR, that the rapid population growth is a law of socialist development, and analysed the contradictions between rapid population growth and national economic development.

The second stage was the early-middle period of reform and "opening-up" (1979–2012). Under the guiding principles of economic reform and opening-up, emancipation of the mind and pragmatism, the dogma that the greater the population the better was overturned. The negative critiques of Ma Yinchu's "new population theory" was rectified in this period. The population problem induced by the contradiction between shortage economy and surplus population was recognized, thus the one-child policy as the pillar of China's population policy was made and thereupon strictly implemented for the following three decades. Most politicians and scholars in China approved of the policy and supported it. However, with an aging population, gradual adjustments were made to this policy. A new voice opposing this policy was released and heard criticising the prevailing strategy as wrong, and arguing for its abolition.

The third stage (2013–) has been the period since a comprehensive deepening of the reforms. In this period, a series of adjustments to population policy were made and one adjustment refers to people's autonomy over reproductive rights. According to my investigations, most officials and scholars approve of this policy. However, others worry that it will trigger a rebound in the fertility rate and thus advocate a continuation of strict population controls so as to guarantee some population stability despite this new fertility policy. Still, there are few arguments about the policy and these have not had any significant influence. I will explain this in further detail later.

I have consulted discussions of population policy by scholars, officials, and other experts at home and abroad. In those works that favour China's population policy, I find much convincing empirical research supporting both its theory and practice. For example, Professor Yuan Huarong (Zhejiang University, Population and Development Institute, Western China Development Institute of Environment and Resources and Non-traditional Security and Peace Development Research Centre) has published a series of books, which provide systematic explanations of policy changes, the relationship between population and natural resources and environment, and in particular the reality of China's population policy from the macro perspective of population and development. In the following works he provides detailed arguments: *Theory and Case Studies: Population, Environment and Development* (Zhejiang University Press, 2014), *"Multhus Revolution" and the "End" of "Optimum Population"* (China Environmental Science Press, 2013), *Succession, Implications, and Challenges of Civilizations and Our Response* (China Environmental Science Press, 2013), *"Ecological Purpose" and Environmental Ethics* (China Environmental Science Press, 2013), and *"Salvage", "Return" and "Human Revolution"* (China Environmental

Science Press, 2013). At a seminar on "Research on Population Issues in Social Change and Transformation" held in Beijing in July 2018, Mr. Yuan presented a paper titled *Family Planning: A Contemporary Benevolent Rule in China*" that can be regarded as the summary of Mr. Yuan's viewpoints and discussions on population policy in the books mentioned above.[1] In particular, the statements he made about the new changes in population, economy, society and environment during the 40 years of reform and opening-up not only express his own consistent views, but also add new content. They are representative of the basic views of the advocates of China's family planning policy. Mr. Yuan's main arguments are summarized below according to the four-part structure of his paper. Revolving around the titles of each part, the paper clearly expounds each theme and thus no further analysis is made.

The first part is entitled "*Family planning: a fundamental strategy in the global crisis*". The explanations to this conclusion are as follows.

a. "Population explosion: the root of global crisis and global change".
b. "The enlightenment of Chinese history: dynastic collapse—correlation between recession and the peak of population".
c. "Warning from nature: the link between biological extinction and species' saturation in geological history".

Combined with the relationship between population fluctuations and changes on the earth, discussions at various levels are made.

The second part is "*Family planning: a benevolent rule in contemporary China*".

a. Family planning is the safeguard of reform and opening-up, and a guarantee of long-term social stability. First, family planning is the safeguard of reform and opening-up. Second, family planning guarantees food security. Third, family planning improves the situation of the poor and sick because of having too many children through "green poverty alleviation" by breaking the vicious circle of having too many children and poverty. Fourthly, family planning is a driving force for social harmony, which has been proven by sociology and bio-ecology.
b. Family planning is at the root of improving the cultural fitness of the Chinese nation and realizing the "Chinese Dream". Firstly, the size of population should be controlled in order to improve its capacities. Secondly, family planning promotes the development of a powerful country, a prosperous and wise nation.
c. Family planning is a contemporary benevolent rule in China from the perspective of hierarchy theory. First, from this perspective, family planning safeguards the right to subsistence and thus ensures social justice. Second, low-level transgressions cannot deny the fact that family planning can be recognized as the benevolent rule in *contemporary* China. There are two reasons. One is that the nature of a system is determined by its upper level. The other is that the interplay between the upper level and the lower level is not equal—the upper level has a strong effect on the lower level, which has barely any influence on the upper level.

[1] The Editorial Board (2018).

The third part is "Fertility autonomy, the springboard to full freedom of fertility".

a. The concepts of fertility willingness, natural fertility and fertility autonomy are proposed and expounded here. Fertility willingness refers to the choices about how many children to have, of which gender, and when to have them. Natural fertility means taking the natural process of bearing children, and not controlling the number or gender of children through abortion or contraception but through environment and diet. Fertility autonomy refers to making decisions about medical interventions in the birthing process, understanding of the issues involved, and self-determination in fertility choices.
b. This section is about family planning, including firstly the theory of population explosion and practical solutions to it: Malthusianism and Restraint on Population Growth and secondly, family planning and fertility autonomy: hard and soft restraints, imposed and voluntary regulation.
c. Fertility autonomy is the springboard to freedom of fertility. Firstly, people have children according to their fertility wishes. Secondly, Mr. Yuan discusses misunderstandings of the freedom of fertility, which has been misinterpreted by liberals. Some self-willed couples have children without a reasonable plan. Thirdly, people enjoy fertility autonomy taking into consideration of their financial situations.

The fourth part is "over-interpretation of policy: the study of population policy".

a. Over-zealous interpretation of policy is the key cause of behavioural errors.
b. Who is the "stick" and who is the "ghost" in the two versions of "Zhong Kui[2] busts a ghost"?
c. The decline of population theory research and the rise of policy interpretation research. It asks whether the relationship between population research and family planning is one of "consultation" or "service". "Consultation" emphasizes the independence and scholarship of research, while "service" emphasizes the explanatory and problem-orientated nature of research. In fact, both are necessary elements of research. Perhaps "consultation" should take priority over service, and independent research is more important than research at the service of a policy.

10.1.2 Negative and Opposing Evaluations

The first phase of change in population policy followed the founding of the People's Republic of China. Disapproval of and opposition to population control and family planning was first given full expression in criticizing Ma Yinchu's "New Population Theory". However, a new theoretical dogma emerged from the criticism, arguing that the greater the population, the more labourers and products there are, and the faster development happens. Thus, more people are better. Mr Ma's opponents

[2] Zhong Kui was a powerful man who can use magic powers to drive away ghosts in a Chinese tale.

even describe the rapid growth of population as the "law of socialist population development". This idea was even written into textbooks of political economy.

In the second stage of disapproval and opposition, some voices were published and others were not published but were printed by individuals, or posted online. Some views were expressed not only about population policy in the second stage, but also about the third stage of policy adjustments. Yi Fuxian's views are representative of critiques in this stage and are discussed here for the following reasons.

First, his views are clearly expressed, reflecting his frankness about what he is for and what he is against.

Second, although many scholars (including myself) disagree with, even fiercely oppose Mr. Yi's stand, he raises a number of questions worth considering, and tries his best to support his argument.

Third, I notice that the book *Big Country with an Empty Nest: A Reflection on China's Family Planning Policy* published in 2013 is a revision of the book *Big Country with an Empty Nest: Misguided Family Planning in China* published in 2007. Both content and language have been revised. For example, there is less extreme sarcasm and more positive elaboration. On this point, no one who has similar opinions presents the case better than he does. Indeed, some of them are lack of intellectual integrity, relying on subjective assumptions, and even making false charges against others, criticizing or rejecting objective facts, and asserting that certain documents do not exist, or simply leaving them off their reference lists. No assessments of right and wrong can be obtained through debating with such critics. So, only Mr. Yi is chosen as our interlocutor.

Mr. Yi wrote: "Nature doesn't let the trees grow beyond the sky, nor does it allow the population to increase indefinitely." And "wealth and offspring cannot both flourish… offspring are the source of wealth and the flourishing offspring brings prosperity …you won't be poor for long if you have sons and you won't be rich for long if you don't have sons. But financial prosperity leads to a decline in offspring." It seems that there is an invisible hand controlling population numbers, through the rises and falls that promote social mobility and shifts of civilization.

This is Mr. Yi's basic view of population. The flourishing of offspring brings prosperity, and after prosperity, the offspring declines, controlled by the "invisible hand" that promotes a shift of civilization. It is the foundation on which he evaluates population policy. He takes this basic view into multi-perspective and multi-level elaborations.

Mr. Yi has sharply criticized China's family planning policy. He thinks that had the policy not been implemented in 1980, or even had it been stopped and replaced by one that encouraged fertility, the population would not have increased indefinitely. Today, the population would be just over 1.4 billion (not over 1.5 billion) and the population would have peaked at about 1.7 billion. But the demographic composition will be reasonable. In 1980, the family planning policy should have been stopped completely rather than replaced by the one-child policy. In one word, Mr. Yi directly rejects China's population policy and the basic state policy on family planning.

In the 2013 version of *Big Country with an Empty Nest: Reflection on China's Family Planning Policy*, Mr. Yi reiterates his original views and makes some additions and modifications at the same time. But at the end of the book, he writes: "in the past, all problems were put in the 'too many people' basket without any real and active search for a solution. In the future, we should not just pass the responsibility to family planning and then walk away."[3] In fact, any existing social problem contains solutions to itself.

Regarding the third stage population and fertility policy, some people are against it, but open opposition is rare. Since the publication of my article *Review and Prospects of Population Policies in People's Republic of China* in the People's Daily in 2009, few voices of opposition have been raised compared with the voices of approval both at home and abroad. However, there is opposition and it is expressed. It is reported, for example, that one official who is also a scholar (to be fair, a real scholar with great achievements) once submitted an internal report calling for a continuation of the current population policy and the basic state policy of family planning to maintain a low birth rate. One reason for not approving or even opposing the changes to the fertility policy is concern about the consequences of a fertility rebound and new population growth.

I have expressed my views frankly on this question in media interviews: a small rebound is possible, but a large rebound will not happen. The achievement of the ultimate population development strategic goals will not be impacted greatly, so the adjustment is necessary and beneficial.

10.1.3 The Substance of the Population Policy Debate

What are the origins of the different views of China's population policies and different evaluations of their effects? As mentioned before, they must be traced back to China's population development strategy. Where does the population development strategy come from and what is the key to it? To answer this, we must go back to the nature of the population problem. Different views on China's population policy are ultimately to be found in the nature of China's population problem. On the basis of the different understandings of the problem, quite different population policies are advocated.

Those who believe that the nature of China's population problem is that population numbers were holding back productive forces give central place to the control of population growth, believing that other contradictions will be solved if we grasp this principal contradiction. This approach thus pays little attention to questions of the age and gender structures of the population, including problems associated with an aging population, such as old-age security, employment, intergenerational relationships, the gender imbalance and urban–rural structural changes.

On the other hand, those who understand China's population problem as the productive forces holding back the population believe that "the flourishing of offspring brings prosperity" but later, this prosperity, "leads to a decline in offspring".

---

[3] Yi (2013).

It is thus essential to maintain "the flourishing of offspring" by encouraging population growth, maintaining the momentum of population expansion and reproduction, and seeking a corresponding population policy, one aimed at increasing fertility and birth rates.

Throughout the history of development at home and abroad, two voices and two sets of views have often co-existed and it is difficult to reconcile them. What is it that prevents the two sides from being reconciled? Epistemologically or theoretically speaking, the key to the problem is that attention has not been paid to the nature of the population and its evolution; everything in the world, including policy, can be turned in the opposite direction under certain conditions. When the People's Republic of China was founded, the nature of the population problem was that population was limiting the productive forces. On the one hand, the social productive forces were underdeveloped and on the other hand, the fertility and birth rates were too high. When the two factors are combined, the population problem is mainly manifested in an insufficient supply of the means of livelihood and production. These problems must be solved, by the provision of food, clothing, housing, transport, employment, schooling and medical care. And the contradiction between shortage economy and surplus population remains. The solution is, on the one hand, to vigorously develop the economy and increase supply, but on the other hand, population growth should be controlled so as to relieve the pressures of demand. Focusing on both supply and demand is a wise choice and it undoubtedly contributes to solving the problems. Fortunately, especially since the reform and opening-up era began, the national economy has continuously and rapidly developed, and the contradiction between the shortage economy and surplus population has changed greatly. With the socio-economic development, science and technology has progressed, modernization has been promoted, and the shortage economy is changing constantly. This change accelerated especially after the implementation of market system reforms, and the situation of shortage economy has disappeared. At the same time, the population transformation has progressed to the post-demographic transition stage in which the population growth is declining, quality of population is greatly improved, and the age, gender, urban–rural and regional population structures have changed to varying degrees, which leads to the emergence of new population situations. The simultaneous changes to economy and population show that China is stepping out of the "development trap" of the shortage economy and surplus population in the original sense. In coastal areas, where the economy has developed rapidly and reached a higher level than in inland areas, the nature of the contradictions in which the transformation has occurred has changed completely.

This does not mean that the contradiction between the shortage economy and overpopulation no longer exists. Rather, it is the nature of the problem that has changed: the contradiction has shifted from quantitative to qualitative. That is, the contradiction is the lack of accommodation between high-quality economic development and improvement in the quality of the population. This point has been elucidated in the previous chapters. The key point is that after strengthening economic efficiency, improving quality and efficiency, changing the mode of economic development, transformation and upgrading, and so on, China has achieved high-quality economic development, which is a fundamental leap. The population development strategy during the first stage focused on quantity control and achieved a low fertility rate within a generation; it then entered the second stage, of post-demographic transition, which focused on comprehensively improving population capacities and achieved a fundamental change in the history of China's population development. Thus the contradiction between a shortage economy and surplus population changed from quantitative to qualitative, that is, to a contradiction between high-quality economic development and overall improvement in the quality of the population. High-quality economic development should be based on the overall improvement in the quality of people's physical health, intelligence, scientific and education levels, and cultivation of good habits. The population development strategy and population policy should shift their focus to the overall improvement people's capacities because this is an objective law of the coordinated development of population and economy. Moreover, the overall improvement of population quality is conducive to maintaining a relatively stable low birth rate, which contributes to the rational adjustment of the age, gender, urban–rural distribution and regional distribution structures of the population. Thus, it is the second step of the population development strategy, the adjustment of population and fertility policies that plays the key role in the transition stage.

10.2 The Test of Practice

It is inevitable that there have been differing and even opposing understandings and evaluations of population policy since the founding of the People's Republic of China. But as long as no prejudices about it prevail, and the materialist principle is respected, different opinions can converge on a consensus through discussion and negotiation. The materialist principle is the principle of practice. And practice is the sole criterion for testing truth. Based on this criterion, this author believes that the population policies implemented since 1949, especially in the past 40 years of reform and opening-up, have been basically correct. However, there have been many problems, too, from which we should draw lessons. Based on personal experience and research, the author presents a summary of his views here in order to promote a greater consensus.

To evaluate China's population policy since the founding of the PRC, especially the last 40 years, particular attention should be directed to the process of its development, implementation and adjustment, with a focus on the benefits resulting from it. We must also analyse the benefits of the policy for the control of population numbers and quality, and the structural changes to the population, as well as for economic and social development, the allocation of resources, and the environment.

10.2.1 The Benefit for Control of Population Numbers

The "three-step" population development strategy set a population target, the realization of which was dependent on economic development and social progress which would bring about a decline in the fertility rate. Second, it relied on a fertility policy to control population growth, to complete the transformation of the population reproduction pattern from high birth rate, low mortality rate and high natural increase rate to one of low birth rate, low mortality rate, and low natural increase rate, and to promote a "post demographic transition" after the realization of this "three low" mode. Through the first three decades after the founding of the PRC, there were fluctuations in the policy that aimed to control population growth, during the reform and opening-up era, its strict implementation of the policy, with the one-child policy as the main element, greatly promoted the acceleration of the demographic transition and resulted in a continuous decline in the fertility, birth and population increase rates. How to evaluate the effect of population policy on the decline of fertility? As is noted in existing research on the question, there is some controversy about the effect that this policy had on population decline, with estimates that the population would have been anywhere between 400 and 250 million larger without it. However, all studies recognize the effect of the population policy and family planning on population growth. The author of the current book contends that though the effect of fertility policy on the reduction of the birth rate is difficult to determine due to the dual effects of socio-economic development and the fertility policy, the total fertility rate, women's net reproduction rate, and the related variable indicators such as birth rate, death rate, natural increase rate, and so on, are specific and measurable. Since 1980, the birth rate in 2017 dropped from 18.21 per 1000 people to 10.94 per 1000, a drop of 7.27 per 1000. The natural increase rate dropped from 11.87 per 1000 people to 3.81 per 1000, a drop of 8.06 per 1000.[4] Meanwhile, as to the change of population age structure, which is decisive for future population changes, the proportion of the population aged 0 to 14 dropped by half from 33.6% in 1982 to 16.8% in 2017. The proportion of adults aged 15 to 64 rose from 61.5% to 71.8%, an increase of 10.3%. The proportion of the elderly, aged 65 or above, rose from 4.9% to 11.4%, an increase of 6.5%.[5] The long-term continuous decline in the birth and increase rates shows that population growth has slowed down greatly, and the day of zero population growth is within sight. The age structure of the population has changed continuously from the original young population to the adult and elderly profile. It has now entered the middle stage of the elderly population, and the speed of aging is relatively fast. From traditional population transformation to post-demographic population transition, the potential energy of population growth has declined greatly. This is closely related to the one-child policy, vigorously promoted from 1980. These changes can be seen as the result of the "two-wheel drive" of socio-economic development and the population control policy; that is the concentrated manifestation of the effect of fertility policy.

[4] National Statistics Bureau of the PRC (2018).

[5] National Statistics Bureau of the PRC (2018).

In 1994, 182 countries participated in the International Conference on Population and Development held in Cairo, Egypt. *The Program of Action on International Population and Development* adopted by the conference declared that "the center of sustainable development is human", which is an important supplement to the *Rio de Janeiro Declaration on Environment and Development* and *Agenda for the 21st Century* adopted by the United Nations Conference on Environment and Development in 1992. The ideas that evolved from these include the following: the fundamental purpose of development is to meet the needs of all-round development of people, including that of future generations; the fundamental driving force of development is human capital, and the transformation from output capital to human capital should be gradually realized; the establishment of a people-oriented framework for sustainable development because resources are the prerequisite for sustainable development, and all development can be attributed to the material transformation of resources; environmental protection is the ultimate goal of sustainable development, and the goal of sustainable development is to create a natural and social environment suitable for the all-round development of humankind; economic and social development are the essential means for the achievement of widespread development, including sustainable development. Observing current trends of global population changes, and those in China in particular, in accordance with these sustainable development strategic goals, control of population growth is still necessary. Controlling population growth is conducive to protecting resources and the environment, and the weighting effect of population growth on resources and the environment is more prominent today. Controlling population growth is also conducive to improving labour productivity and social efficiencies, and to the realization of modernization. In working towards these goals, especially in the past 40 years, China has vigorously controlled its population growth and reduced the number of births by hundreds of millions. Population reproduction has undergone a fundamental change. In the middle stage of the post-demographic transition, the benefits of China's population policy have been vividly demonstrated.

Despite overall success, problems persist. One serious problem is that some population targets were set too high and demands too urgent, resulting in some coercion. For example, during the 1980 population symposium, there was a view that the national population growth rate should be reduced to 5 per 1000 people by 1985 and zero in 2000. This goal was too stringent to be mentioned in the report. In another example of these tight demands, the population of the country should be controlled so that it remain under 1.2 billion in 2000. This target was officially announced to the public and became the goal for action in the *An Open Letter to All Party Members and Communist Youth League Members from the Central Committee of the Communist Party of China about the Control of Population Growth in China.* Members of the Communist Party, members of the Communist Youth League and people all over the country were exhorted to strive to achieve this goal. Yet the *Report on China's Population and Employment in 2000* eased these goals, suggesting a control target of about 1.25 billion. In the field of demography and other relevant disciplines, many proposed a revision of this target to keep the population within 1.2 billion. In the end, the official document reflected the adoption of the recommendation that the population remain at 1.2 billion. Over-zealous and excessive targets were implemented, resulting in damage to the image of family planning. However, as far as this author knows, the policy was always opposed to forced orders, excessive slogans, and extreme practices. The relevant authorities also stepped in to stop illegal acts, ensuring that the administration operated within the law, with an emphasis on publicity and education, contraception and birth control to meet targets. This is the mainstream and dominant aspect of the policy, and we should not condemn the entire population policy and family planning system because of problems in its implementation.

The introduction of a new policy allowing a second child for couples in rural households whose first child was girl, played a significant role in boosting the discrepancy in the gender ratio of new-borns. Historically, China has had a strong preference for male heirs; customs and traditions favour boys over girls and men over women. In 1980, the Central Committee Population Symposia drew attention to this issue, and in a report to the Secretariat of the Central Committee stated: "due to the influence of feudal ideology, 'son preference' leads to the drowning and abandonment of female infants, which deserves attention and should be prohibited by law." That is, clear legal stipulation was required to prevent gender selection when advocating the one-child policy. As mentioned above, the policy allowed single-daughter households in rural areas to have two children, which meant that if a boy were the first-born, then the household should remain a one-child household, but if a girl were the first-born, the couple could have another child. As a result, many people used every possible means to ensure that the second child was a boy. This was a mistake. The decision whether to have another child based on gender greatly promoted the increase in the ratio of males to females of new-borns.

10.2.2 The Benefits of Enhancing Population Capacity

As the saying goes, what is rare is precious. With the implementation of the one-child policy, children became scarce and thus valuable. They are cherished by families and parents who are willing to invest a great deal in their health, intellectual development and education. This highlights the effect of the population policy on population quality. The comprehensive implementation of a national population policy itself includes the enhancement of the quality of the population, with such policies as safe and healthy births and good care, late marriages and late/planned childbirth. More than 40 years of practice of the population policy has proved that the improvement of the population has worked.

a. *The benefits of enhancing physical health*

First, the population policy includes improving physical health. First, it involves fewer and safer births. Thus, for example, on February 9th, 1982, the CPC Central Committee and the State Council issued the *Instructions on Making More Efforts to Carry out Family Planning Work* stating that: "We will keep encouraging late marriage, late childbirth, fewer births, and better births." Second, it indicates a clear incentive. The key method was to reward compliance with by "distributing a bonus to the only child" and "appropriately extending paid maternity leave without affecting salary adjustment and promotion." "In rural areas and units where the production responsibility system is implemented, the contracting of production for one-child families is lower, or more productive farms are contracted." "In rural areas and units where the production responsibility system is implemented, families with only one child should have a lower quota of output or contract more productive farmland." The third widespread benefit was that the policy strengthened scientific and technological research, and guidance on family planning, which led to the establishment of a family planning network, and enhanced family planning services. In general, the quality of maternal and child health care services was greatly improved, and significant progress was made in maternal and child health. The effect of the implementation of these population and fertility policies has been clear.

Second, there is a substitution effect between quantity and quality of children. According to G.S. Becker's theory, when the number of children is reduced, parents will transfer part of the quantity cost to other children, so as to improve the quality cost of other children. The implementation of the one-child fertility policy (and having a second child only in certain circumstances) reduced the number of children born, thus maternal and child health care have gained greater attention, and the idea of a good birth and good care has been taken up in thousands of families. These initiatives, starting with baby care and reproductive health, demonstrate a fundamental improvement in physical health. Statistics show that the infant mortality rate decreased from 34.7 per 1000 people in 1981 to 6.8 per 1000 in 2017, a drop of nearly 27.9 per 1000, or an 80.4% reduction. life expectancy estimates for both sexes combined increased by 8.8 years, from 67.9 years to 76.7 years, an increase of 11.5%.[6]

b. *The benefits of enhanced education and scientific knowledge*

China has a long tradition of respecting teachers and valuing education. In the "Old China", the vast masses of working people struggled on the treadmill of "being poor—giving birth—being poor". Even if they had every intention of respecting teachers and valuing education, they lacked the capacities to do so. Most of their children were unable to go to school or become educated, and rates of illiteracy and semi-illiteracy were very high. Since the founding of the PRC, most workers' and peasants' children have had access to education. With the development of economic production and the increase in incomes, the national funding for science and education keeps increasing; on the other hand, common families can send their children to school. However, since the family economy has oscillated between poverty and comfort for a long time, when the fertility rate is high most families remain trapped in the dilemma of imbalance: supporting one child, but not the other; supporting boys, but not the girls. According to Becker's theory of quantity and quality substitution of children, a fertility policy promoting one child allows families to concentrate their financial resources on supporting one or two children to receive primary, secondary

[6] National Health Commission (2018).

and higher education, including scientific, technical and vocational education. Thus, China's population policy has had an effect of promoting the quality of science and education.

Statistics show that from 1978 to 2017, the number of undergraduates and junior college students increased from 856,000 to 27,536,000, an increase of 24.88 million or a factor of 28.1, with an average annual increase of 9.3%. The number of students in ordinary high schools increased from 15.531 million to 23.745 million, an increase of 8.214 million or 52.9%, with an average annual increase of 1.1%. The number of students in junior middle and primary schools decreased. The number of junior middle school students went down from 49.952 million to 44.421 million, a decrease of 5.531 million, with an average annual decrease of 0.3%. The number of ordinary primary school students decreased from 146.24 million to 100.937 million, a decrease of 45.303 million, with an average annual decrease of 0.9%. At the same time, the number of full-time teachers in ordinary colleges and universities increased from 206,000 to 1,633,000, with an average annual increase of 5.4%. The number of full-time teachers in ordinary high schools increased from 741,000 to 1,774,000, with an average annual increase of 2.2%. In junior middle schools, the number of full-time teachers grew from 2.441 million to 3.549 million, with an average annual increase of 0.9%. In ordinary primary schools, they increased from 5.226 million to 5.945 million, with an average annual increase of 0.3%.[7] By comparison, the number of students in colleges and universities grew the most rapidly, with an average annual increase of 9.3%. The average annual increase of teachers in colleges and universities is 5.4%, ranking first in the growth rate of teachers. The next biggest increase was in ordinary high schools, with an average annual increase of 1.1% for students and 2.2% for teachers. A different set of trends is seen in junior middle schools and primary schools: students numbers showed a small decline, and teacher numbers a small growth. Our task is to find out whether these changes in the number of teachers and students in schools is linked to the population policy, and how exactly the policy enhanced education.

First, the decline in fertility rate and the reduction in the average number of children in a family, increased the opportunities for education, especially higher education. As is known to all, rapid social and economic development since the reform era began, people's material and cultural lives have continuously improved, and household incomes have been increasing over the past 40 years. Originally, it was not easy for ordinary families to send their children to school, especially college. However, as the country has vigorously promoted the one-child policy, family transfer payments have increased. There are two reasons for this. First, the number of children has been reduced to one or two, who are treasured and cared for by the family. At the same time, with higher aspirations for their children, parents are willing to pay a higher cost for their development. Second, not only middle- and high-income groups, but also low-income groups are now willing and able to pay their children's education costs and to support their children to complete higher education. If conditions permit, they will go to graduate school and may even study abroad. Moreover, with the

[7] National Statistics Bureau of the PRC (2018).

improvement in overall incomes, the capacity of ordinary families to pay for their children's education has increased. Some parents, though they might have difficulty paying the costs of their child's education, are willing to work from morning to night, live frugally and save as much as they can in order to support their children, the "only seedling", to go to school and to develop his or her talents, to contribute to the country and to win glory for the family. The children of such families are often aware of how hard-won are these opportunities. Generally, students from such highly motivated families study harder, have better grades and demonstrate a higher capacity to repay their parents in the future.

Second, the decline in fertility and birth rates has led to a slight decrease in the number of junior middle school and primary school students and a slowdown in the growth in teacher numbers. The birth rate has continued to decline, falling to 14.03 per 1000 people in 2000, resulting in 17.78 million births, in contrast to more than 20 million in the 1980s and 1990s. In the same period, the natural increase rate also dropped to 7.58 per 1000 people, or 9.6 million, below the previous level of more than 10 million. By 2018, the birth rate had decreased to 10.94 per 1000 people, and the number of births to 15.23 million, while the natural increase rate had further dropped to 3.81 per 1000 people, or 5.32 million. It can be seen that by the end of the twentieth century, and especially since the beginning of the twenty-first century, the overall trend of the birth rate, the number of births, the natural increase rate, and the number of natural growth continued to decline, and the inertia resulting in these falls is still strong. This development has had a direct impact on the number of students in primary and secondary schools and led to a small reduction in school sizes. In addition, the reduction has generated some new problems, the main one being a decrease in the working-age population in the future. However, from the perspective of improving the quality of the population's intelligence, science and education, the decrease in the juvenile population has been conducive to the growth of population investment, which is a force for economic and social development.

The effect of the population policy on scientific and technological literacy is like that of education. From 2000, expenditure on research and experimental development increased from 8.957 million RMB to 176.061 million in 2017, an 18.66-fold increase, with an average annual increase of 19.1%. The number of patent applications for new inventions increased from 13,000 to 420,000, an increase by a multiple of 31.32, with an average annual increase of 22.7%.[8] The rapid growth of expenditure on science and technology and patent development can be seen as linked to the substantial improvements in the scientific, technological and education standards of the population. The number of years of formal education has increased rapidly. The average schooling has risen to nearly 12 years. High school education has been popularized, and the number of disciplines offered in basic university degrees and above has grown more rapidly. In line with this growth, employment in scientific research and technical service industries has also expanded significantly. Between 2005 and

[8] National Statistics Bureau of the PRC (2018).

2017, the number of employees in this sector went from 2.277 million to 4.204 million, an increase of 84.6%, which can be attributed largely to the enhancement of the scientific and education attainment levels of the general population.

10.2.3 The Benefits of Structural Adjustment for the Population

As a result of China's population policies over the past 70 years, great changes and new trends have occurred to the country's demographics. Changes have occurred in areas such as age, gender, urban and rural populations, and geographical distribution across the country. But the effects have been both positive and negative for the population, and for economic and social development.

a. *The "Golden Age" of changes in the age structure*

In 1983, the current author published a proposition about demographic change in *Suggestions on Using the Change of Population Age Structure to Promote Modernization* (*People's Daily*, June 5th, 1983), predicting that under the current fertility policy, a "golden age" of about 40 years was coming, when there would be a lower dependency ratio among the population (that is, the ratio of juvenile population + elderly population to the working-age population), as well as a relatively high proportion of people of working age. This period would provide corresponding demographic profits and demographic dividends, supporting sustained and high economic growth. The prediction proved correct for the period from 1990 to 2010, during which time the population dependency ratio dropped from 49.8% to its lowest value of 34.2%, and the proportion of the working-age population (aged 15 to 64) rose from 66.7% to a maximum value of 74.5%. After the turning point in 2010, the dependency ratio began to go up, rising to 39.2% in 2017, as the proportion of the working-age population began to go down, falling to 71.8% in 2017.

It should be clarified that although the dependency ratio has increased, the proportion of working-age population has decreased, and the demographic profits and demographic dividends have declined, they have not come to an end. In fact, the forecasts show that the demographic dividends will not disappear until around 2030. If the economic effect of the population policy increased year by year until 2010, it has decreased year by year after 2010, forming an inverted U-shaped curve; for at least the next decade, that trend will continue, but more gradually. This is the economic effect of the population policy, an effect of supporting national high-speed or medium–high-speed economic growth. Academic debate has generally accepted that the population policy's contribution to GDP growth is between 25 and 30%. Indeed, there is no doubt that China's sustained and rapid economic growth over the past four decades is mainly the result of adopting the reform and opening-up policy. However, the vigorous implementation of population policy, the effect and function of demographic profits and dividends resulting from the "golden age" of the age structure are also undeniable and objective.

b. *Lessons from changes in gender structure*

As mentioned above, China used to be a country of strong gender preference, with the ratio of males to females at birth being slightly higher than normal for thousands of years. After the founding of the PRC, great efforts were made to promote gender equality and equal pay for equal work. A great many women were liberated from housework and took up paid work outside the home. To a large extent, gender discrimination decreased. This is also reflected in the gender ratio at birth, which showed a downward trend in the 1960s and 1970s. The one-child policy generally permits each couple to have only one child, with an expansion to an upper limit of two children if the parents qualified, and thus increased the risk of a higher gender ratio at birth. In particular, the policy stipulating that single-daughter families in rural areas could have a second child provided a legal platform for the increase of the gender ratio at birth, which has remained high for a long time. It fell slightly in recent years, but it is still above a normal value. This is a lesson worthy of serious study, a negative effect of improper policy formulation.

c. *Historic changes in the urban–rural population structure*

In a narrow sense, the population policy, which focuses on fertility, does not encompass the urban–rural distribution structure. However, in terms of population management and regulation, there are policies directly or indirectly related to this question. For example, in the 1960s and 1970s, the population was seen as divided into agricultural and non-agricultural populations. The former was mainly a rural population and the latter mainly an urban population. Shifting from agriculture to non-agriculture was difficult, because the non-agricultural population was mainly engaged in secondary and tertiary industries, living on a grain entitlement supplied by the government according to quantitative standards. The agricultural population was not supplied with commercial grain in the same way; instead, each people's commune distributed grain according to labour contributed, and if this was not sufficient then a person could buy surplus grain released to the market. After the reform period, the grain ration was gradually abolished, and grain entitlement stamps were also gradually cancelled, entering the museum of history, but there was still a distinction between the registered population and the permanent resident population. Reform has not stopped in the twenty-first century. Some provinces and cities have abolished the division of the two population classifications and in its place implemented a single resident ID card system. However, about 240 million migrant workers in cities still fall into an indefinite category—they are registered in the countryside but work in cities, members of a veritable amphibious population. In terms of permanent residents, the proportion of the urban population rose to 51.27% in 2011, while the rural population dropped to 48.73%. For the first time, the urban population surpassed the rural population, representing a historic change to China's urban–rural population structure. Urbanization is not only the result of industrialization and modernization, which have broken the urban–rural duality, but also an effect of the continuous reform of the population policy and the increased mobility. For the future, changing migrant workers into citizens must be an aim of further reforms to the population policy and a concentrated effect of the urbanization policy.

10.3 The Chinese Style Family Planning

Catchphrases like "Chinese style" or "with Chinese characteristics" seldom appear in my books and papers, because I am unable to clarify the terms. For that reason, it is necessary to elucidate the term "Chinese style" as it is used in this section. First, if we compare "Chinese-style family planning" with current practices of family planning around the world, it is indeed a type of "family planning" with a similar basic structure to that elsewhere. Second, Chinese-style family planning is different from other models in some respects, for example, some of its characteristics are reflected in the goals of birth control, social participation, delivery of service, the way of individual participation and so on. Since China is in transition from a highly centralized and unified family planning policy to an autonomous family planning policy, it is reasonable to retain some features of the former policy for a certain period.

10.3.1 China's Population Policy at a Crossroads for a Second Time

In the three decades following the foundation of the People's Republic of China in 1949, population policy went through changes and encountered many challenges. The one-child policy was finally established as the main pillar of China's population policy in 1980. Slight adjustments were made in order to keep the whole Chinese population under a targeted 1.2 billion by the year 2000. The policy remained firmly in place until 2010, when the one-child policy was no longer viable. In the academic community, some colleagues and I had made several appeals for adjustments to the policy, but with no positive responses. Therefore, in 2009, I published an article in *People's Daily*, an influential newspaper in China, in order to make an appeal that the promise should be kept and that timely adjustments be made to the one-child policy. My suggestions included permitting a couple of whom either the man or the woman was an only child to have two children, and restricting permissions granted for three children in one family. A couple of years passed before the policy was adapted to allow married couples to have two children if one parent is a single child in 2013, and permission for all married couples to have two children was finally granted in 2015. That period was a crossroads in history. Indeed, this metaphor was chosen as the title of an interview with me by a certain media source. Now we are in the middle-to-late stage of post-demographic times. With the percentage and the absolute number of working-age people passing their peak for several years, population policy is once again at a crossroads. Questions have arisen: what will be the next adjustment to population policy? What are the second and third objectives of the population development strategy? These questions must be addressed after careful argument and discussion, based on summarizing our own experiences, as well as what we can learn from the experiences of other countries.

Table 10.1 Natural changes in the Chinese population since 2000

Year	Total population (million)	Birth rate (per 1000 people)	Mortality rate (per 1000)	Population increase rate (per 1000)
2000	1267.43	14.03	6.45	7.58
2002	1284.53	12.86	6.41	6.45
2004	1299.88	12.29	6.42	5.87
2006	1314.48	12.09	6.81	5.28
2008	1328.02	12.14	7.06	5.08
2010	1340.91	11.90	7.11	4.79
2012	1354.04	12.10	7.13	4.93
2014	1367.82	12.37	7.16	5.21
2016	1382.71	12.95	7.09	5.86
2018	1395.38	10.94	7.13	3.81

Fertility policy adjustments must take into consideration the needs of future development and demographic change, the reality of social development, and the new situation of the post-demographic period, especially after the introduction of the policy permitting all couples to have two children. Currently, new demographic realities that are closely related to population policy can be seen in the following phenomena.

First, there is the low fertility rate. According to the 2010 national census, the total fertility rate was 1.18, extremely low—a figure that is widely doubted, though the skeptics have produced no convincing counter evidence. Before the census, the fertility rate publicized by the National Population and Family Planning Commission was around 1.8, which was generally believed by the academics to be quite different from the real rate. The United Nations estimated the median total fertility rate of the whole world to be 2.4 between 2015 and 2020. The median fertility rate of developed countries was estimated to be 1.66, of developing countries 2.5, and China's median total fertility rate was estimated at 1.84.[9] In recent years, the fertility rate figure released by the relevant authorities in China is around 1.6, with slight growth. Unfortunately, my colleagues and I cannot obtain the first-hand data for this figure. Based on data obtained from sporadic surveys, the real total fertility rate in China is thought to be a little higher than 1.6. The general understanding is that the fertility rate has remained below the replacement level since the early 1990s. This trend is reflected in Table 10.1, showing the natural changes of population since the twenty-first century.[10]

Table 10.1 shows that the Chinese birth and population increase rates follow a weak "downward–upward–downward" trend and generally speaking, are now at a rather low level. It is not difficult to see that after the introduction of policies like those permitting a couple within which either the man or woman is an only child, to have two children (in 2013), and permitting all couples to have two children (in

[9] U. N. Population Division (2009).

[10] National Statistics Bureau of the PRC (2018).

2015), people's enthusiasm for having more children was slightly stimulated. The birth rate rose from 12.10 per 1000 people in 2012 to 12.95 per 1000 in 2016, an increase of 0.85 per 1000 people. Within the same period, the natural population increase rate rose from 4.93 per 1000 people to 5.86 per 1000. However, further changes have taken place in these few years. In 2017, the birth rate fell to 12.43 per 1000 people, a decrease of 0.52 per 1000 over 2016. In 2018, the decrease in the birth rate continued, going down to 10.94 per 1000 people, a decrease of 1.49 per 1000 over 2017. Meanwhile, the natural increase rate also showed a downward trend. In 2017, it declined to 5.32 per 1000 people, and in 2018, it went on declining, down to 3.81 per 1000, a decrease of 1.51 per 1000 people over 2017. Will this downward trend continue? Observations must be made of the reality of population changes before the question can be answered.

Second, the trend towards an aging society continues in China. There is now a consensus on the reality of this aging society, but opinions about the speed and degree of aging diverge. Some believe that the society is aging too quickly and that China already had a high percent of elderly. Consequently, they believe that urgent measures must be taken to curb this trend. Others believe that China is at an early stage of the aging society, and that the percentage of elderly is not yet high, therefore there is no need for concern. As far as I am concerned, evaluation of the situation should begin with serious investigation of the realities of Chinese demographics.

It is generally believed that there are three stages in the transformation to an aging society. In the early stage, the rate of aging, the percentage of those above 65 years old, is relatively low, with senior citizens constituting seven to 12% of the total population. In the second stage, the rate of aging is intermediate, in transition from a relatively low level to a relatively high level, with senior citizens constituting 12 to 17% of the total population. In this second stage, the trend towards an aging society and resultant problems are aggravated. In the third stage, the aged population has reached a high level, with senior citizens constituting more than 17% of the total population. As a result, serious problems such as a high social dependency ratio will emerge.

China has just entered the second stage, the intermediate level of an aging society, which should be recognized and evaluated correctly. If we wish to fulfill the strategic task of keeping population growth under control, aim for zero growth in the overall population and maintain an acceptable number of people in the long run, then an aging society is unavoidable to some extent. However, when I say that a certain degree of aging in the population is necessary and positive, I do not mean it is better to have a society with a growing percentage of older people or to become an aging society at a faster rate. On the contrary, only a moderate aging population percentage is desirable. Indeed, a higher percentage of old people, or heading towards that at a faster speed should be avoided. A moderately aging population means maintaining an acceptable rate of aging and a sustainable social dependency ratio when it is time to implement the second and the third steps of the Three-Step Population Development Strategy. One objective of this strategy is to keep the total population under control and achieve an all-around optimum population step by step.

Another objective of the strategy is not to let the aging problem go beyond what the development of population, economy and society can bear. There is no gain without loss. No matter how hard we try, it is impossible to fulfill the strategy of keeping population growth under control and simultaneously to have a young structure population. It must be clarified that no effort should be spared to accomplish the population development strategy target since this is our top priority. Meanwhile, the adverse effects of an aging population should be restricted and minimized. Our analysis shows that adjusting the speed towards an aging society is compatible with our objective of the second step of the population strategy. It is inadvisable either to let the aging population problem take its own course or to be so worried by it that we call for some immediate solution to the problem. It is better to develop scientific policy and then implement it step by step.

Third, the proportion of working-age people is generally declining. The working-age population (those aged between 15 and 64) has shown a downward trend in recent years, after peaking in 2010 and 2013. The median prediction scheme indicated that the percentage of working-age population would decline from 74.5% in 2010 to 69.0% in 2020 and it is predicted to continue to decline, to 67.4% in 2030, a decrease of 1.57% over 2020 level. It is predicted to continue declining to 62.9% by 2040, a decrease of 4.5% over 2030. In 2050, it is predicted to reach 61.2%, a decrease of 1.7% over the 2040 prediction. The absolute number of working-age people is predicted to decline from 1005.82 million in 2013 to 996.52 million in 2020, a decrease of 0.9% and to 987.97 million in 2030, another decrease of 0.9%. The working-age population is predicted to fall to 912.33 million in 2040, a 7.7% decrease over the 2030 level, and then continue to decline, down to 857.52 million in 2050. In total, then, the percentage of working-age population is predicted to decline by 13.3% from 2010 to 2050, with an average decrease of 0.33% per annum. This is a fairly fast rate of decline but not too rapid. The absolute number of working-age people will decrease by 148.3 million, an average decrease of four million per annum. This is quite a large reduction, but not too large. In general, the situation is acceptable, but the resultant problems must be addressed. What is worthy of our attention is the predicted quite substantial decrease in working-age people in both percentage terms and absolute numbers from 2030 to 2050. The situation is deemed worrying when it reaches a bottom limit that should not be touched.

Fourth, citizens should have autonomy over reproductive decisions. The international community has long objected to the Chinese government's implementation of its family planning policy. The reasoning is that the right to have children has been controlled by the state, so Chinese citizens of childbearing age are deprived of their reproductive rights. Since reproductive rights are basic human rights, when people cannot enjoy their reproductive rights, they are believed to have lost part of their human rights. The across-the-board implementation of the two-child policy means that the majority of citizens of childbearing age can start to enjoy their full reproductive rights, as more than 75% of the people in question are willing to have two children with no restriction on the spacing between the first and second children. Before this new policy, whether a person conformed to the family planning policy also determined the accomplishments of the individual and their working unit. If someone broke the one-child policy, he or she could not be promoted at work, even if he or she had fulfilled their tasks brilliantly. This was the so-called "veto power", an implicit punishment for a breach of the family planning policy. As people begin to exercise reproductive autonomy, this so-called "veto power" has disappeared. Now, the population policy in China is once again at a crossroads, after the first crossroads when the one-child policy was issued in 1980, and the second, when the population policy was adjusted in the first decade of the twenty-first century. In other words, China is now faced with a decision whether a certain population policy is still needed and what kind of policy is needed.

10.3.2 Take the Path of Chinese-Style Family Planning

Information from an array of sources indicates that a great majority of people think that a population policy is still needed. When there is a strategic population development objective, there will be certain appeals and a corresponding policy orientation. In my view, to follow the trend of times, China's future population policy should be family-planning oriented. As China is the country with the biggest population in the world, its future population policy is confronted with complicated situations. Factors that contribute to the complicated situations include the dominance of the one-child policy for several decades, the dramatic changes to Chinese society following the reform and opening-up policies, and the transformation of basic aspects of Chinese demographics because of fierce clashes between traditional Chinese culture and ideas of modernization. Countries in the western world had already completed transformations of their population and undergone long periods of post-demographic transition. In some countries, family planning has been practiced for more than half a century. Thus, the US, Japan and some European countries have accumulated some experiences that could be enlightening for China. In the meantime, social economic development in China has laid new foundations for population reproduction. With the preliminary establishment of a market economy system and appreciation of exchange value, the marginal costs and benefits of rearing children have changed profoundly, and a growing number of people are choosing to have fewer children and thus to offer them better care. In these circumstances, the future family planning-oriented population policy will draw lessons from the good practices and useful experience of western countries and simultaneously assimilate the achievements of both Chinese post-reform social economic development and certain elements of traditional Chinese culture. Finally, a sound population policy should be formulated to serve Chinese-style family planning.

If we are to adopt a family planning approach and implement the population policy that serves it, then interpretation of western family planning models is needed. In the traditional view, population control objectives and corresponding policy should resolve the twin problems of overpopulation and a population shortage, by encouraging or restraining population growth and promoting internal migration. Since Thomas Robert Malthus's proposal in *An Essay on the Principle of Population* was made, the negative inhibitive measures evolved into fertility rate controls and contraception methods. Thus, family planning was created and was strongly dynamic, laying a new path to the solution of population problems. Family planning refers to planning intended to determine the number and spacing of children through birth control, taking into account the family's financial situation, the mother's health, and so on. It was first introduced in Britain in the 1930s and gradually gained popularity in other countries, with its denotations and connotations being enriched in the process.

When family planning is discussed, the efforts and contribution made by Margaret Sanger should be acknowledged. In 1915, thanks to the efforts of Sanger and others, the Planned Parenthood Federation of America was founded in New York, with an official publication entitled *Birth Control Review*. Subsequently, she traveled to Japan, India, and China to promote her birth control proposals. In 1923, she established a sizable birth control research laboratory, which was the first to offer consultancy services on birth control. In 1927, she organized the first World Birth Control Conference, drawing the world's attention to the issue of contraception and launching a practical action plan. After the end of World War II, family planning associations in eight countries and regions (the US, the Federal Republic of Germany, the Netherlands, Sweden, India, Singapore, UK and Hong Kong) held a founding conference for the International Planned Parenthood Federation, the largest nongovernmental voluntary international organization of birth control, in Bombay (Mumbai today). Presently, its headquarters are based in London, and it has over 180 member countries and affiliates. This organization has played an active international role in areas of population, fertility and other related fields. It advocates the view that family planning is an indispensable component of human rights and that it is necessary to strike a balance between world population, resources and environment in order to achieve human happiness and world peace and prosperity. The International Planned Parenthood Federation (IPPF), encourages every country to establish a national association of family planning and urges governments to shoulder the responsibility for implementing family planning programs.

The purpose of the IPPF is to facilitate family planning in each country, protect both the physical and psychological well-being of children and their parents, to deliver population education, and to promote research on human reproduction, its adjustments and the application of new research findings. The goals of IPPF member associations include promoting and protecting human rights, including teenagers' human rights, providing people with the information necessary to make informed sexual health decisions, enabling them to make their own choices about the number of children they will have and the spacing between them, and making available high quality sexual and reproductive health services. The goals also include paying special attention to the care of women and children, providing family planning and safe abortion services, eliminating unsafe abortion, fighting for equal rights for women, and enabling them to engage fully in social and economic development.

The *International Planned Parenthood Federation (IPPF) Charter on Sexual and Reproductive Rights* published in 1997 identified a list of 12 sexual and reproductive health rights. They include the Right to Life, the Right to Liberty and Security of Person, the Right to Privacy, the Right to Equality and to be free from All Forms of Discrimination, the Right to Freedom of Thought, the Right to Information and Education, the Right to Choose Whether or Not to Marry and to Found and Plan a Family, the Right to Health Care and Health Protection, the Right to the Benefits of Scientific Progress, the Right to Freedom of Assembly and Political Participation, the Right to be Free from Torture and Ill Treatment. More detailed and explicit acts have been specified to guarantee that legal services be provided to protect teenagers' sexual and reproductive rights. What is clear is the understanding of and respect for the dignity, rights, personal wishes and preferences of its clients, an emphasis on service first, as well as taking action according to law. The charter has set standards and criteria to help member associations to understand and solve their population problems.

I have been to the UK, France, the Netherlands, Belgium, the US, Canada, Japan and some other countries, as well as the regions like Taiwan and Hong Kong of China for academic exchanges, so I am acquainted the implementation of family planning in those places. Their family planning models all have similar structures and some common features can summarized as follows:

Form:	nongovernmental and non-profit organizations.
Mission:	to provide care to women and children, to provide reproductive health services and related technical support, and to promote scientific research and publicity.
Action:	service first. Delivering all-around service to clients to protect all their rights.
Organization:	management personnel plus volunteers.
Funding:	donations. Mainly from developed countries, and some financial support from individuals and nongovernmental foundations.

If China's population policy is to take the path of family planning in the future, it will be necessary to make fundamental changes to its forms, tasks, action, organization and funding, and align with the practices of the international community. As a member of IPPF, China has already established similar organizations and made some progress in keeping pace with IPPF. However, there are still obvious gaps and even fundamental differences in the above five basic aspects. Therefore, to follow the path of western family planning, China must implement reforms and establish a standardized family planning organization from top to bottom, according to these five basic principles and norms. There will definitely be Chinese-style family planning or family planning with Chinese characteristics. A key goal must be that those characteristics be kept within the ambit of the basic international principles and norms of family planning. In other words, Chinese-style family planning is the application and development of those principles and norms in China, and not letting China's family planning go beyond those principles and norms. For example, what is the relationship between government and family planning organizations, and what is the government's role? It is fundamentally different from the organization, leadership and working methods of the current state-led system. Another example is that much of the basic work will be undertaken by volunteers, which is different from the previous situation where "top leaders took overall responsibility". Furthermore, while funding for family planning is generally sponsored by social organizations and individuals in the west, financial support from government is indispensable in China to some extent, at least for a certain period. In view of the current state of social organizations in China, research and pilot projects should be carried out. The transition from family planning as basic national policy and government-led family planning to a service-oriented, non-governmental family planning model requires fundamental changes and adjustments that cannot be made overnight. We must grasp the overall problems that affect transformation and adjustment of family planning, and follow a route of making breakthroughs in key areas, trying out new measures and applying them carefully.

Practically, Chinese-style family planning should be carried out step by step. The direction and focus of reform in family planning should be to abandon the strict control of fertility indicators, to grant people autonomy over their reproductive rights and to create a service-oriented family planning model.

First, the regime of strict control over fertility should be abandoned. For a long time, the one-child stipulation, the planned fertility rate, the unplanned fertility rate, contraceptive efficiency, total fertility rate, population increase rate, and so on, were the main quantitative indicators of excellence against which the implementation of the population policy was measured and family planning work was assessed. Forms of organization at all levels, cadre allocation, focus of work, etc., were established according to the indicators. Cadre selection aimed to choose competent people who would visit thousands of households, speak thousands of words, and try thousands of means to introduce the fertility indicators to every household and every person. It is time to come up with better management methods, applying both rewards and punishments to achieve better results. The ideal breakthrough will be to reform family planning and abandon the regime of strict control using fertility indicators; this means that the systems and methods of the original target scheme will withdraw from the stage of history.

Second, we must hand over autonomy to people to exercise their reproductive rights. A review of the whole of human history might suggest that the right to reproduce is the most basic of all rights, and that it predates all other laws and regulations. *Tehran Declaration*, adopted by the UN International Conference on Human Rights in 1968, stated that parents should be free to determine the number of children and the spacing between them, the most basic human right. Yet, reproductive rights are constantly evolving. *World Population Program of Action*, adopted by the UN World Population Conference in Bucharest in 1974, incorporated the right to relevant information, education and methods in reproductive rights on the basis of the *Tehran Declaration*. *Convention on the Elimination of All Forms of Discrimination against Women*, adopted by the UN General Assembly in 1979, incorporated reproductive rights, and China was one of the earliest signatories. The International Conference on Population and Development held in Cairo in 1994 reiterated this spirit, emphasizing that the principle of informed free choice was essential to the long-term success of family planning programs and that no coercion of any form should be used. In the whole process of childbirth and rearing, women have special functions and responsibilities, which are irreplaceable. Reproductive rights therefore concern not only parents but also all couples and individuals. Reproductive rights involve not only the choice to have children and when to have them, but also access to relevant knowledge, education, contraception methods, and so on, which are related to the maintenance of gender equality in reproductive rights. Reproductive rights are defined in the international context similarly to how they are defined and interpreted in China, yet there are differences. To give people autonomy over their reproduction can facilitate exchanges with the international community and pave the way to more consensuses.

Third, a service-oriented system and mechanisms should be created. The family planning slogan could be summed up as "Service First". That is to say, for people of childbearing age, all-round, high-quality services should be provided, including maternal and child health care, reproductive health care, technical support, the dissemination of scientific research findings, and so on. To highlight the characteristics of a general service, it is necessary to respect the right to life, equality, freedom, security, privacy, education, marriage, childbirth, assembly, participation in politics, and so on, and also to add some specific provisions to the rights that the clients of such services should enjoy, including the sexual and reproductive health of adolescents. The reason why western family planning has remained so popular for so long is that it does not dictate what rules the reproductive group must follow and how to follow them, but focuses rather on providing high-quality services. In China, control of population size and the family planning implementation, has always raised issues of coercion, which have been the focus of opponents' critiques. To be fair and objective, the Chinese government has consistently opposed coercion in the implementation of the family planning policy, and it has taken up many cases of coercion that had adverse effects. At the same time, however, it must be admitted that a number of cases in which coercion has been used have cast a pall over the policy. Reform of family planning rules, driven by an idea of a family planning agency, will cut off the sources of coercion. When the family planning agency is a non-governmental organization, it has no power to impose coercive orders. Cutting this link between power and capital is conducive to maintaining an upright and moral society.

References

National Health Commission. (2018). *China Health Statistical Yearbook 2018.* China Union Medical University Press. p. 235.

National Statistics Bureau of the PRC. (Ed.). (2018). *China Statistical Yearbook of 2018.* China Statistical Publishing House. pp. 10–11; 32, 33.

The Editorial Board (Ed.). (2018). *Research on Population Issues in Social Change and Transformation.* China Social Sciences Press. pp. 89–107.

Tian, X. Y. (2009, December 4). Review and Prospects of Population Policies in People's Republic of China. *People's Daily.*

U. N. Population Division. (Ed.). (2009). *World Population Prospects: The 2008 Revision (Vol.1)*. New York. p. 48, 52, 54, 184.

Yi, F. X. (2013). *Big Country with an Empty Nest: Reflection on China's Family Planning Policy*. China Development Press. p. 358.

Yuan, H.R. (2014). *Theory and Evidence: Population, Environment, and Development*. Zhejiang University Press.

MIX
Papier aus verantwortungsvollen Quellen
Paper from responsible sources
FSC® C105338

If you have any concerns about our products,
you can contact us on
ProductSafety@springernature.com

In case Publisher is established outside the EU,
the EU authorized representative is:
Springer Nature Customer Service Center GmbH
Europaplatz 3, 69115 Heidelberg, Germany

Printed by Libri Plureos GmbH
in Hamburg, Germany